Springer Climate

Series Editor

John Dodson, Institute of Earth Environment, Chinese Academy of Sciences,
Xian, Shaanxi, China

Springer Climate is an interdisciplinary book series dedicated to climate research. This includes climatology, climate change impacts, climate change management, climate change policy, regional climate studies, climate monitoring and modeling, palaeoclimatology etc. The series publishes high quality research for scientists, researchers, students and policy makers. An author/editor questionnaire, instructions for authors and a book proposal form can be obtained from the Publishing Editor.

Now indexed in Scopus® !

More information about this series at https://link.springer.com/bookseries/11741

Seema Rani · Rajesh Kumar
Editors

Climate Change

Impacts, Responses and Sustainability in the Indian Himalaya

Springer

Editors
Seema Rani
Department of Geography
Institute of Science
Banaras Hindu University
Varanasi, Uttar Pradesh, India

Rajesh Kumar
Department of Geography
School of Human Sciences
Sikkim University
Gangtok, Sikkim, India

ISSN 2352-0698 ISSN 2352-0701 (electronic)
Springer Climate
ISBN 978-3-030-92781-3 ISBN 978-3-030-92782-0 (eBook)
https://doi.org/10.1007/978-3-030-92782-0

This Springer imprint is published by the registered company Springer Nature Switzerland AG
The registered company address is: Gewerbestrasse 11, 6330 Cham, Switzerland

To our students

Preface

The word Himalaya is derived from the Sanskrit words *Hima* means "snow", and *Alaya* means "abode". It is the highest mountain system in the world and forms part of high Asia. It is a young mountain which is about 2500 km long, from the Nanga Parbat (8126 m) (Gilgit-Baltistan, Pakistan) to the Namcha Barwa (Assam Himalaya) with an average width of approximately 240 km. It is covering an area of about 0.5 million km^2 in India that is about 16.2% of the country's total geographical area. There are threefold axes in the Indian Himalayan Region (IHR), namely Himadri (Greater Himalaya), Himachal (Lesser Himalaya) and the Shiwaliks (Outer Himalaya). The IHR is spreading in nine states (Himachal Pradesh, Uttarakhand (UK), Sikkim, Arunachal Pradesh, Nagaland, Manipur, Mizoram, Tripura, Meghalaya), hilly parts of two states (Assam and West Bengal), and two union territories (Jammu and Kashmir and Ladakh) of India. Apart from the three folded axes or longitudinal sub-divisions, the IHR can be broadly divided into three regions based on regional characteristics, namely western Himalaya (Kashmir and Himachal Himalaya), central Himalaya (UK Himalaya) and eastern Himalaya (Darjeeling–Kalimpong–Sikkim–Arunachal Pradesh and Assam Himalaya and Purvanchal in Nagaland, Manipur, Mizoram, Tripura, and Meghalaya states).

The IHR has a complex climate and unique physiological conditions which support different types of biodiversity. Many large rivers of the world (like the Ganga, Indus, and Brahmaputra) flow from this region and are providing water to a large part of the Indian subcontinent. It is directly or indirectly influencing the livelihood of a large population in the Himalayan and great plain regions over the several centuries. The region is also characterized by various environmental, demographic, economic and social systems. Approximately, 50 million people depend on the resources of the IHR.

The concern of climate change in the region has become a talk all around the world at different spatial scales. Recently, the intergovernmental panel on climate change (IPCC) in its special report (2019) on the Ocean and Cryosphere in a Changing Climate (SROCC) highlighted the impacts of shrinking of the cryosphere mass loss from ice sheets/glaciers and reduction in snow cover on the resources and aesthetic/cultural aspects of the IHR. It is resulting in extreme weather events (warm winter and frequent cloud bursts) in the IHR. The importance of the Himalaya in the Indian monsoon system has also been recognized by many scientists over several decades. This region is undergoing several physical changes over the years in forms of natural disasters (glacial lake outburst floods, landslides, flash floods due to cloud burst and earthquake) those are also impacting the lives of millions of populations. As per the assessment of scientists in the International Centre for Integrated Mountain Development (ICIMOD) (Kathmandu city of Nepal) an intergovernmental body, this region shows signs of warming which is higher than the global average rate. The report mentioned an increase in warm nights and days by 1.7 and 1.2/decade which would have serious implications on the physical aspects of the area such as snow/glacier and river discharge. The rising climate anomalies will indeed impact the livelihoods of millions of people in the region. Hence, it is imperative to study the impacts and responses of this mountainous region towards climate change for sustainable planning and adaptability. The Government of India (GoI) launched National Action Plan on Climate Change (NAPCC) in 2008 to achieve sustainable development of the region while addressing climate change issues. Among the missions of the NAPCC, the National Mission for Sustaining the Himalayan Ecosystem (NMSHE) is one of the most important missions that aim to contribute to the country's sustainable development through improving understanding of climate change, its anticipated consequences, and adaptation efforts for the Himalayas.Based on the recommendations of five thematic Working Group Reports, in 2018, the National Institution for Transforming India (NITI) Aayog has constituted "Himalayan State Regional Council" for Sustainable Development in IHR to review the implementation of the identified action points (Inventory and Revival of Springs in the Himalayas for Water Security, Sustainable Tourism in Indian Himalayan Region, Shifting Cultivation: Towards Transformational Approach, Strengthening Skill and Entrepreneurship (E and S) Landscape in the Himalayas and Data/Information for Informed Decision Making) based on the reports. There is a prerequisite of huge data, to make plans under the mentioned themes, which is also a major concern due to the rugged terrain and extreme weather conditions of the region.

Thus, an attempt is made to comprehend available data in an integrated edited book volume *Climate Change: Impacts, Responses and Sustainability in the Indian Himalaya.* This book intends to integrate the existing research on Himalayan dynamics and its implications on physical systems at different spatio-temporal scales. Especially, this work includes many relevant parts of weather/climate, snow/glacier hydrology, and ecology associated with the IHR. The book is mainly suitable for academic researchers, scientists, planners, and policymakers working in the field of geography and geosciences. We strongly believe that this book would be very useful and informative for those working on the physical aspects of the IHR. We also wish that the coming researchers will enhance and contribute to the provided findings in this book by using updated fine quality data and recent evolving geospatial technologies.

Varanasi, Uttar Pradesh, India
Gangtok, Sikkim, India
July 2021

Seema Rani
Rajesh Kumar

Acknowledgements

Expert Reviewers

Dr. Atul Kumar Singh, Inter-University Accelerator Centre (IUAC), New Delhi, India
Dr. Jaishri Sanwal, Geodynamic Unit, Jawaharlal Nehru Centre for Advanced Scientific Research (JNCASR), Jakkur, Bangalore, Karnataka, India
Dr. Manish Kumar Pandey, University Centre for Research and Development (UCRD), Chandigarh University, Mohali, Punjab, India
Dr. Prem C. Pandey, Centre for Environmental Sciences and Engineering (CESE), School of Natural Sciences (SoNS), Shiv Nadar University, Greater Noida, Uttar Pradesh, India
Dr. Ruby Ghosh, Birbal Sahni Institute of Palaeosciences, Lucknow, Uttar Pradesh, India
Dr. Surya Narayan Pradhan, Sambalpur University, Sambalpur, Odisha, India
Mr. Muneer Ahmad Mukhtar, Geological Society of India, Srinagar, Jammu and Kashmir (UT), India
Mr. R. K. Gautam, Sikkim State Council of Science and Technology, Vigyan Bhawan, Deorali, Gangtok, Sikkim, India

Expert Guidance

Dr. Suraj Mal, Department of Geography, Shaheed Bhagat Singh College, University of Delhi, New Delhi, India. Presently, Chair of International Geographical Union (IGU) Commission on Biogeography and Biodiversity.

Chapter's Coordinating and Lead Authors

Omvir Singh, Mayank Shekhar, Ipsita Roy, Parminder Singh Ranhotra, Rupendra Singh, A. A. Shah, R. K. Sharma, Riyaz Ahmad Mir, Amit Kumar Tripathi, Prem Chandra Pandey, Hukum Singh, K. K. Vikrant, Damanjeet Kaur, Saurabh Gupta, Shashi Shekhar, and Prachita Arora

Springer Team

Ms. Doris Bleier, Springer Dordrecht, Van Godewijckstraat 30, P.O. Box 17 3300 AA Dordrecht, The Netherlands and her team

Contents

Contributors

S. Nawaz Ali Birbal Sahni Institute of Palaeosciences, Lucknow, Uttar Pradesh, India

Prachita Arora Birbal Sahni Institute of Palaeosciences, Lucknow, Uttar Pradesh, India

Navakanesh M. Batmanathan Southeast Asia Disaster Prevention Research Initiative, Universiti Kebangsaan Malaysia, Bandar Baru Bangi, Malaysia

Amalava Bhattacharyya Birbal Sahni Institute of Palaeosciences, Lucknow, Uttar Pradesh, India

D. S. Chauhan Department of Forestry, Hemvati Nandan Bahuguna Garhwal University, Srinagar, Pauri Garhwal, Uttarakhand, India

Rupesh Dhyani Amos Campus, Forest Research Institute (IRF), Université du Québec en Abitibi-Témiscamingue (UQAT), Rouyn-Noranda, Canada

Saurabh Gupta Department of Microbiology, Mata Gujri College (Autonomous), Punjab, India

Sanjay K. Jain Water Resource System Division, National Institute of Hydrology, Roorkee, Uttarakhand, India

Supinder Kaur University School of Law, Sri Guru Granth Sahib World University, Punjab, India

Damanjeet Kaur Department of Microbiology, Mata Gujri College (Autonomous), Fatehgarh Sahib, Punjab, India;
Department of Biotechnology, Punjabi University, Patiala, Punjab, India

Manoj Kumar GIS Centre, Forest Research Institute, Dehradun, Uttarakhand, India

Rajesh Kumar Department of Geography, School of Human Sciences, Sikkim University, Tadong, Gangtok, Sikkim, India;
Department of Environmental Sciences, School of Earth Sciences, Central University of Rajasthan, Ajmer, Rajasthan, India

Syed Umer Latief Post-Graduate Department of Geography, Amar Singh College, Srinagar, Jammu and Kashmir (UT), India

Pyarimohan Maharana School of Ecology and Environment Studies, Nalanda University, Rajgir, Bihar, India

Riyaz Ahmad Mir Department of Earth Sciences, Indian Institute of Technology, Roorkee, Uttarakhand, India;
Geological Survey of India, Srinagar, Jammu and Kashmir (UT), India

P. Morthekai Birbal Sahni Institute of Palaeosciences, Lucknow, Uttar Pradesh, India

Prem Chandra Pandey School of Natural Sciences, Center for Environmental Sciences & Engineering, Shiv Nadar University, Greater Noida, Uttar Pradesh, India

Pranay Pradhan Sikkim State Council of Science and Technology, Deorali, Gangtok, Sikkim, India

Asiya Qadir Srinagar, India

Parminder Singh Ranhotra Birbal Sahni Institute of Palaeosciences, Lucknow, Uttar Pradesh, India

Seema Rani Department of Geography, Institute of Science, Banaras Hindu University, Varanasi, Uttar Pradesh, India

R. H. Rizvi Central Agroforestry Research Institute, Jhansi, Uttar Pradesh, India

Ipsita Roy Birbal Sahni Institute of Palaeosciences, Lucknow, Uttar Pradesh, India

A. A. Shah Department of Physical and Geological Sciences, Universiti of Brunei, Darussalam, Brunei

Arpan Sharma Department of Computer Applications, Sikkim Manipal Institute of Technology, Majitar, Sikkim, India

Jyoti Kumar Sharma School of Natural Sciences, Center for Environmental Sciences & Engineering, Shiv Nadar University, Greater Noida, Uttar Pradesh, India

Milap Chand Sharma Centre for the Study of Regional Development, Jawaharlal Nehru University, New Delhi, India

Yogesh K. Sharma Department of Botany, University of Lucknow, Lucknow, Uttar Pradesh, India

Mayank Shekhar Birbal Sahni Institute of Palaeosciences, Lucknow, Uttar Pradesh, India

Shashi Shekhar Himachal Pradesh State Pollution Control Board, Shimla, Himachal Pradesh, India

R. K. Sharma Department of Environmental Science, School of Basic Sciences and Research, Sharda University, Greater Noida, Uttar Pradesh, India;
Sikkim State Council of Science and Technology, Deorali, Gangtok, Sikkim, India

Ayushi Singh Birbal Sahni Institute of Palaeosciences, Lucknow, Uttar Pradesh, India

Omvir Singh Department of Geography, Kurukshetra University, Kurukshetra, Haryana, India

Shaktiman Singh School of Geosciences, University of Aberdeen, Aberdeen, UK

Hukum Singh Plant Physiology Discipline, Genetics and Tree Improvement Division, Forest Research Institute, Dehradun, Uttarakhand, India

Rupendra Singh CPS, School of Social Sciences, Jawaharlal Nehru University, New Delhi, India

Prashant K. Srivastava Remote Sensing Laboratory, Institute of Environment and Sustainable Development, Banaras Hindu University, Varanasi, Uttar Pradesh, India; DST-Mahamana Centre of Excellence in Climate Change Research, Banaras Hindu University, Varanasi, Uttar Pradesh, India

Anupreet Singh Tiwana Department of Geography, Mata Gujri College (Autonomous), Punjab, India

Nidhi Tomar Birbal Sahni Institute of Palaeosciences, Lucknow, Uttar Pradesh, India

Dimitris Triantakonstantis Department of Soil Science of Athens, Institute of Soil and Water Resources, Hellenic Agricultural Organization DIMITRA, Attiki, Greece

Amit Kumar Tripathi School of Natural Sciences, Center for Environmental Sciences & Engineering, Shiv Nadar University, Greater Noida, Uttar Pradesh, India

K. K. Vikrant Forestry Department, Himalayan College, Roorkee, Haridwar, Uttarakhand, India

Abbreviations

Aa	Average Annual
Aab	Average Ablation
AAbD	Average Ablation Discharge
Aac	Average Accumulation
AAcD	Average Accumulation Discharge
AAD	Average Annual Discharge
AAI	Absorbing Aerosol Index
ADP	Ad Hoc Working Group on Durban Platform for Enhanced Action
AHP	Analytic Hierarchy Process
AL	Additive Rescaling Factor
Am	Average Monsoon
AMD	Average Monsoon Discharge
AMJJAS	April to September
AMRUT	Atal Mission on Rejuvenation and Urban Transformation
AMSL	Above Mean Sea Level
AP	Arunachal Pradesh
Apo	Average Postmonsoon
APoD	Average Postmonsoon Discharge
Apr	Average Premonsoon
APrD	Average Premonsoon Discharge
ARSTAN	Autoregressive Standardization
AS	Arabian Sea
ASTER	Advanced Spaceborne Thermal Emission and Reflection Radiometer
Aw	Average Winter
AWD	Average Winter Discharge
AWG-KP	Ad Hoc Working Group on KP
AWG-LCA	Ad Hoc Working Group on LCA
AWM	Area Weighted Mean
BA	Bølling-Allerød
BBMB	Bhakra Beas Management Board

BD	Biodiversity
BGA	Bhojbasa Glacial Advance
BGS	Bhagirathi Glacial Stage
BJS	Bhojbasa Stage
BKD	Bhojbasa Kame Deposit
BoB	Bay of Bengal
BOP	Bhojbasa Outwash Plain
BP	Before Present
BR	Bayesian Regression
BSIP	Birbal Sahni Institute of Palaeosciences
BSRA	Birbal Sahni Research Associate
C.E.	Common Era
C3S	Copernicus Climate Change Service
CBDR&RC	Common, But Differentiated, Responsibilities, and Respective Capabilities
CC	Climate Change
CCAP	Climate Change Action Programme
CDB	Convention on Biological Diversity
CDM	National Clean Development Mechanism Authority
CE	Coefficient of Efficiency
CGIAR-CSI	Consultative Group on International Agricultural Research, Consortium for Spatial Information
CH	Central Himalayas
CMS	Centre for Media Studies
CO	Carbon Monoxide
COFECHA	Quality Control Software for Tree-Ring Width Measurement
COP	Conference of Parties
COVID 19	Corona Virus Disease
CPCB	Central Pollution Control Board
CRN	Cosmogenic Radionuclide
CRU	Climatic Research Unit
CVR	Cell Value Radiance
CWC	Central Water Commission
DEM	Digital Elevation Model
DJF	December, January, February
DN	Digital Number
DST	Department of Science and Technology
DTM	Digital Terrain Model
E. Kumaon	Eastern Kumaon
E. Nepal	Eastern Nepal
EASM	East Asian Summer Monsoon
EB	Eastern Brahmaputra
ECMWF	European Centre for Medium-Range Weather Forecasts
EDW	Elevation-Dependent Warming
EEFP	Energy Efficiency Financing Platform

ELA	Equilibrium Line Altitude
ENSO	El-Nino Southern Oscillation
EPS	Expressed Population Signal
ERA	ERA-Interim Reanalysis
ESA	European Space Agency
ESIP	Ecosystems Services Improvement Project
ET	Evapotranspiration
ETM+	Enhanced Thematic Mapper Plus
EWS	Early Warning System
EWW	Early Wood Width
FEEED	Framework for Energy Efficient Economic Development
FGD	Focus Group Discussion
FSI	Forest Survey of India
FTT	Fourier Transform
GBPIHED	Govind Ballabh Pant Institute of Himalayan Environment and Development
GD	Growth Disturbances
GDP	Gross Domestic Product
GHGs	Greenhouse Gases
GIM	National Mission for Green India
GIS	Geographic Information System
GIZ	German Agency for International Cooperation
GLOFs	Glacial Lake Outburst Floods
GoI	Government of India
GPCC	Global Precipitation Climatology Centre
GPLs	Glacier Protection Laws
G-SHE	Governance for Sustaining Himalayan Ecosystem
GSI	Geological Society of India
GW	Gigawatts
HADP	Hill Area Development Programme
HKH	Hindu Kush Himalaya
HP	Himachal Pradesh
ICFRE	Indian Council of Forestry Research and Education
ICIMOD	International Centre for Integrated Mountain Development
IGP	Indo-Gangetic Plain
IH	Indian Himalaya
IHCAP	Indian Himalayas Climate Adaptation Programme
IHR	Indian Himalayan Region
IKI	International Climate Initiative
IMD	India Meteorological Department
INC	Intergovernmental Negotiating Committee
INCCA	Indian Network for Climate Change Assessment
INDCs	Intended Nationally Determined Contributions
IPBES	Intergovernmental Science-Policy Platform on Biodiversity and Ecosystem Services

IPCC	Intergovernmental Panel on Climate Change
ISM	Indian Summer Monsoon
ISRO	Indian Space Research Organisation
ITCZ	Inter Tropical Convergence Zone
IUCN	International Union for Conservation of Nature
J-A	January-April
JA	July, August
JJAS	June, July, August, September
JK	Jammu and Kashmir
JR	Jhelum River
ka BP	Kilo Annum Before Present
KB	Kashmir Basin
KP	Kyoto Protocol
LCA	Long Term Cooperative Action
LGM	Last Glacial Maxima
LH	Late Holocene
LIA	Little Ice Age
LISS	Linear Imaging Self Scanning Sensor
LOOCV	Leave One Out Cross Validation
LST	Land Surface Temperature
LULC	Land Use/Land Cover
LWW	Late Wood Width
MAM	March, April, and May
MAP	Mean Annual Precipitation
MAT	Mean Annual Temperature
MEA	Millennium Ecosystem Assessment
MEF	Major Economies Forum on Energy and Climate
MELM	Maximum Elevation of Lateral Moraine
MERRA	Modern-Era Retrospective Analysis for Research and Applications
MGNREGA	Mahatma Gandhi National Rural Employment Guarantee Act
MH	Mid Holocene
MIS	Marine Isotope Stage
MK	Mann–Kendall
ML	Multiplicative Rescaling Factor
MnAD	Minimum Annual Discharge
MNRE	Ministry of New and Renewable Energy
MoEF	Ministry of Environment and Forests
MoEFCC	Ministry of Environment, Forests, and Climate Change
MSL	Mean Sea Level
MTEE	Market Transformation for Energy Efficiency
MWP	Mediaeval Warm Period
MxAD	Maximum Annual Discharge
NABARD	National Bank for Agriculture and Rural Development
NACs	Non-Attainment Cities
NAPCC	National Action Plan on Climate Change

NASA	National Aeronautics and Space Administration
NBAP	National Biodiversity Action Plan
NCAR	National Centre for Atmospheric Research
NCEP	National Centres for Environmental Prediction
NCEPP	National Council for Environmental Policy and Planning
NDCs	Nationally Determined Contributions
NDJF	November, December, January, and February
NE	North East
NEP	National Environment Policy
NFP	National Forest Policy
NGOs	Non-Government Organizations
NJHEP	Nathpa Jhakri Hydro Electric Project
NMEEE	National Mission for Enhanced Energy Efficiency
NMHS	National Mission on Himalayan Studies
NMSA	National Mission for Sustainable Agriculture
NMSH	National Mission on Sustainable Habitat
NMSHE	National Mission for Sustaining the Himalayan Ecosystem
NMSKCC	National Mission on Strategic Knowledge for Climate Change
NOAA	National Oceanic and Atmospheric Administration
NPP	Net Primary Productivity
NRSC	National Remote Sensing Centre
NSA	National Solar Mission
NWIH	North Western Indian Himalaya
NWM	National Water Mission
OD	Older Dryas
OECD	Organisation for Economic Co-operation and Development
OLS	Ordinary Least Square
ONDJFM	October to March
OSL	Optically Stimulated Luminescence
PAT	Perform Archive and Trade
PDSI	Palmer Drought Severity Index
PET	Potential Evapotranspiration
PM	Particulate Matter
PMCCC	Prime Minister Council on Climate Change
PMT	Product Mean Test
PSp	Pollen-Spores
RBAR	Routine Bar
RCPs	Representative Concentration Pathways
RDCC	Research and Development Committee
RE	Reduction of Error
Riv DIS v1.0	River Discharge Database, Version 1.0
RMSE	Root Mean Square Error
Sa	Sum Annual
Sab	Sum Ablation
SAC	Space Applications Centre

Sac	Sum Accumulation
SAPCC	State Action Plan on Climate Change
SBSTA	Subsidiary Science and Technology Advisory Agency
SCA	Snow Cover Area
SDC	Swiss Agency for Development and Cooperation
SDGs	Sustainable Development Goals
SDI	Standardized Discharge Index
SGA	Shivling Glacial Advance
SI	Standardized Indices
SICB	State Institute of Capacity Building
SK	Sikkim
SLL	South Lhonak Lake
Sm	Sum Monsoon
SNAP	Sentinel Application Platform
SO	September October
SoI	Survey of India
SPEI	Standardized Precipitation-Evapotranspiration Index
Spo	Sum Postmonsoon
Spr	Sum Premonsoon
SRCCL	Special Report on the Ocean and Cryosphere in a Changing Land
SRI	Standardized Rainfall Index
SROCC	Special Report on the Ocean and Cryosphere in a Changing Climate
SRTM	Shuttle Radar Topography Mission
SSCCC	Sikkim State Climate Change Cell
ST	Sign Test
STI	Standardized Temperature Index
Sw	Sum Winter
THAR	Terminus to Headwall Altitude Ratio
TL	Thermo Luminescence
TM	Thematic Mapper
TP	Tapovan Palaeolake
TP	Tibetan Plateau
TRD	Traumatic Resin Ducts
TRMM	Tropical Rainfall Measuring Mission
TRW	Tree-Ring Width
UK	Uttarakhand
UNEP	United Nations Environment Programme
UNFCCC	United Nations Framework Convention on Climate Change
UNSC	United Nations Security Council
USGS	United States Geological Survey
UTs	Union Territories
UV	Ultraviolet
W. Himalaya	Western Himalaya
W. Tibet	Western Tibet
WB	West Bengal

WD	Western Disturbance
WHO	World Health Organisation
WIH	Western Indian Himalaya
WMO	World Meteorological Organization
YD	Younger Dryas
δ13C	Carbon Isotope
δ18O	Oxygen Isotope

Chapter 1
Climate Change, Its Impacts, and Sustainability Issues in the Indian Himalaya: An Introduction

Seema Rani, Rajesh Kumar, and Pyarimohan Maharana

Abstract Climate change poses threats to humans and brings the toughest challenges for economic development in the twenty-first century. The scientific communities warn the world leaders regarding the threats of climate change and its inevitable impacts on the physical and cultural environment. The Intergovernmental Panel on Climate Change (IPCC) reported global warming of 1.5 °C which is a matter of great concern for all the stakeholders around the world. The manifestation of recent climate change is increasing flooding events, shrinking of the cryosphere (mass loss from ice sheets/glaciers, reductions in snow cover), vegetation changes, and loss of biodiversity which are having adverse effects on available resources and aesthetic/cultural aspects of the Indian Himalayas (IH). Hence, it is imperative to study the impacts and responses of the mountainous region towards climate change for sustainable planning and adaptability. This chapter aims to review the scientific works on emerging trends in climate change, its impacts, and sustainability issues in the IH. The review work suggests the need for more research on innovative ideas for better adaptation and to combat the increasingly adverse impacts of climate change.

Keywords Indian Himalaya · Climate change · Sustainability · Planning · Adaptability

S. Rani (✉)
Department of Geography, Institute of Science, Banaras Hindu University, Varanasi, Uttar Pradesh, India
e-mail: seemarani@bhu.ac.in

R. Kumar
Department of Geography, School of Human Sciences, Sikkim University, Tadong, Gangtok, Sikkim, India

P. Maharana
School of Ecology and Environment Studies, Nalanda University, Rajgir, Bihar, India

S. Rani and R. Kumar (eds.), *Climate Change*, Springer Climate,
https://doi.org/10.1007/978-3-030-92782-0_1

Introduction

The United Nations has adopted the 2030 Agenda for Sustainable Development in 2015, towards sustainable development of the planet, now and in the future (Assembly 2015). Climate action is the thirteenth goal of the Sustainable Development Goals (SDGs) to take immediate actions to combat climate change and its impacts on the physical conditions of the Earth's surface as well as on human beings. Climate change poses threats to human beings and brings the toughest challenges for economic development in the twenty-first century. The scientific communities warn the world leaders regarding the threats of climate change and also about its inevitable impacts on the physical and cultural environment (Mal et al. 2019; Schickhoff and Mal 2020; Shugar et al. 2021; Schickhoff et al. 2022). The Intergovernmental Panel on Climate Change (IPCC) reported global warming of 1.5 °C which is a matter of great concern for all the agencies around the world (IPCC 2018). IPCC (2018) reported that approximately 1.0 °C of global warming has been caused by humans and as a result, global warming may reach 1.5 °C between 2030 and 2052 if it continues to increase at the current rate. The report pointed that the mountainous region would be worst affected due to warming of 1.5 °C or 2 °C. Recently, IPCC has released Special Report on the Ocean and Cryosphere in a Changing Climate (SROCC) and Land (SRCCL) and the Intergovernmental Science-Policy Platform on Biodiversity and Ecosystem Services (IPBES) (IPCC 2019). SROCC has highlighted the impacts of shrinking of the cryosphere (mass loss from ice sheets and glaciers and reductions in snow cover) on the resources and aesthetic/cultural aspects of the Indian Himalaya (IH). Hence, it is imperative to study the impacts and responses of the mountainous region particularly Himalayas towards climate change for sustainable planning and adaptability. To assess climate change, an interdisciplinary study is required to engage several experts from geosciences, ecology, and social sciences to produce a research work explicitly focusing on the impacts of climate change and its related issues in the IH. The present chapter will cover climate change-related studies on the Indian Himalayan Region (IHR) from west to east. This chapter will focus more on the scientific methodologies such as trend analysis of instrumental/gridded data of elements of climate, tree-ring-based estimation of climatic and hydrological variables of past, geospatial technology for assessing the impact of climate change on the forest ecosystem, assessment of glacial recession and glacial lake outburst floods (GLOFs) induced by climate change, the impact of flash floods on a channel and adjoining human activities, air quality monitoring and its assessment techniques. Besides, this will also provide a separate section on planning and policies related to climate change and the impacts of climate change on local as well as the regional economy.

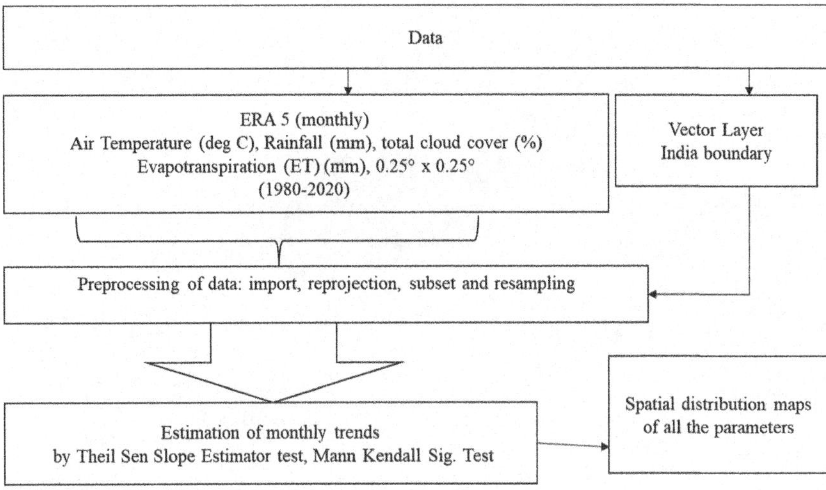

Fig. 1.1 Data and methods of processing the data used in the present study

Material and Methods

ERA 5 monthly air temperature, rainfall, cloud cover, and evapotranspiration (ET) at 0.25° × 0.25° horizontal resolution for the period 1980–2020 was obtained for climate variability assessment of the IHR (https://cds.climate.copernicus.eu/) (Hersbach et al. 2018) (Fig. 1.1). The administrative boundary of India is corrected as per the instructions of the Government of India (Census of India 2011) and used as a study area (India) for the present study. Shuttle Radar Topography Mission (SRTM) 30 m data was obtained from the Consultative Group on International Agricultural Research, Consortium for Spatial Information (CGIAR-CSI) to show the elevation distribution of the Himalayas. Total flood-affected areas (in km^2) (1953–2016) of the region were obtained from the Central Water Commission (CWC), New Delhi (India) (CWC 2018). The non-parametric Theil Sen slope estimator test (Theil 1950) has been carried out for reporting the trends in climate parameters and its significance has been tested by the Mann Kendall significance test (Mann 1945; Kendall 1975; Neeti and Eastman 2011) (sig is tested at 0.10 level). Spatial distribution maps of trends in climate parameters over the IH have been prepared in the GIS environment.

Indian Himalayas

Asia has the highest and most populated mountains in the world. These mountains (the Himalayas, Karakoram, and Hindu-Kush) ranges are extending from the east of India (Namcha Barwa) to the west in Afghanistan. These highest mountain ranges are formed between 40 and 50 million years ago, by plate movement. The Himalayas

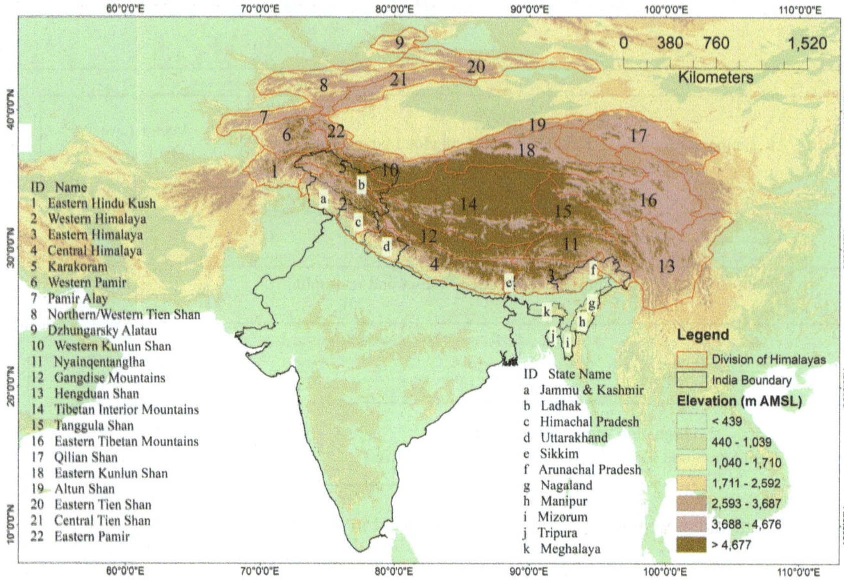

Fig. 1.2 Location and extent of the Indian Himalayan region. The elevation detail in the map is derived from the SRTM DEM 250 m (twenty two regions shown in the map are divided according to Bolch et al. 2019)

have about 30 highest peaks in the world whose height varies from 7620 m to more than 8500 m above the mean sea level (AMSL) (Mount Everest in Nepal). These are often known as Hindu-Kush-Himalayan (HKH) mountains covering an area of about 4.3 million km^2 including eight countries (Afghanistan, Pakistan, India, China, Nepal, Bhutan, Bangladesh, and Myanmar) (Fig. 1.2).

These ranges in the Indian part are known as the IHR which spread in 10 states of India, namely Jammu and Kashmir (presently it is divided into two union territories (i) Jammu and Kashmir (JK) and (ii) Ladakh), Himachal Pradesh (HP), Uttarakhand (UK), Sikkim (SK), Arunachal Pradesh (AP), Meghalaya, Nagaland, Manipur, Mizoram, Tripura, and hilly regions of Assam and West Bengal (WB). This region extends from Shivaliks to the Tibetan Plateau (TP) in the north (China). The region comprises about 16.2% of the total geographical area of India, and the majority area has snow-clad peaks including glaciers, dense forests with the highest diversity compared to other parts of the country. It has a low and dispersed human population compared to other parts due to rugged terrain, extreme weather conditions, and poor infrastructure. This region is considered fragile because of its limited capacity to deal with the ongoing impacts of climate change. These regions are highly vulnerable parts of India to natural and man-made disasters (floods, dam breach, landslides, debris flow, permafrost degradation, avalanche, earthquake, etc.).

Emerging Climate Change Issues

Air Temperature

Changes in the global air temperature are widely studied over several decades (IPCC 1996, 2001, 2007, 2013; Tank et al. 2006) because it influences other weather parameters such as evaporation, relative humidity, precipitation, wind speed, and direction. These are used as input in the modeling framework to study the climate change impact at varying spatial scales (Boyer et al. 2010; Bhatt et al. 2013; Luo et al. 2013). According to the IPCC (2013), an increase of about 0.89 °C was observed in global combined land and sea surface temperature during 1901–2012. An impact of a 1.5 °C rise in temperature has been taken into consideration on natural and human systems (IPCC 2018). Changes in air temperature have also been reported in India in the last decades (Gadgil and Dhorde 2005; Arora et al. 2005; Dash et al. 2007, 2013; Dash and Hunt 2007; Singh et al. 2008, 2013; Jhajharia and Singh 2011; Pal and Al-Tabbaa 2011; Duhan et al. 2013) (Table 1.1). A review on changes in air temperature over India pointed out significant warming in annual temperature (mean, maximum, and minimum) that varied between 0.027 °C and 0.072 °C/decade, during 1901–2007 (Jain and Kumar 2012). According to the annual climate summary of the India Meteorological Department (IMD 2019), warming trend at the rate of 0.061 °C/decade has been observed in the annual mean air temperature with significant warming in the maximum air temperature (0.1 °C/decade) and relatively lower warming in the minimum air temperature (0.022 °C/decade) during 1901–2019. Many regions across India have shown warming at the rate of 0.044 °C/decade (average temperature), 0.051 °C/decade (maximum temperature), and 0.019 °C/decade (minimum temperature) during 1941–2012 (Rani and Sreekesh 2018; Saxena and Mathur 2019; Ray et al. 2019). The higher elevations of the TP (>4000 m) show even stronger warming which is attributed to elevation-dependent warming (EDW) (Liu et al. 2009; Krishnan et al. 2019b).

Changes in air temperature in the Western Indian Himalaya (WIH) are quite different from other parts of India as mentioned by Dash et al (2007), Bhutiyani et al. (2007), and Dimri and Dash (2012). Dash et al (2007) indicate an increase of 0.9 °C in annual mean maximum air temperature in the region (1901–2003) with a substantial rise after 1972. However, this study also observed a decline of 1.9 °C in the annual minimum temperature in the region. Similarly, Bhutiyani et al. (2007) found an annual warming rate of 0.16 °C/decade and mentioned that the rise of annual maximum temperature is faster than the minimum temperature during 1901–2002. In the present study, a significant warming trend (at sig level of 0.10) has been found in the IHR at an average rate of 0.36 °C/decade during 1980–2020 that varied from 0.25 to 0.53 °C/decade (Fig. 1.3) indicating the regional variations. Further, the warming is particularly noteworthy in winter's minimum and maximum temperature (0.17 °C/decade). A warming trend was also found in the mean maximum temperature (1.1–2.5 °C) during winter months (Dec–Feb) in the WIH during 1975–2006 (Dimri and Dash 2012). Dimri et al. (2018a), found significant warming in minimum air

Table 1.1 Changes in air temperature and rainfall in the Indian Himalayan region based on previous studies

Region/State/Basin	Air temperature (°C/year)	Rainfall (mm/year)	Sources
Shimla (Himachal Pradesh) and Jammu and Kashmir regions (1901–2003)	$A = 9$		Dash et al. (2007)
Shimla (Himachal Pradesh) and Jammu and Kashmir (1901–2002)	$A = 0.02$		Bhutiyani et al. (2007)
Western Indian Himalaya (1876–2006)	$A = 0.11$		Bhutiyani et al. (2010)
Western Himalaya (1979–2007)	$A = 2.6$		Gautam et al. (2010)
Siachen Glacier (1984–2006)	$W = 0.012–0.0317$*		Dimri and Dash (2012)
Dehradun, Uttarakhand (1967–2007)	$A = 0.47$		Singh et al. (2013)
Uttarakhand, Himalaya (1957–2007)		$A = -0.324 - -0.389$	Singh and Mal (2014)
Upper Beas basin (1969–2010)	$A = 0.06^{**}$, $W = 0.03–0.07^{**}$, Pr-M $= 0.06^{*}$ M $= 0.06^{**}$ Ps-M $= 0.09^{**}$	$A = 4.66^{**} - -10.81^{*}$ $W = -5.97^{**}$	Rani (2017)
Sichuan Basin (WH) (1979–2010)	$A = <1$		Forsythe et al. (2017)
Over the eastern side of the HKH range (1901–2014)	$A = 2$		Ren et al. (2017)
Uttarakhand Himalaya (1983–2008)		$A = -122.6 - -328.2$* $W = -25.6$* $M = -74.6 - -223.7$* Ps-M $= -8.2 - -8.8$*	Banerjee et al. (2020)
Satluj River basin (1901–2013)		$A = -2.637 - 4.029$ W $= -0.512 - 1.305$ Pr-M $= -0.878 - 0.331$ $M = -2.326 - 1.742$ Ps-M $= -0.054 - -0.194$	Tirkey et al. (2020)
Gangtok, Sikkim Himalaya (1981–2010)	$A = 0.035$		Kumar et al. (2020)

Note Significant at 0.01^{**} level and 0.05^{*} level; *W* winter, *Pr-M* pre-monsoon, *M* monsoon, *Ps-M* post monsoon, *A* Annual

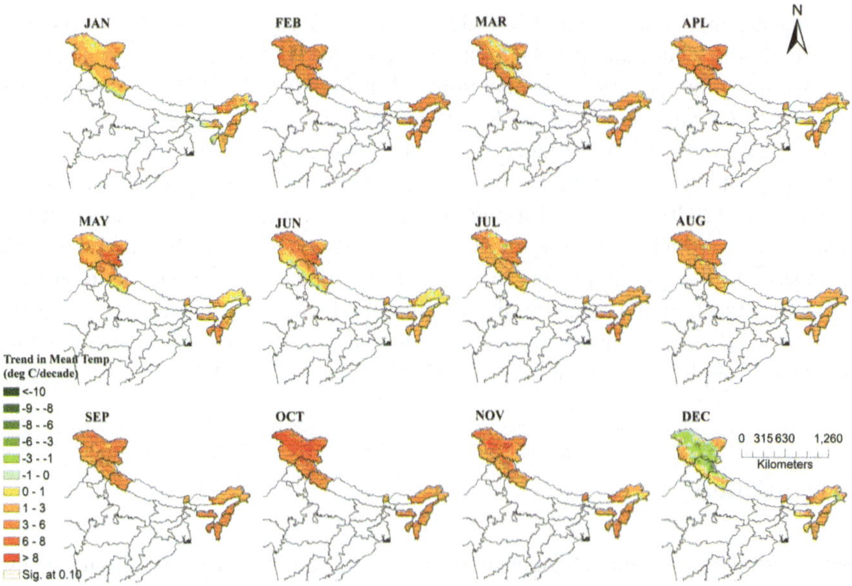

Fig. 1.3 Trend in mean air temperature over the Indian Himalayan region during 1980–2020

temperature compared to the maximum air temperature in the WIH. The highest rate of warming (up to 0.42 °C/decade) has been observed during the winter months over the WIH during 1980–2020 (Fig. 1.3). Similarly, warming trends are observed by dendroclimatic studies over the eastern Himalaya (state of Sikkim in India) in recent decades (Yadava et al. 2015; Borgaonkar et al. 2018). The warming trends in minimum air temperatures were recorded at Gangtok (0.36 °C/decade during 1961–2017) and Tadong (0.65 °C/decade from 1981 to 2010) stations, while no trend was found in the maximum temperatures at Tadong station of Sikkim in the eastern IH (1981–2010). However, a negative trend was observed for the overall assessed period (−0.27 °C/decade during 1961–2017) in Gangtok city of Sikkim (Kumar et al. 2020).

Schickhoff et al. (2016) stated that warming in most Himalayan regions is faster than the global mean trend (0.85 °C) during the period 1880–2012, with a general high in winter season temperature trends compared to other seasons. Some studies found a significant increasing/decreasing trend in summer temperature and rainfall based on tree-ring based reconstructions of the IH and Karakoram region of Himalaya during the past three to four centuries (Hughes 2001; Borgaonkar et al. 1994, 1996; Pant et al. 1998; Yadav et al. 1999). A statistically significant projection of warming (0.3–0.9 °C/decade) is reported across all seasons under representative concentration pathway (RCPs) over the IHR (Dimri et al. 2018b). Temperatures in the mountain are increasing at a faster rate than the global average (Pepin et al. 2015). Ren et al. (2017) found both rise and fall in temperature trends over the HKH region using long historical data of over a century (moderate rising trend in mean temperature during 1901–1940 and a falling trend during 1940–1970). The annual mean temperature has shown warming of more than 0.3 °C/decade in the TP region during the 1901–2014 and about 0.2 °C/decade over the eastern side of the HKH range (Ren et al. 2017). The warming rates in annual temperature is less than 0.10 °C/decade over northern India (the Sichuan Basin) and the Karakoram range during summer (Forsythe et al. 2017). You et al (2017) attributed this observed warming signal to the increasing anthropogenic activities which are contributing to greenhouse gas concentrations. Gupta et al. (2020a) observed the highest rate of warming at the highest altitude station (Kaza) (0.84 °C/decade) in RCP 8.5 over the Satluj River Basin in the WIH. Compared to the global average, the rate of warming in the Himalayas is greater, indicating the region's vulnerability to climate change and its impacts (Shrestha et al 2012). However, uncertainties exist in both observational studies and modeling projections of climate change in the HKH (Mayewski et al. 2020).

Precipitation

Several studies evaluated changes in inter-annual and intra-seasonal rainfall around the world over the several decades (Diaz et al. 1989; IPCC 1996, 2001, 2007, 2013; Tank et al. 2006; Westra et al. 2013) though these changes vary from region to region. Decreasing trends in global precipitation were found in the tropics since the 1970s whereas summer rainfall is increasing in South Asia (IPCC 2013). Rainfall has been showing both increasing and decreasing trends in different parts of the world (Westra et al. 2013). It is important to understand the changes in rainfall in India because river water mainly depends on the monsoon and the majority of the population depends on agriculture that depends on rainfall. Consequently, the variability in rainfall including the rainy days, are extensively studied in the Indian context over the several decades (Naidu et al. 1999; Dash and Hunt 2007; Guhathakurta and Rajeevan 2008; Ghosh et al. 2009; Pal and Al-Tabbaa 2010; Bhutiyani et al. 2010; Kumar et al. 2010; Kumar and Jain 2011; Pal and Al-Tabbaa 2011; Rana et al. 2012; Ratna 2012; Jain and Kumar 2012; Jhajharia et al. 2012; Dash et al. 2013; Banerjee et al. 2020; Mal et al. 2022). These studies found regional variations in the rainfall trend. However, no significant trend was found for rainfall at the Indian level (Kumar et al. 2010). Guhathakurta and Rajeevan (2008) and Bhutiyani et al. (2010) found increasing pre-monsoon precipitation and decreasing monsoon precipitation in the WIH. Consistent change in rainfall has not been reported over India (Babar and Ramesh 2013; Kumar et al. 2018). In the present study, a significant decreasing trend (−0.32 mm/decade) in rainfall was observed in the IHR during the last four decades that varied between −1.13 and 0.14 mm/decade (Fig. 1.4). The highest decline in rainfall has been found in the winter months (up to −1.10 mm/decade) over the IH (Fig. 1.4). At a seasonal scale, rainfall is showing rising and falling trends during spring and monsoon seasons, respectively, in the WIH (Bhutiyani et al 2010). Singh and Mal (2014) reported a declining trend in annual rainfall in the high altitudes during 1957–2007 in the state of Uttarakhand (located in the WIH). However, the monsoon rains increased in the low altitudes, while winter rainfall show mixed trends. Summer precipitation show rather

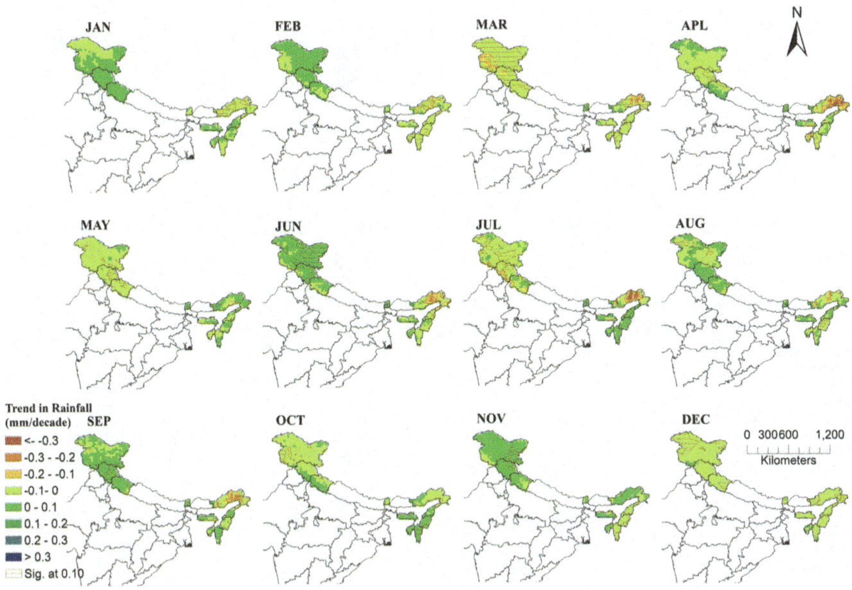

Fig. 1.4 Trend in rainfall over the Indian Himalayan region during 1980–2020

decreasing trends in the Himalayas (Schickhoff et al. 2016). In contrast, parts of the Karakoram Himalayas have shown a rising trend in wintertime frozen precipitation in the recent decades, which is termed as the "Karakoram Anomaly" (Kapnick et al. 2014; Kääb et al. 2015; Krishnan et al. 2019a).

A similar trend was also reported over the Karakoram region of northwest Himalayas. However, a falling trend in precipitation was found over Central Himalayas (CH) (Cannon et al. 2015; Madhura et al. 2015; Krishnan et al. 2019a). Zhan et al. (2017) observed a significant change in the intensity of light precipitation in the northern Hindu-Kush and central India. Gupta et al. (2020a) found the highest rate of decline in precipitation (−6.362 mm/year) at low altitude station (Kasol) (under RCP 8.5 scenarios) in the Satluj River basin in the WH. However, the same study found an increasing trend in precipitation under RCP 2.6. There is a need to be looked more comprehensively at such changes under different RCPs in different elevations zones using the orographic processes. In the present study, a significant decreasing trend (at sig level of 0.10) was observed in total cloud cover over the IH at a mean rate of −0.0004%/decade since 1980 that varied from −0.03 to 0.03%/decade (Fig. 1.5). All the winter months (Dec-Mar) show a significant decline in total cloud cover over the region (Fig. 1.5). There is a need to understand the cause and effect relationship of rainfall and cloud cover in detail over the region using advanced spatial technology.

Land Surface Temperature (LST) and evapotranspiration (ET) changes are also observed at a basin scale. A decline in annual mean clear-sky daytime LST at the rate −0.17 °C $year^{-1}$ in the Ganga basin during the period 2001–2019 (Mal et al.

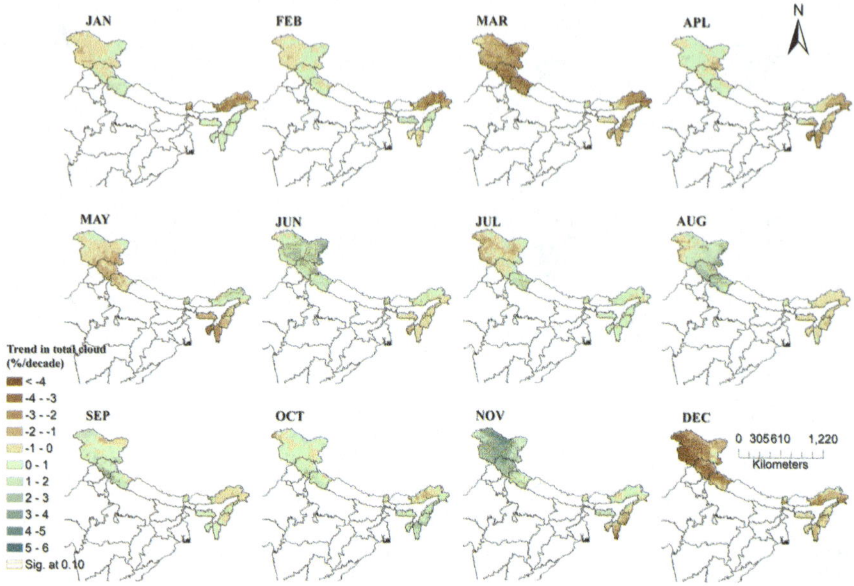

Fig. 1.5 Trend in total cloud cover over the Indian Himalayan region during 1980–2020

2021a). The study further shows warming in LST during monsoon (average rate of 0.26 °C/decade). However, they have shown the highest declining rate in LST during winter (−0.78 °C/decade) in the basin. Elevation-wise, the study observed warming trends in LST in post monsoon (0.03–0.7 °C/decade) in more than 3000 m AMSL indicating EDW in the study area (Mal et al. 2021a). Rising air temperature and LST can lead to higher ET in any area. The global surface temperature was 1.09 °C higher in the period 2011–2020 compared to the period 1850–1900 (IPCC 2021). A significant declining trend (at sig level of 0.10) in mean ET was observed over the IH with a mean rate of −0.002 mm/decade over the last 40 years that varied from −0.12 to 0.06 mm/decade (Fig. 1.6) suggesting the need to understand their relationship and impact at different spatial scales.

Impacts of Climate Change

As studies have shown warming in the IH, their impacts on different aspects such as climate extremes, hydrology, hazardous events, vegetation, etc. are also increasing with time (Fig. 1.7). Sign of rapid warming has been observed in the HKH during the global warming period, which is also influencing the climate extremes conditions affecting the hydrological cycles in the region (You et al. 2017). Another study shows a significant decline in the frequency of extreme cold events and an increase in the frequency of extreme warm events over the entire region of the HKH during the study

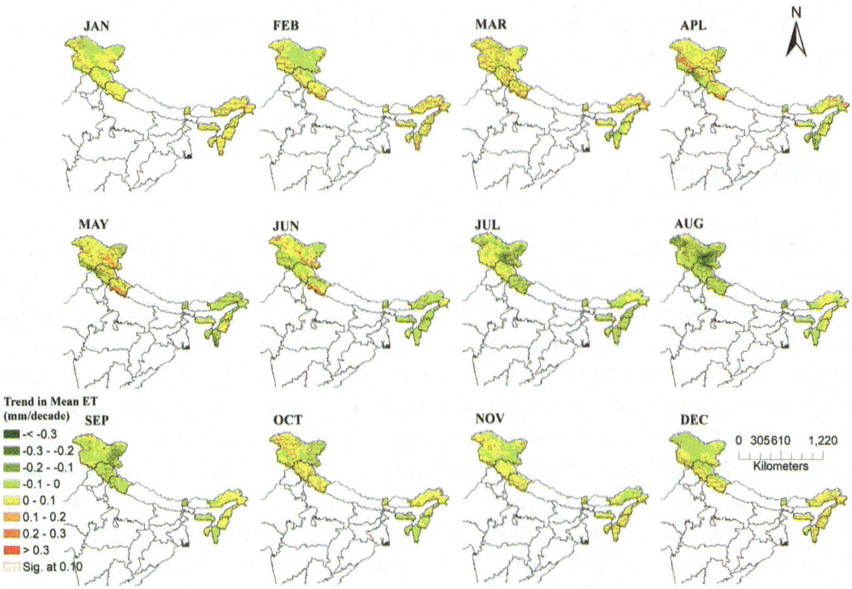

Fig. 1.6 Trend in mean evapotranspiration (ET) over the Indian Himalayan region during 1980–2020

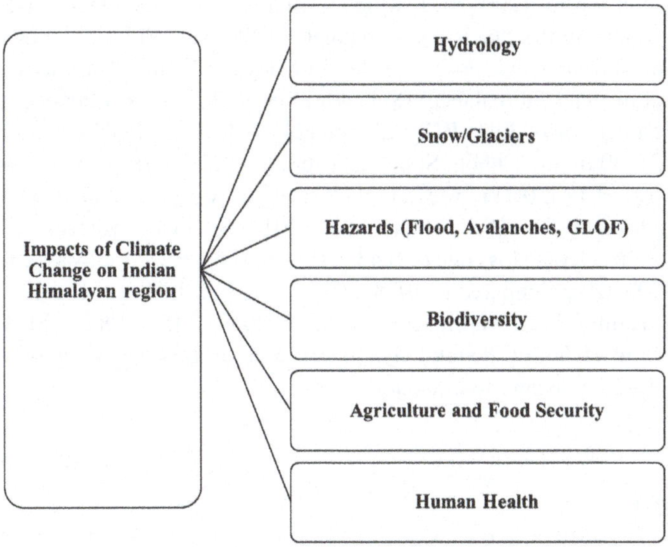

Fig. 1.7 Possible impacts of climate change on the Indian Himalayan region

period of 1961–2015 (Sun et al. 2017). Succeeding sections will cover the impact of climate change on different aspects of the IH.

Snow/Glaciers

Several studies have also been done on the IH to analyze the changes in the snow cover area (SCA) by using remote sensing data (Jain et al. 2001; Kulkarni et al. 2002; Negi et al. 2009; Kulkarni 2010; Gurung et al. 2011; Rani 2018; Banerjee et al. 2021) because about 50% of the total glaciated area in the world is located in the central Asian mountains and its largest part drains into the Indian sub-continent. They found evidence of a decline in SCA in the IH, though the declining rate varies over time and space. The warming trend in the upper Beas river basin is negatively influencing the SCA during 2000–2010 (winter =−0.247%/year; pre-monsoon = −0.525%/year) (Rani and Sreekesh 2016). A significant decline in SCA is observed in the pre-monsoon season, resulting in the early melting of snow during the season (Rani 2014). Sood et al. (2020) have shown a shift of one month of snow accumulation and snowmelt during the SCA variability since the past decade over the IH including the Karakoram mountain ranges.

According to Kulkarni et al. (2005), a retreat of 578 m has been estimated in the Pārbati glacier during 1990–2001, which is almost 52 m/year. On average, negative trends are found in the glacier mass budgets in the past 50 years, with a regional disparity in glacier's mass losses in the Himalayas (distinct mass losses), and in the Karakoram (close to balance) (Schickhoff et al. 2016). Engelhardt et al. (2017) estimated a decrease of 35% (RCP 4.5 scenario) to 70% (RCP 8.5 scenario) by 2099 relative to 2000 on the Chhota Shigri Glacier area (WIH). Its average annual mass balance was −0.4 (± 0.3) m w.e. a (−1) (meters of water equivalent per year) over the study period (1951–2099). Maurer et al. (2019) found a consistent ice loss and a doubling of the average loss rate (-0.43 ± 0.14 m w.e. year^{-1}) across the Himalayas during 2000–2016, compared to 1975–2000 (-0.22 ± 0.13 m w.e. year^{-1}). This is attributed to atmospheric warming (Maurer et al. 2019). Majeed et al. (2014) reported a loss of about 45.6% of the total Gya glacier area of the Ladakh region during the period 1969–2019 using satellite data.

Hydrology

Changes in precipitation and snow/glacier will indeed influence the hydrology of the region. Singh and Bengtsson (2005) studied the impact of a warmer climate on rainfall fed, snow-fed, and glacier-fed river basins of HP and found that warmer climate would result in the reduction of water availability in the snow-fed basins due to a compound effect of an increase in evaporation and decrease in the melt (due to availability of the lesser amount of snow in the basin) whereas the same from the glacier-fed basin has increased. Pārbati River shows a significant decline in annual runoff during 1980–2005. The rate of decline was found to be about −64.5 cusecs/decade (winter = 3.7 cusecs/decade, spring = −3.0 cusecs/decade, monsoon = −27.7 cusec/decade and autumn = −30.0 cusecs/decade) (Rani 2014). Rani et al (2019) found fluctuations in the potential evapotranspiration (PET) in the upper Beas basin, situated in the WIH, under future climate change scenarios (by the mid-twenty-first century) due to rising air temperature. Such changes would influence the basin hydrology because it is more sensitive to climate change. Consequently, the mean annual discharge would increase by 0.31 and 9.65% by 2050 (Rani and Sreekesh 2019, 2020).

Singh et al. (2016), highlighted that changes are more evident in glacio-hydrological proxies (glaciers, glacier mass balance, and streamflow in downstream areas) for changing climate, in recent decades over the IHR. Engelhardt et al. (2017) found that simulated average annual runoff does not differ substantially in the catchment area containing the Chhota Shigri Glacier (Western Himalaya) for the period 1951–2099. However, it may shift towards earlier snowmelt onset in 2040 that would increase the runoff in summer months (May and June), and a reduction in glacier melt might decrease runoff in monsoon months (August and September). Tiwari, Kar, and Bhatla (2018) examined the mid-twenty-first century climate projections over the WH under two scenarios (RCP 4.5 and RCP 8.5) in the Satluj River basin and found the impact of warming on streamflow in the basin by mid-century. They estimated that total annual discharge from Satluj will be less especially in peak discharge season (JJAS) in the future concerning the present climate. Jasrotia et al. (2021) found strong influences of precipitation projections on runoff of Jhelum catchment in the WH that results in an increase in runoff gradually from 2020 to 2080, then decreases afterward. Sen and Kansal (2019) revealed that the communities believe that climate change was responsible for worsening the spatio-temporal inequities in the Himalayas because it changes precipitation and temperature conditions. They also highlighted changes in the availability of seasonal water due to the shift in melting. These problems can be solved by several strategies in these areas by involving all stakeholders starting from the local communities to the government.

Geo-Environmental Hazards

The Indus, Ganga, and Brahmaputra basins are at risk of enhanced flooding in the future due to the absence of added adaptation and disaster risk mitigation measures (Lutz et al. 2014). As per CWC (2018), the total flood-affected area during 1953–2016 is 194,620 km^2 that is distributed in JK (10.1%), HP (47.8%), UK (0.4%), SK (9.4%), AP (7.4%), Meghalaya (1.4%), Nagaland (0.1%), Manipur (12.1%), Mizoram (3.8%), and Tripura (7.6%) states of the IH. Dimri et al. (2016) pointed out that extreme rainfall situations in the recent decade may lead to various flash flooding in and around the Himalayan region. Dimri et al. (2017) advocated that most of the cloudbursts in and around the southern rim of the IH are occurring in elevation zone ranging from 1000 to 2500 m AMSL. These cloudbursts are often associated with flash floods that cause damage to public and private properties. A 50-year return flood occurred in the Kashmir valley on September 10, 2014, due to the combined effects of low-pressure systems (originated from Bay of Bengal and Saurashtra and Kutch regions) and mid-latitude westerlies and affected 488.2 km^2 area of the valley (Kumar and Acharya 2016; Kumar 2016). An ongoing encroachment of high-value land use (built-up area and arable land) on the floodplain and channel belts of the Jhelum River is one of the main factors for increasing flood-induced damage (Kumar 2016). For example, a massive flash flooding in the Rishiganga and Dhauliganga River basins occurred on February 07, 2021, and destroyed Rishiganga and Tapovan hydroelectricity projects (Sain et al. 2021). Dimri et al. (2021) found extreme flood as most frequent natural disaster in the IHR.

Permafrost (permanently frozen ground) in the IH is highly sensitive to climate change-induced rise in temperature. The increase in temperature in IH is almost double compared to the global average (Ali et al. 2018). Such rise in temperature is causing permafrost thawing which in turn increases the instability in the steep rocky slopes of the IH (Ali et al. 2018). Landslides in the IH region is mainly caused by seismic activities, permafrost degradation, extreme rainfall (cloud bursts), extension and expansion of road networks, dams, and construction of tunnels (Dikshit et al. 2020).

A total of 35 lakes were mapped in Nanda Devi Biosphere Reserve (NDBR) of IH which has risen in damage potential (Mal and Singh 2014). A total of the 362 glacial lakes were identified in Uttarakhand Himalaya and eight glacial lakes are critical considering the outburst potential (Raj and Kumar 2016). Some of them are growing due to the recession of the glaciers. Prakash and Nagarajan (2017) found an increase of 47% in all glacial lakes' areas (size >5000 m^2) during 2002–2014, (an increase of 57% in moraine-dammed lakes) in the Chandra–Bhaga basin of the WIH using multi-temporal satellite and modeling. The eight lakes out of 16 moraine-dammed lakes were classified as very high and high outburst susceptibility using the analytic hierarchy process (AHP), emphasizing the need for further investigation in detail (Prakash and Nagarajan 2017). In 2015, a total of 4950 glacial lakes covering a total area of 455.3 ± 72.7 km^2, (between 4000 and 5700 m AMSL) were found in the Himalayas (Nie et al. 2017). An expansion of approximately 14.1% was observed in Himalayan glacial lakes during 1990–2015 particularly in the southern slopes of the CH. This may be primarily attributed to increasing glacier meltwater by global warming (Nie et al. 2017).

Sattar, Goswami, and Kulkarni (2019) illustrated the worst-case GLOF scenario in the South Lhonak lake of Sikkim, during an overtopping failure of the moraine. This can produce a peak flood of 6064.6 $m^3\ s^{-1}$ that would release a total water volume of 25.7×10^6 m^3. Veh et al. (2020) found that the 100-y outburst flood "has an average volume of $33.5^{+3.7}/_{-3.7} \times 10^6$ m^3 with a peak discharge of $15{,}600^{+2000}/_{-1800}$ $m^3\ s^{-1}$. The estimated 100-year return period GLOF discharge ($\sim$14,500 $m^3\ s^{-1}$) is more than three times that of the adjacent Nyainqentanglha Mountains, indicating a magnitude higher than in the HKH". Projections of future hazards of meteorological floods need to take into consideration for the extreme runoffs during lake outbursts, because of an increasing human population, infrastructure, and hydropower projects in the Himalayan region (Veh, Korup, and Walz 2020). Dubey and Goyal (2020) identified 329 glacial lakes (area >0.05 km^2) with a total of 23 lakes at very high risk and 50 at high risk in the IH. Mal et al. (2020) estimated a total of 1532 glacial lakes, covering an area of 93.7 km^2 in AP of eastern Himalayas.

Majeed et al. (2014) estimated a peak discharge of 470 m^3 s^{-1} by 25% of the Gya Glacier lake that may lead to inundating an area of ~4 km^2 around Gya village. However, the study also indicated that breaching of the terminal moraine. It would be 5.5 times larger than the 2014 GLOF. Mal et al. (2021 b) estimated the danger level of GLOF and demonstrated that potentially, JK is the most threatened region in terms of a total of 556 lakes at very high and high danger, followed by AP ($n = 388$) and SK ($n = 219$). In terms of sector-wise, JK faces the greatest GLOF threat to sectors like roads and population, whereas the threat to agriculture and hydropower generation is highest in AP and SK, respectively. There is a need to do local investigation and potential risk reduction measures of lakes in SK (13 lakes), HP (5 lakes), JK (4 lakes), UK (2 lakes), and AP (1 lake) in the IH (Mal et al. 2021b).

Biodiversity

Singh and Pusalkar (2020) showed that the angiosperms are represented by ca. 8700 taxa in the IHR, followed by bryophytes (1955 taxa), pteridophytes (766 taxa), and gymnosperms (51 taxa). More than 75% of families of Ranunculaceae, Brassicaceae, Rosaceae, Asteraceae, and Orchidaceae have their Indian taxa represented in the region. About 12% of the Himalayan angiosperms are in endemics, and threatened/vulnerable. India accounted for 6.45% of its recorded faunal species on its 2.20% of the world's area (Venkataraman et al. 2020). It covers only two-thirds of the total area of the country and the remote islands with such ecosystems are still unexplored. Chandra et al. (2018) mentioned two biogeographic zones including seven biotic provinces in the IH namely: (i). Trans Himalayas (Ladakh Mountains, TP, SK: 1), and (ii) Himalayas (North-West, West, Central, and East). According to them, the IH has about 27.6% of the total vertebrate diversity of the country (280 species mammals, 940 birds, 316 fishes, 200 reptiles, and 80 amphibians). However, about 133 known species of vertebrates from the IH region are listed as threatened under different categories in the International Union for Conservation of Nature (IUCN) and natural resources Red data book. There is a need of specific attention for understanding the effect of various environmental pressures such as climate change, habitat disintegration, and deforestation on biodiversity.

The elevation range shifts of species, intense recruitment of tree species in tree line ecotones, and shifts in phenology biotic responses towards the current climate change are resulting in modified structure, composition, and functioning of Himalayan ecosystems (Schickhoff et al. 2016). Tewari, Verma, and von Gadow (2017) suggested the need to examine the climate change impacts on insects and pathogens in the Himalayas to develop an effective strategy for their adaptation and mitigation. Hamid et al. (2019) attempted to understand the niche dynamics of B. utilis, and the shifting in its climatic niche in future scenarios (by 2050). The study highlighted that the species environmental niche will not remain identical under future climate conditions. Vegetation distribution (grasslands and tropical deciduous forests) in the Kashmir Himalayas shall altogether vanish from the region by 2085, whereas other vegetation (savannah, boreal evergreen forest, temperate evergreen broadleaf forest, and the mixed forest types) shall colonize the areas under polar desert/rock/ice. Some studies suggested a substantial reduction in area under permanent snow and ice by the end of the century that may result in a serious impact on agriculture productivity, streamflow, and biodiversity (Rashid et al. 2015; Rashid and Romshoo 2020). It will be affecting the population's livelihoods and food security in the region. Kumar et al. (2021) examined the vulnerability of forests in the WIH region (JK, HP, and UK) and found that alpine forest of higher altitudes is less vulnerable than lower sub-alpine forests. Climate change has differential impacts on different flora and fauna species, suggesting the need for detailed investigation on this.

Agriculture and Food Security

Agriculture provides direct or indirect livelihood to about 70% of the Hindu-Kush Himalayas which is a significant contributor to the national economy (Tiwari 2000). Changing climate parameters such as temperature extremes, rainfall pattern, snowfall reduction, snow cover area depletion, reduced runoff in rivers are putting pressure on the available resources which are directly or indirectly affecting agricultural production and influencing the food security of the region. Immerzeel et al. (2010) concluded that the Himalayan mountains are threatened by climate change, but the effects of climate change on water availability and food security in Asia differ substantially among basins and cannot be generalized. Agriculture in the Himalayan region is vulnerable to these climate changes. Researchers expected a serious rise in temperature in the coming years that can be a serious issue to deal with the emerging food requirements of rising population of the region. Tulachan (2001) pointed out changes in the production of food grain crops, horticultural and cash crops, and livestock in the five HKH countries, though there are several causes of these changes. Hussain et al. (2016) pointed to low agricultural production and income in Eastern Brahmaputra (EB) (India) in almost all cash and staple crops, which is resulting in very low farm income in the HKH region, EB (India).

Tiwari and Joshi (2015) pointed the out-migration of a huge adult male population (specifically an overall increase of 25–36% during 2001–2013) from the mountain for better livelihoods options because of the negative impact of climate change on agriculture in the Himalayas. They further highlight other issues of recurrent crop failures, declining irrigation potential (25%), agricultural productivity (26%), and rural livelihoods loss (34%) in traditional rural sectors due to climate change in the Himalayas. Shukla et al. (2019a) stated that, to reflect the different and numerous adaptation demands and restrictions of farming families in the Himalayan area, proponents of successful adaptation programmes in the Himalayan region must be aware of the subtleties within agricultural communities.

Hussain et al. (2016) highlighted new strategies in farming practices adopted by farmers to mitigate the climate change impact on agricultural production in several ways such as introducing new resilient crops, replacing water consuming crops and certain livestock, which are vulnerable to scarcity of water and fodder in the Indian part of the EB. The study further suggested some strategies in agricultural practices to cope up with climate change:

> "Governments need to establish separate food security policies for mountains and plains because mountains are different from plains in terms of nature, type, and magnitude of vulnerabilities.
>
> Government and non-government experts in HKH countries need to identify the specific zones within the sub-basins with higher agro-ecological potential for specific high-value crops such as fruits, nuts, vegetables, tea, tobacco, and other cash crops.
>
> Areas having less agro-ecological potential and that are highly vulnerable to hazards may be encouraged not to pursue agricultural activities.
>
> Governments should encourage private investment in production and post-harvest facilities".

There is a need of involving the local farming communities' knowledge for management and sustainable use of bioresources in rural areas to cope up with climate change (Negi et al. 2017). Shukla et al. (2019b) examined farmer perceptions of climate change and emphasized the need of applying flexible adaptation options among the farmers to avoid inequalities in fulfillment of the needs of the diverse farming communities in UK of the WIH. Ogra et al. (2020) observed that local community participation is lacking in effectively adopting or planning to cope up with climate/agro-ecological change because of poverty and other structural barriers in the Himalayan region.

Human Health

Climate change can affect human health directly (e.g. death/injury in floods, cloud burst, storms, and thermal stress) and indirectly through disease vectors (e.g. mosquitoes), water-borne pathogens, air quality, water quality, and food quantity/quality. Majra and Gur (2009) mentioned the threats to public health security by present climate change through different ways including disasters related to extreme weather to vector-borne diseases as dengue and malaria) in India. The study pointed out that any further increase (as projected) in these disasters may cripple the already insufficient public health setup in the country. World Health Organization (2006) reported some specific human health risks due to climate variability/change in the Himalayan mountain regions and proposes some strategies for integration of health with relevant sectors in the region. Ebi et al. (2007) identified the climate change-related health risks (include the spread of vector-borne diseases, diarrheal diseases) and vulnerable populations in the HKH. A report of the International Centre for Integrated Mountain Development (ICIMOD) mentioned the need for suitable food and nutrition, avoidance of disease, hygiene, clean and safe drinking water, clean air, and energy for the well-being of the population in the time of climate change (Chettri et al. 2009).

Policies/Plans on Climate Change and Issues of Sustainability

Barua et al. (2014) suggested that it is essential to intervene in ways to deal with the multidimensional poverty in the region. It will enhance the community's potential to adapt to current as well as future climate risk because rural communities are facing several issues other than climate change such as lack of education facilities, health services, livelihood opportunities, resources, etc. These issues constraints their ability to adapt to the climate change impact. The study has summarized some international initiatives and issues of sustainability (Mal et al. 2018). According to Huggel et al. (2020), the construction of future scenarios with a diverse range of stakeholders is one of the concepts of knowledge co-production as an approach to an inclusive and sustainable adaptation process.

Gupta et al. (2020b) calculated the socio-environmental vulnerability in different altitude zones in the Garhwal Indian Himalayas. Their results indicated that communities in the middle and high altitude zones (1000–2000 m AMSL) were more vulnerable than those in the low and very high zones. Negi et al. (2021) mentioned that the observations of local communities on climate change can be utilized to develop strategies in mitigation and adaptation in the Himalayan region. Vulnerability assessment of the Himalayan region is one of the most important strategies to plan policies for the region to mitigate climate change.

There are several initiatives taken in the past for sustainable development of the Himalayan region such as Hill Area Development Programme (HADP), Wildlife Protection Act (WPA), Project Tiger, Forest Conservation Act (FCA), The National Forest Policy (NFP) 1988, The Environment (Protection) Act, The Biological Diversity Act (BDA) 2002, National Environment Policy 2006 (NEP), Forest Rights Act 2006 (FRA), National Action Plan on Climate Change 2008, Integrated Watershed Development (IWD), National Biological Diversity Action Plan (NBDAP), Indian Network for Climate Change Assessment (INCCA), National Mission For Sustaining The Himalayan Ecosystem (NMSHE), National Water Policy, and Indian Himalayas Climate Adaptation Programme (IHCAP). Badola et al. (2015) suggested the need for sector-wise coordination of a mixed governance approach to sustain ecosystem services in the region. The impact of climate change is not uniform in the IH and there are inter and intra-regional variations. Further investigation is needed in this regard. The present work is suggesting the establishment of more meteorological stations for regular and wide data collection as the availability of climate data is limited to a few stations. Available data should be provided to researchers at an affordable price. There is a need to encourage coordinated and interdisciplinary research on climate change/variability and their effects on the availability of resources at varying levels for effective planning and management of the resources.

Scope of the Book

The book aims to provide a detailed overview of the scientific literature on climate change, its impacts, and sustainability issues over the IHR. It provides a platform for readers and researchers to identify the areas that need further research. It covers the physical and anthropogenic aspects of climate change. This book has a total of 16 chapters including the introduction. Chapter 2 covers the observed changes in the air temperature, rainfall, and discharge in a Himalayan watershed. Chapter 3 presents a review of the available information of tree-ring-based streamflow records and flood events from some of the Himalayan rivers. Chapter 4 discusses the paleoclimate studies considering the biotic (pollen-spore) and abiotic proxies through sedimentary and speleothem archives to understand the role of Indian Summer Monsoon (ISM), Western Disturbances (WD), and East Asian Summer Monsoon (EASM) in deciding the climatic fluctuations over the Himalayan and Tibetan highlands. Chapter 5 focuses on the recession of a WIH glacier. Chapter 6 discusses the previously established glacial stages within the Gangotri Valley (WIH) concerning climatic changes during Holocene. Chapter 7 aimed to review the previous works on the structural setting of the basin with the main focus on the flood hazards in the Jhelum River. Chapter 8 critically reviews the glaciological studies and examines the scientific gaps and challenges in controlling the research and management of glaciers and glacial lakes in Sikkim Himalaya.

Chapter 9 assessed the trends in the observed discharges of the Satluj River in the WIH over five decades. Chapters 10 and 11 discuss the impacts of climate change on the forest and wildlife of the IHR. Chapter 12 presents the estimation of agroforestry area and carbon storage potential in the Garhwal Himalayas. Chapter 13 discusses the causes and consequences of continually varying climate conditions on the behavior and survival of the species in the IHR. Chapter 14 lessons learned from the COVID 19 lockdown to improve the air quality in north-WIH and how it can be used as a strategy to adapt to climate change impacts in the region. Chapter 15 mentions the methodological approaches to understand the people's perception of climate change. Chapter 16 provides a summary of various actions taken by the Indian Government towards controlling and adapting to climate change in the Himalayan region.

Acknowledgements Authors are thankful to scientific members of the European Centre for Medium-Range Weather Forecasts (ECMWF) and Copernicus Climate Change Service (C3S) for providing ERA 5 reanalysis data. The authors also acknowledge the Central Water Commission (CWC) of India for providing total flood-affected areas of the states of the Indian Himalayas.

References

Ali SN, Quamar MF, Phartiyal B, Sharma A (2018) Need for permafrost researches in Indian Himalaya. J Clim Change 4:33–36. https://doi.org/10.3233/JCC-180004

Arora M, Goel N, Singh P (2005) Evaluation of temperature trends over India. Hydrol Sci J 50(1):81–93

Assembly G (2015) Sustainable development goals. SDGs Transform Our World, 2030

Babar SF, Ramesh H (2013) Analysis of southwest monsoon rainfall trend using statistical techniques over Nethravathi basin. Int J Civ Eng Technol 2(3):130–136

Badola R, Hussain SA, Dobriyal P, Barthwal S (2015) Assessing the effectiveness of policies in sustaining and promoting ecosystem services in the Indian Himalayas. Int J Biodivers Sci Ecosyst Serv Manag 11(3):216–224. https://doi.org/10.1080/21513732.2015.1030694

Banerjee A, Chen R, Meadows M, Singh RB, Mal S, Sengupta D (2020) An analysis of long-term rainfall trends and variability in the Uttarakhand Himalaya using google earth engine. Remote Sens 12(4):709. https://doi.org/10.3390/rs12040709

Banerjee A, Chen R, Meadows ME, Sengupta D, Pathak S, Xia Z, Mal S (2021) Tracking 21st century climate dynamics of the third pole: an analysis of topo-climate impacts on snow cover in the central Himalaya using Google Earth Engine. Int J Appl Earth Obs Geoinf 103:102490

Barua A, Katyaini S, Mili B et al (2014) Climate change and poverty: building resilience of rural mountain communities in South Sikkim, Eastern Himalaya, India. Reg Environ Change 14:267–280. https://doi.org/10.1007/s10113-013-0471-1

Bhatt A, Joshi G, Joshi G (2013) Impact of climate changes on catchment hydrology and rainfall-runoff correlations in Karjan Reservoir basin, Gujarat. In: 2012 international SWAT conference proceedings. Indian Institute of Technology, Delhi, pp 118–130. Available via http://swat.tamu.edu/media/69009/swat-proceedings-2012-india.pdf. Accessed 20 May 2020

Bhutiyani M, Kale V, Pawar N (2007) Long-term trends in maximum, minimum and mean annual air temperatures across the north western Himalaya during the twentieth century. Clim Change 85(1–2):159–177

Bhutiyani M, Kale V, Pawar N (2010) Climate change and the precipitation variations in the northwestern Himalaya: 1866–2006. Int J Climatol 30(4):535–548

Bolch T, Shea JM, Liu S, Azam FM, Gao Y, Gruber S, Immerzeel WW, Kulkarni A, Li H, Tahir AA, Zhang G (2019) Status and change of the cryosphere in the extended hindu kush himalaya region. In: Wester P, Mishra A, Mukherji A, Shrestha AB (eds) The hindu kush himalaya assessment: mountains, climate change, sustainability and people, pp 209–255. Springer. https://doi.org/10.1007/978-3-319-92288-1_7

Borgaonkar HP, Gandhi N, Somaru R, Krishnan R (2018) Tree-ring reconstruction of late summer temperatures in northern Sikkim (eastern Himalayas). Palaeogeogr Palaeoclimatol Palaeoecol 504(1):125–135

Borgaonkar HP, Pant GB, Rupa KK (1994) Dendroclimatic reconstruction of summer precipitation at Srinagar, Kashmir, India since the late 18th century. Holocene 4(3):299–306

Borgaonkar HP, Pant GB, Rupa KK (1996) Ring-width variations in Cedrusdeodara and its climatic response over the western Himalaya. Int J Climatol 16(12):1409–1422

Boyer C, Chaumont D, Cha I (2010) Impact of climate change on the hydrology of St Lawrence Tributaries. J Hydrol 384:65–83

Cannon F, Carvalho LMV, Jones C, Bookhagen B (2015) Multi-annual variations in winter westerly disturbance activity affecting the Himalaya. Clim Dyn 44:441–455. https://doi.org/10.1007/s00382-014-2248-8

Census of India (2011) Administrative Atlas of India, Office of the Registrar General & Census Commissioner, India Ministry of Home Affairs, Government of India, Available via. http://www.censusindia.gov. https://censusindia.gov.in/2011census/maps/administrative_maps/Final%20Atlas%20India%202011.pdf. Accessed 20 April 2021

Chandra K, Gupta D, Gopi KC, Tripathy B, Kumar V (2018) Faunal diversity of Indian Himalaya. Director, Zoological Survey of India, Kolkata, pp 1–872

Chettri N, Eriksson M, Jing F, Mool PK, Sharma E, Shrestha AB, Tse-ring K, (2009) Climate change impacts and vulnerability in the Eastern Himalayas. International Centre for Integrated Mountain Development (ICIMOD)
CWC (2018) State-wise flood damage statistics. Central Water Commission, Govt. of India, New Delhi
Dash S, Hunt J (2007) Variability of climate change in India. Curr Sci 93(6):782–788
Dash S, Jenamani R, Kalsi S, Panda S (2007) Some evidence of climate change in twentieth-century India. Clim Change 859(3–4):299–321
Dash S, Saraswat V, Panda S, Sharma N (2013) A study of changes in rainfall and temperature patterns at four cities and corresponding meteorological subdivisions over coastal regions of India. Glob Planet Change 108:175–194
Diaz H, Bradley R, Eischeid J (1989) Precipitation fluctuations over global land areas since the late 1800s. J Geophys Res 94(D1):1195–1210
Dikshit A, Sarkar R, Pradhan B, Segoni S, Alamri AM (2020) Rainfall induced landslide studies in Indian Himalayan region: a critical review. Appl Sci 10:2466. https://doi.org/10.3390/app10072466
Dimri AP, Chevuturi A, Niyogi D, Thayyen RJ, Ray K et al (2017) Cloudbursts in Indian Himalayas: a review. Earth Sci Rev 168:1–23. https://doi.org/10.1016/j.earscirev.2017.03.006
Dimri AP, Kumar D, Choudhary A, Maharana P (2018a) Future changes over the Himalayas: maximum and minimum temperature. Glob Planet Change 162:212–234
Dimri AP, Kumar D, Choudhary A, Maharana P (2018b) Future changes over the Himalayas: mean temperature. Glob Planet Chang 162:235–251
Dimri AP, Thayyen RJ, Kibler K, Stanton A, Jain SK, Tullos D, Singh VP (2016) A review of atmospheric and land surface processes with emphasis on flood generation in the Southern Himalayan rivers. Sci Total Environ 15(556):98–115. https://doi.org/10.1016/j.scitotenv.2016.02.206
Dimri AP, Dash S (2012) Wintertime climatic trends in the Western Himalayas. Clim Change 111(3–4):775–800
Dimri AP, Allen S, Huggel C, Mal S, Ballesteros-Canovas JA, Rohrer M, Shukla A, Tiwari P, Maharana P, Bolch T, Thayyen RJ (2021) Climate change, cryosphere and impacts in the Indian Himalayan Region. Curr Sci 120(5). https://doi.org/10.18520/cs/v120/i5/774-790
Dubey S, Goyal MK (2020) Glacial lake outburst flood (GLOF) hazard, downstream impact, and risk over the Indian Himalayas. Water Resour Res 56(4):e2019WR026533. https://doi.org/10.1029/2019wr026533
Duhan D, Pandey A, Gahalaut K, Pandey R (2013) Spatial and temporal variability in maximum, minimum and mean air temperatures at Madhya Pradesh in central India. C RGeosci 345(1):3–21
Ebi KL, Woodruff R, von Hildebrand A et al (2007) Climate change-related health impacts in the Hindu Kush-Himalayas. EcoHealth 4:264–270. https://doi.org/10.1007/s10393-007-0119-z
Engelhardt M, Leclercq P, Eidhammer T, Kumar P, Landgren O, Rasmussen R (2017) Meltwater runoff in a changing climate (1951–2099) at Chhota Shigri Glacier, Western Himalaya Northern India. Ann Glaciol 58:47–58. https://doi.org/10.1017/aog.2017.13
Forsythe N, Fowler HJ, Li X-F, Blenkinsop S, Pritchard D (2017) Karakoram temperature and glacial melt driven by regional atmospheric circulation variability. Nat Clim Change 7(9). https://doi.org/10.1038/nclimate3361
Gadgil A, Dhorde A (2005) Temperature trends in twentieth century at Pune, India. Atmos Environ 39:6550–6556
Gautam R, Hsu NC, Lau KM (2010) Premonsoon aerosol characterization and radiative effects over the Indo-Gangetic Plains: implications for regional climate warming. J Geophys Res 115:1383e1392
Ghosh S, Luniya V, Gupta A (2009) Trend analysis of Indian summer monsoon rainfall at different spatial scales. Atmos Sci Lett 10:285–290
Guhathakurta P, Rajeevan M (2008) Trends in the rainfall pattern over India. Int Int J Climatol 28:1453–1469

Gupta A, Dimri AP, Thayyen R et al (2020a) Meteorological trends over Satluj River Basin in Indian Himalaya under climate change scenarios. J Earth Syst Sci 129:161. https://doi.org/10.1007/s12040-020-01424-x

Gupta A, Negi M, Nandy S, Kumar M, Singh V, Valente D, Petrosillo I, Pandey R (2020b) Mapping socio-environmental vulnerability to climate change in different altitude zones in the Indian Himalayas. Ecol Indic 109:105787

Gurung D, Kulkarni A, Giriraj A, Aung K, Shrestha B, Srinivasan J (2011) Changes in seasonal snow cover in Hindu Kush-Himalayan region. Cryosphere Discuss 5:755–777

Hamid M, Khuroo AA, Charles B et al (2019) Impact of climate change on the distribution range and niche dynamics of Himalayan birch, a typical treeline species in Himalayas. Biodivers Conserv 28:2345–2370. https://doi.org/10.1007/s10531-018-1641-8

Hersbach H, Bell B, Berrisford P, Biavati G, Horányi A, Muñoz Sabater J, Nicolas J, Peubey C, Radu R, Rozum I, Schepers D, Simmons A, Soci C, Dee D, Thépaut J-N (2018) ERA5 hourly data on pressure levels from 1979 to present. Copernicus Climate Change Service (C3S) Climate Data Store (CDS). https://doi.org/10.24381/cds.bd0915c6. Accessed 01 May 2020

Huggel C et al (2020) An integrative and joint approach to climate impacts, hydrological risks and adaptation in the Indian Himalayan Region. In: Dimri A, Bookhagen B, Stoffel M, Yasunari T (eds) Himalayan Weather and Climate and their Impact on the Environment. Springer, Cham. https://doi.org/10.1007/978-3-030-29684-1_26

Hughes MK (2001) An improved reconstruction of summer temperature at Srinagar, Kashmir since 1660 AD based on tree-ring width and maximum latewood density of Abiespindrow (Royle) Spach. Palaeobotanist 50:13–19

Hussain A, Rasul G, Mahapatra B et al (2016) Household food security in the face of climate change in the Hindu-Kush Himalayan region. Food Sec 8:921–937. https://doi.org/10.1007/s12571-016-0607-5

Immerzeel W, van Beek L, Bierkens M (2010) Climate change will affect the Asian water towers. Science 328:1382–1385

India Meteorological Department (IMD) (2019) Annual climate summary. Available via. https://imdpune.gov.in/Links/annual_summary_2019.pdf. Accessed 01 February 2021

IPCC (2018) Global Warming of 1.5°C: an IPCC special report on the impacts of global warming of 1.5°C above pre-industrial levels and related global greenhouse gas emission pathways, in the context of strengthening the global response to the threat of climate change, sustainable development, and efforts to eradicate poverty. Available via. https://www.ipcc.ch/site/assets/uploads/sites/2/2019/06/SR15_Full_Report_High_Res.pdf. Accessed 20 Jan 2021

IPCC (1996) Climate change 1995: the IPCC scientific assessment, contribution of working group I to second assessment report of intergovernmental of climate change (IPCC). Cambridge University Press, Cambridge

IPCC (2001) Climate change 2001: the scientific basis contribution of working group I to the third assessment report of the intergovernmental panel on climate change. Cambridge University Press, Cambridge

IPCC (2007) Climate change 2007: impacts, adaptation and vulnerability. In: Parry ML, Canziani OF, Palutik JP, van der Linden PJ, Hanson CE (eds) Contribution of working group II to the 4th assessment report of the intergovernmental panel on climate change. Cambridge University Press, Cambridge

IPCC (2013) Climate change 2013: the physical science basis, contribution of working group I to the 5th assessment report of the intergovernmental panel on climate change (IPCC). Cambridge University Press, Cambridge

IPCC (2019) IPCC special report on the ocean and cryosphere in a changing climate. Available via. https://www.ipcc.ch/srocc/ (SROCC) 19 Summary for Policymakers, SROCC. Accessed 20 June 2021

IPCC (2021) Summary for policymakers. In: Climate change 2021: the physical science basis. In: Masson-Delmotte V, Zhai P, Pirani A, Connors SL, Péan C, Berger S, Caud N, Chen Y, Goldfarb L, Gomis MI, Huang M, Leitzell K, Lonnoy E, Matthews JBR, Maycock TK, Waterfield T,

Yelekçi O, Yu R, Zhou B (eds) Contribution of working group I to the 6th assessment report of the intergovernmental panel on climate change. Cambridge University Press. In Press. Accessed 20 Oct 2021

Jain S, Kumar V (2012) Trend analysis of rainfall and temperature data for India. Curr Sci 102(1):37–49

Jain, S, Rathore, D, Chaudhry A (2001) Snow cover estimation in Himalayan Basins using remote sensing. A Report by the National Institute of Hydrology, Roorkee

Hydrological modeling to simulate stream flow under changing climate conditions in Jhelum catchment, western Himalaya. JHydrol 593:125887. https://doi.org/10.1016/j.jhydrol.2020.125887

Jhajharia D, Yadav B, Maske S, Chattopadhyay S, Kar A (2012) Identification of trends in rainfall, rainy days and 24 h maximum rainfall over subtropical Assam in northeast India. CR Geosci 344(1):1–13. https://doi.org/10.1016/j.crt.e.2011.11.002

Jhajharia D, Singh V (2011) Trends in temperature, diurnal temperature range and sunshine duration in Northeast India. Int J Climatol 31:1353–1367

Kääb A, Treichler D, Nuth C, Berthier E (2015) Brief communication:contending estimates of 2003–2008 glacier mass balance over the Pamir–Karakoram–Himalaya. Cryosphere 9(2):557–564. https://doi.org/10.5194/tc-9-557-2015

Kapnick SB, Delworth TL, Ashfaq M, Malyshev S, Milly PCD (2014) Snowfall less sensitive to warming in Karakoram than in Himalayas due to a unique seasonal cycle. Nat Geosci 7:834–840

Kar SC, Mali P, Routray A (2014) Impact of land surface processes on the South Asian monsoon simulations using WRF modeling system. Pure Appl Geophys 171(9):2461–2484. https://doi.org/10.1007/s00024-014-0834-7

Kendall MG (1975) Rank correlation methods. Oxford University Press, New York

Krishnan R, Sabin TP, Madhura RK, Vellore R, Mujumdar M, Sanjay J, Nayak S, Rajeevan M (2019a) Non-monsoonal precipitation response over the Western Himalayas to climate change. Clim Dyn 52:4091–4109. https://doi.org/10.1007/s00382-018-4357-2

Krishnan R et al (2019b) Unravelling climate change in the Hindu Kush Himalaya: rapid warming in the mountains and increasing extremes. In: Wester P, Mishra A, Mukherji A, Shrestha A (eds) The Hindu Kush Himalaya assessment. Springer, Cham

Kulkarni A (2010) Monitoring Himalayan cryosphere using remote sensing techniques. J Indian Inst Sci 90(4):457–469

Kulkarni A, Mathur P, Rathore B, Alex S, Thakur N, Kumar M (2002) Effect of global warming on snow ablation pattern in Himalaya. Curr Sci 83(2):120–123

Kulkarni A, Rathore B, Mahajan S, Mathur P (2005) Alarming retreat of Parbati glacier. Curr Sci 88(11):1844–1850

Kumar M, Kalra N, Sing, H, Sharma S, Rawat P, Singh R, Gupta A, Kumar P, Ravindranath NH (2021) Indicator-based vulnerability assessment of forest ecosystem in the Indian Western Himalayas: an analytical hierarchy process integrated approach. Ecol Indic 125:107568

Kumar N, Tischbein B, Beg MK (2018) Multiple trend analysis of rainfall and temperature for a monsoon-dominated catchment in India. Meteorol Atmos Phys 131(4):1019–1033

Kumar P, Sharma MC, Saini R et al (2020) Climatic variability at Gangtok and Tadong weather observatories in Sikkim, India, during 1961–2017. Sci Rep 10:15177. https://doi.org/10.1038/s41598-020-71163-y

Kumar R (2016) Flood hazard assessment of 2014 floods in Sonawari sub-district of Bandipore district (Jammu and Kashmir): an application of geoinformatics. Remote Sens Appl Soc Environ 4:188–203. https://doi.org/10.1016/j.rsase.2016.10.002

Kumar R, Acharya P (2016) Flood hazard and risk assessment of 2014 floods in Kashmir Valley: a space-based multisensor approach. Nat Hazards 84(1):437–464. https://doi.org/10.1007/s11069-016-2428-4

Kumar V, Jain S, Singh Y (2010) Analysis of long-term rainfall trends in India. Hydrolog Sci J55(4):484–496

Liu X, Cheng Z, Yan L et al (2009) Elevation dependency of recent and future minimum surface air temperature trends in the Tibetan Plateau and its surroundings. Glob Planet Change 68:164–174

Luo Y, Ficklin D, Liu X (2013) Assessment of climate change impacts on hydrology and water quality with a watershed modeling approach. Sci Total Environ 450–451:72–82

Lutz AF, Immerzeel WW, Shrestha AB, Bierkens MFP (2014) Consistent increase in High Asia's runoff due to increasing glacier melt and precipitation. Nat Clim Chang 4:587–592. https://doi.org/10.1038/nclimate2237

Madhura RK, Krishnan R, Revadekar JV, Mujumdar M, Goswami BN (2015) Changes in western disturbances over the Western Himalayas in a warming environment. Clim Dyn 44:1157–1168. https://doi.org/10.1007/s00382-014-2166-9

Majeed U, Rashid I, Sattar A, Allen S, Stoffel M, Nüsser M, Schmidt S (2014) Recession of Gya Glacier and the 2014 glacial lake outburst flood in the Trans-Himalayan region of Ladakh, India. Sci Total Environ 756:144008. https://doi.org/10.1016/j.scitotenv.2020.144008

Majra JP, Gur A (2009) Climate change and health: Why should India be concerned? Indian J Occup Environ Med 13(1):11–16. https://doi.org/10.4103/0019-5278.50717

Mal S, Singh RB (2014) Changing glacial lakes and associated outburst floods risks in Nanda Devi Biosphere Reserve, Indian Himalaya. Proc Int Assoc Hydrol Sci 364:255–260

Mal S, Kumar A, Bhambri R, Schickhoff U, Singh RB (2020) Inventory and spatial distribution of glacial lakes in Arunachal Pradesh, Eastern Himalaya, India. J Geol Soc India 96(6):609–615

Mal S, Singh RB, Huggel C, Grover A (2018) Introducing linkages between climate change, extreme events, and disaster risk reduction. In: Mal S, Singh R, Huggel C (eds) Climate change, extreme events and disaster risk reduction. Sustainable development goals series. Springer, Cham. https://doi.org/10.1007/978-3-319-56469-2_1

Mal S, Mehta M, Singh RB, Schickhoff U, Bisht MP (2019) Recession and morphological changes of the debris-covered Milam glacier in Gori Ganga valley, Central Himalaya, India, derived from satellite data. Front Environ Sci 7:42

Mal S, Rani S, Maharana P (2021a) Estimation of spatiotemporal variability in land surface temperature over the Ganga River Basin using MODIS data. Geocarto Int. https://doi.org/10.1080/10106049.2020.1869331

Mal S, Allen SK, Frey H, Huggel C, Dimri AP (2021b) Sectorwise assessment of glacial lake outburst flood danger in the Indian Himalayan region. Mt Res Dev 41(1):R1–R12. https://doi.org/10.1659/MRD-JOURNAL-D-20-00043.1

Mal S, Arora M, Banerjee A, Singh RB, Scott CA, Allen SK, Karki R (2022) Spatial variations and long-term trends (1901–2013) of rainfall across Uttarakhand Himalaya, India. In: Schickhoff U, Singh R, Mal S (eds) Mountain landscapes in transition. Sustainable development goals series. Springer, Cham. pp 163–183. https://doi.org/10.1007/978-3-030-70238-0_3

Mann HB (1945) Nonparametric tests against trend. Econometrica 13:245–259. https://doi.org/10.2307/1907187

Maurer JM, Schaefer J M, Rupper S, Corley A (2019) Acceleration of ice loss across the Himalayas over the past 40 years. Sci Adv 5: eaav7266

Mayewski PA, Perry LB, Matthews T, Birkel SD (2020) Climate change in the Hindu Kush Himalayas: basis and gaps. One Earth 3(5):551–555. https://doi.org/10.1016/j.oneear.2020.10.007

Naidu C, Rao B, Rao D (1999) Climatic trends and periodicities of annual rainfall over India. Meteorol Appl 6(4):395–404

Neeti N, Eastman R (2011) A contextual Mann-Kendall approach for the assessment of trend significance in image time series. Trans GIS 15(5):599–611

Negi VS, Thakur S, Dhyani R, Bhatt ID, Rawal RS (2021) Climate change observations of indigenous communities in the Indian Himalaya. Weather ClimSoc 13(2):245–257

Negi H, Kulkarni A, Semwal B (2009) Estimation of snow cover distribution in Beas basin, Indian Himalaya using satellite data and ground measurements. J Earth Syst Sci 118(5):525–538

Negi VS, Maikhuri RK, Pharswan D et al (2017) Climate change impact in the Western Himalaya: people's perception and adaptive strategies. J Mt Sci 14:403–416. https://doi.org/10.1007/s11629-015-3814-1

Nie Y, Sheng Y, Liu Q, Liu L, Liu S, Zhang Y, Song C (2017) A regional-scale assessment of Himalayan glacial lake changes using satellite observations from 1990 to 2015. Remote Sens Environ 189:1–13. https://doi.org/10.1016/j.rse.2016.11.008

Ogra M, Manral U, Platt R, Badola R, Butcher LM (2020) Local perceptions of change in climate and agroecosystems in the Indian Himalayas: a case study of the Kedarnath Wildlife Sanctuary (KWS) landscape, India. Appl Geogr 125:102339

Pal I, Al-Tabbaa A (2010) Long-term changes and variability of monthly extreme temperatures in India. Theor Appl Climatol 100:45–56

Pal I, Al-Tabbaa A (2011) Assessing seasonal precipitation trends in India using parametric and non-parametric statistical techniques. Theor Appl Climatol 103(1–3):1–11

Pant GB, Borgaonkar HP, Rupa Kumar K (1998) Climatic signals from tree-rings: a dendro climatic investigation of Himalayan spruce (piceasmithiana). Himalayan Geol 19(2):65–73

Pepin N et al (2015) Elevation-dependent warming in mountain regions of the world. Nat Clim Change 5:424–430

Prakash C, Nagarajan R (2017) Outburst susceptibility assessment of morainedammed lakes in Western Himalaya using an analytic hierarchy process. Earth Surf Process Landf 42:2306–2321. https://doi.org/10.1002/esp.4185

Raj KBG, Kumar KV (2016) Inventory of glacial lakes and its evolution in Uttarakhand Himalaya using time series satellite data. J Indian Soc Remote Sens 44:959–976. https://doi.org/10.1007/s12524-016-0560-y

Rani S (2014) Assessment of the influence of climate variability on the snow cover area of the upper Beas river basin. Unpublished M. Phil. dissertation, Centre for the Study of Regional Development, Jawaharlal Nehru University, New Delhi, India

Rani S (2017) Effect of climate and land cover changes on flow regime in the upper Beas River basin. Unpublished PhD thesis, Centre for the Study of Regional Development, Jawaharlal Nehru University, New Delhi, India

Rani S, Sreekesh S (2016) An analysis of pattern of changes in snow cover in the upper Beas River Basin, Western Himalaya. In: Raju N (eds) Geostatistical and geospatial approaches for the characterization of natural resources in the environment. Springer, Cham. https://doi.org/10.1007/978-3-319-18663-4_139

Rani S, Sreekesh S (2020) Flow regime changes under future climate and land cover scenarios in the Upper Beas basin of Himalaya using SWAT model. Int J Environ Stud. https://doi.org/10.1080/00207233.2020.1811574

Rani S (2018) Evaluating snow cover changing trends of the Western Indian Himalaya. Spat Inf Res 26:103–112. https://doi.org/10.1007/s41324-017-0158-7

Rani S (2015) Assessment of annual, monthly, and seasonal trends in the long term rainfall of the Garhwal Himalayas. IGI Global, 222–242. https://www.igi-global.com/chapter/assessment-of-annual-monthly-and-seasonal-trends-in-the-long-term-rainfall-of-the-garhwal-himalayas/140572

Rani S, Sreekesh S (2018) Variability of temperature and rainfall in the upper Beas basin, Western Himalayas. In: Mal S, Singh RB, Huggel C (Eds) Climate change, extreme events and disaster risk reduction. Sustainable development goals series. Springer Cham, pp 101–120

Rani S, Sreekesh S (2019) Evaluating the responses of streamflow under future climate change scenarios in a Western Indian Himalaya Watershed. Environ Process 6:155–174. https://doi.org/10.1007/s40710-019-00361-2

Rani S, Sreekesh S, Krishnan P (2019) Effect of climate change on potential evapotranspiration in the upper Beas basin of the Western Indian Himalaya. Int Arch Photogramm Remote Sens Spatial Inf Sci XLII-3/W6:51–57. https://doi.org/10.5194/isprs-archives-XLII-3-W6-51-2019

Rashid I, Romshoo SA (2020) Impact of climate change on vegetation distribution in the Kashmir Himalaya. In: Dar G, Khuroo A (eds) Biodiversity of the Himalaya: Jammu and Kashmir State. Topics in Biodiversity and Conservation, vol 18. Springer, Singapore. https://doi.org/10.1007/978-981-32-9174-4_40

Rashid I, Romshoo SA, Chaturvedi RK et al (2015) Projected climate change impacts on vegetation distribution over Kashmir Himalayas. Clim Change 132:601–613. https://doi.org/10.1007/s10584-015-1456-5

Ray LK, Goel NK, Arora M (2019) Trend analysis and change point detection of temperature over parts of India. Theor Appl Climatol 138(1–2):153–167

Ren YY, Ren GY, Sun XB, Shrestha AB, You QL, Zhan YJ et al (2017) Observed changes in surface air temperature and precipitation in the Hindu Kush Himalayan region during 1901–2014. Adv Clim Change Res 8(3). https://doi.org/10.1016/j.accre.2017.08.001

Sain K, Kumar A, Mehta M, Verma A et al (2021) A perspective on Rishiganga-Dhauliganga flash flood in the Nanda Devi Biosphere Reserve, Garhwal Himalaya, India. J Geol Soc India 97:335–338

Sattar A, Goswami A, Kulkarni AV (2019) Hydrodynamic moraine-breach modeling and outburst flood routing—a hazard assessment of the South Lhonak lake, Sikkim. Sci Total Environ 10(668):362–378. https://doi.org/10.1016/j.scitotenv.2019.02.388

Saxena R, Mathur P (2019) Recent trends in rainfall and temperature over North West India during 1871–2016. Theor Appl Climatol 135(3–4):1323–1338

Schickhoff U, Singh RB, Mal S (2016) Climate change and dynamics of glaciers and vegetation in the Himalaya: an overview. In: Singh R, Schickhoff U, Mal S (eds) Climate change, glacier response, and vegetation dynamics in the Himalaya. Springer, Cham. https://doi.org/10.1007/978-3-319-28977-9_1

Schickhoff U, Mal S (2020) Current changes in alpine ecosystems of Asia. In: Goldstein MI, DellaSala DA (eds) Encyclopedia of the world's biomes, Elsevier, pp 589–598. https://doi.org/10.1016/B978-0-12-409548-9.12399-1

Schickhoff U, Bobrowski M, Mal S, Schwab N, Singh RB (2022) The world's mountains in the *Anthropocene*. In: Schickhoff U, Singh RB, Mal S (eds) Mountain landscapes in transition. Springer, Cham. pp 1–144

Sen SM, Kansal A (2019) Achieving water security in rural Indian Himalayas: a participatory account of challenges and potential solutions. J Environ Manage 245:398–408. https://doi.org/10.1016/j.jenvman.2019.05.132

Shrestha UB, Gautam S, Bawa KS (2012) Widespread climate change in the Himalayas and associated changes in local ecosystems. PLoS ONE 7(5):e36741. https://doi.org/10.1371/journal.pone.0036741

Shugar DH, Jacquemart M, Shean D, Bhushan S, Upadhyay K, Sattar A, Schwanghart W, McBride S, de Vries MV, Mergili M, Emmer A, et al (2021) A massive rock and ice avalanche caused the 2021 disaster at Chamoli, Indian Himalaya. Sci 373(6552):300–306

Shukla R, Agarwal A, Gornott C et al (2019a) Farmer typology to understand differentiated climate change adaptation in Himalaya. Sci Rep 9:20375. https://doi.org/10.1038/s41598-019-56931-9

Shukla R, Agarwal A, Sachdeva K et al (2019b) Climate change perception: an analysis of climate change and risk perceptions among farmer types of Indian Western Himalayas. Clim Change 152(1):103–119

Singh DK, Pusalkar PK (2020) Floristic diversity of the Indian Himalaya. In: Dar GH, Khuroo AA (eds) Biodiversity of the Himalaya: Jammu and Kashmir State. Springer, Singapore

Singh O, Arya P, Chaudhary BS (2013) On rising temperature trends at Dehradun in Doon valley of Uttarakhand, India. J Earth Syst Sci 122(3):613–622

Singh RB, Mal S (2014) Trends and variability of monsoon and other rainfall seasons in Western Himalaya. India 15(3):218–226. https://doi.org/10.1002/asl2.494

Singh P, Bengtsson L (2005) Impact of warmer climate on melt and evaporation for the rainfed, snowfed and glacierfed basins in the Himalayan region. J Hydrol 300:140–154

Singh P, Kumar V, Thomas T, Arora M (2008) Basin-wide assessment of temperature trends in northwest and central India. Hydrol Sci J 53(2):421–433

Singh S, Kumar R, Bhardwaj A, Sam L, Shekhar M, Singh A, Kumar R, Gupta A (2016) Changing climate and glacio-hydrology in Indian Himalayan Region: a review. Wiley Interdiscip Rev Clim Change 7:393–410

Sood V, Singh S, Taloor AK, Prashar S, Kaur R (2020) Monitoring and mapping of snow cover variability using topographically derived NDSI model over north Indian Himalayas during the period 2008–19. Comput Geosci 8:100040

Sun XB, Ren GY, Shrestha AB, Ren YY, You QL et al (2017) Changes in extreme temperature events over the Hindu Kush Himalaya during 1961–2015. Adv Clim Chang Res 8(3):157–165. https://doi.org/10.1016/j.accre.2017.07.001

Tank A, Peterson T, Quadi D, Dorji S, Zou X, Tang H et al (2006) Changes in daily temperature and precipitation extremes in central and south Asia. J Geophys Res 111:1–8

Tewari VP, Verma RK, von Gadow K (2017) Climate change effects in the Western Himalayan ecosystems of India: evidence and strategies. For Ecosyst 4:13. https://doi.org/10.1186/s40663-017-0100-4

Theil H (1950) A rank-invariant method of linear and polynomial regression analysis, I. Proc Koninklijke Nederlandse Akademie Wetenschappen Ser Math Sci A53:386–392

Tirkey N, Parhi PK, Lohani AK, Chandniha SK (2020) Analysis of precipitation variability over Satluj Basin, Himachal Pradesh, India: 1901–2013. J Water Clim Change 12:127–135. https://doi.org/10.2166/wcc.2020.136

Tiwari PC (2000) Land-use changes in Himalaya and their impact on the plains ecosystem: need for sustainable land use. Land Use Policy 17(2):101–111

Tiwari PC, Joshi B (2015) Climate change and rural out-migration in Himalaya. Change Adapt SocioeEcol Syst 2:8–25

Tiwari S, Kar SC, Bhatla R (2018) Mid-21st century projections of hydroclimate in Western Himalayas and Satluj River basin. Glob Planet Change 161:10–27

Tulachan PM (2001) Mountain agriculture in the Hindu Kush-Himalaya: a regional comparative analysis. Mt Res Dev 21(3):260–267

Veh G, Korup O, Walz A (2020) Hazard from Himalayan glacier lake outburst floods. Proc Natl Acad Sci USA 117(2):907–912. https://doi.org/10.1073/pnas.1914898117; Venkataraman K, Sharma G, Banerjee D (2020) Faunal diversity of India. In: Dar G, Khuroo A (eds) Biodiversity of the Himalaya: Jammu and Kashmir State. Topics in Biodiversity and Conservation, vol 18. Springer, Singapore. https://doi.org/10.1007/978-981-32-9174-4_4

Westra S, Alexander L, Zwiers F (2013) Global increasing trends in annual maximum daily precipitation. J Clim 26.3904–3918

World Health Organization (2006) Regional Office for South-East Asia. Human health impacts from climate variability and climate change in the hindu kush-himalaya region: report of an inter-regional workshop. WHO Regional Office for South-East Asia. https://apps.who.int/iris/handle/10665/205297

Yadav RR, Park WK, Bhattacharya A (1999) Spring temperature variations in the western Himalayan region as reconstructed from tree-rings: AD 1390–1987. Holocene 9(1):85–90

Yadava AK, Yadav RR, Misra KG, Singh J, Singh D (2015) Tree ring evidence of late summer warming in Sikkim, northeast India. Quat Int 371:175–180

You QL, Ren GY, Zhang YQ, Ren YY, Sun XB, Zhan YJ et al (2017) An overview of studies of observed climate change in the Hindu Kush Himalayan (HKH) region. Adv Clim Chang Res 8(3):141–147. https://doi.org/10.1016/j.accre.2017.04.001

Zhan YJ, Ren GY, Shrestha AB, Rajbhandari R, Ren YY, Sanjay J (2017) Change in extreme precipitation events over the Hindu Kush Himalayan region during 1961–2012. Adv Clim Change Res 8(3). https://doi.org/10.1016/j.accre.2017.08.002

Chapter 2
Variability and Trends in Temperature, Rainfall, and Discharge in a Western Himalayan Catchment

Omvir Singh and Milap Chand Sharma

Abstract Temperature, radiation, precipitation, and stream flow are all critical climatic variables that affect the ecosystem. Understanding the complex mechanisms involved in climate change is a concern for scientists because it is likely to exacerbate current food shortages and issues with irrigated agricultural systems across the world. Therefore, the current research aims to quantify the long-term variability and patterns in temperature, rainfall, and discharge in a catchment of the western Himalayan region in the state of Himachal Pradesh over a four-decade period. Temperature, rainfall, and discharge trends were studied using Mann–Kendall and simple linear regression models. The investigation revealed that the rainfall amount in the catchment has not changed significantly during different seasons as well as annually. However, temperature trends in the basin demonstrated a slight increase during all the seasons but only the winter season temperature has demonstrated a significant positive change. Conversely, these warming reflections on the water discharge have not been observed accordingly and a significant decline was detected in the annual, seasonal as well as monthly streamflow pattern of the catchment. The shifting nature of rainfall, less snow cover in the lower and middle reaches, and thinning of small glaciers and ice patches over the study period can all be blamed for the decreasing discharge. This reduction in streamflow will influence the hydropower production in the upstream parts and agricultural activities in the downstream areas and thus affecting local as well as the national economy.

Keyword Temperature · Rainfall · Streamflow · Trend analysis · Mann–Kendall test · Parbati catchment

O. Singh (✉)
Department of Geography, Kurukshetra University, Kurukshetra, Haryana, India

M. C. Sharma
Centre for the Study of Regional Development, Jawaharlal Nehru University, New Delhi, India

S. Rani and R. Kumar (eds.), *Climate Change*, Springer Climate,
https://doi.org/10.1007/978-3-030-92782-0_2

Introduction

The atmosphere, cryosphere, hydrosphere, biosphere, and geosphere make up the climatic system, which is a very complex system (Qin et al. 2005). Changing climate has not only an effect on the climate that humans depend on for survival but also on diverse issues of socio-economic development. The scientific community has acknowledged climate change because of anthropogenic activities, and it is quickly becoming a major international issue impacting global development (IPCC 2001; Kamga 2001). On a global scale, weather and climate patterns have been reconstructed, showing an increase in temperature of the air to the tune of 0.5–1.1 °C over the previous century (IPCC 2001). The last two decades have seen this increase, with the ten warmest years since 1860, all happening since 1980 (Crowley 2000). In such an alarming situation, water yields may reduce significantly, with adverse effects concerning the regional water availability (Larson et al. 2011). Increased temperature and fluctuating precipitation have a significant impact on the hydrological cycle, which has implications for water supply on a local, regional, and global scale (Oo et al. 2019). As a result, before establishing any long-term water resource project, an assessment of climate change's effects on hydrological resources is required.

More than one-sixth of the world's population gets its water from glaciers and snowmelt (Barnett et al. 2005; Rai et al. 2017; Singh et al. 2021). Surprisingly little is known about shifts in hydrological and climatological attributes concerning the mountain river basins (Murtaza and Romshoo 2017; Minaei and Irannezhad 2018). Changes in climate regimes as a result of the general rise in air temperature have had an impact on their river systems, resulting in the degradation of the present socio-economic structures of the people (Beniston 2003; Mushtaq and Lala 2017). The Himalayas, which serve as an obstacle on the globe where tropical, Mediterranean, and polar forces converge, are critical to the survival and control of the Asian continent's monsoon system (Borgaonkar and Pant 2001). In Himalayan rivers, snow and the melting of glaciers play a vital role in maintaining their perennial flow (Latief et al. 2016; Ahmed et al. 2017; Marazi and Romshoo 2018; Haijun et al. 2019). Therefore, temperature and rainfall shifts in the Himalayan region are expected to have an immediate impact on cryosphere processes and water resources of headwater catchments (Immerzeel et al. 2009; Huai et al. 2018; Normatov and Normatov 2020; Thapa et al. 2021). Because of the inaccessibility, the ruggedness of the landscape, and a sparse network of river gauging sites, studies for river basins fed by snow and glaciers over the Himalayas about the impacts of changing climate on water yields are minimal and have not been studied adequately (Rana et al. 2014). In this context, the current study examines temperature, rainfall, and discharge trends at a regional scale in the Parbati catchment of the upper Beas basin from 1968 to 2005.

Study Area

The present study has been conducted in the Parbati River basin (Himachal Pradesh) located in the western Indian Himalaya. In Himachal Pradesh's Kullu district, the entire area of Kullu tehsil is drained by the Parbati River (Fig. 2.1). It is an independent hydrological system. The basin lies between 31° 50′ to 32 ° 5′ north latitudes and 77 °5′–77 ° 50′ east longitudes. The Parbati River catchment is a hilly and mountainous area having an elevation extent of 1096–6250 m above mean sea level (AMSL). It rises from the Mantalai glacier at an elevation of about 5200 m AMSL, which is positioned on the western slopes of the Great Himalayan ranges. On the Beas River's left bank, the river Parbati is a major tributary at a height of approximately 1096 m AMSL, it joins the Beas River, at Bhuntar. It drains an area of about 1760 km^2. Rocky outcrops with barren surfaces are the basin's most important land uses. In terms of economics, the Parbati catchment is significant since an 800 MW power project is under construction, with a second 520 MW project expected. Additionally, the catchment is a home of several micro and mini-hydropower projects that are either in the planning stages or are already operational. Apart from this, Kullu Tehsil has 44,056 households. Overall, there are 206,716 people in these households, with 106,128 men and 100,588 women.

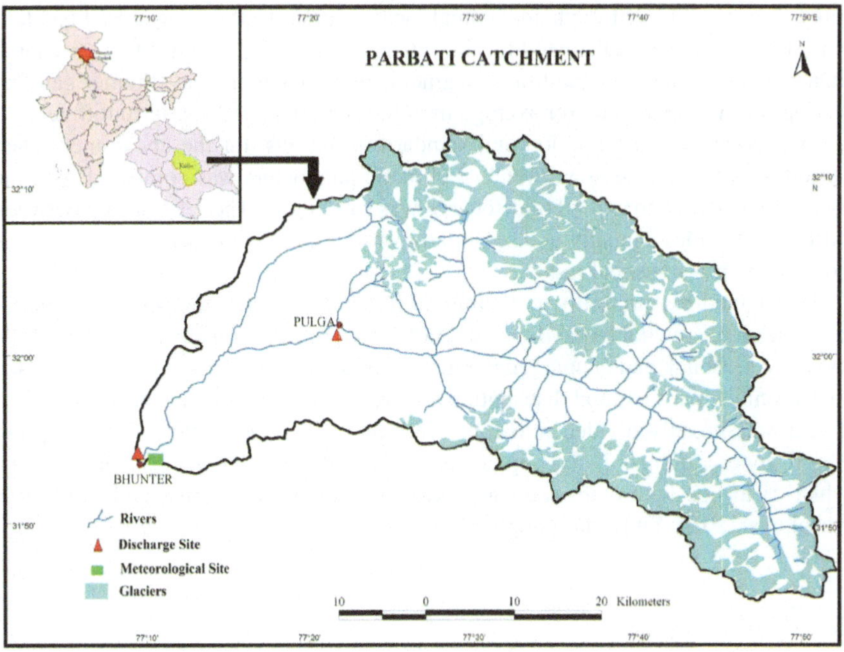

Fig. 2.1 Location of Parbati catchment in Himachal Pradesh

Materials and Methods

Temperature, Rainfall and Discharge Data Collection

Due to its remoteness, ruggedness, and inaccessibility, data on temperature, rainfall, and discharge in the western reaches of the Himalayan region is minimal. For this investigation, daily temperature, rainfall, and discharge data measured at Bhunter station (outlet of Parbati catchment) were collected from Sub Divisional Officer, Bhakra Beas Management Board, Pandoh (Mandi), Himachal Pradesh. Bhunter hydro-meteorological station is located at 1080 m AMSL. The data was collected over 38 years, from 1968 to 2005.

Temperature, Rainfall and Discharge Data Analysis

For the determination of the nature of climatic variation in the Parbati catchment, the meteorological data for temperature and rainfall have been analyzed. The diurnal maximum and minimum temperatures were converted to the mean diurnal temperature, which was then converted to the monthly, seasonal, and annual mean temperatures. The monthly, seasonal, and annual mean rainfall depths, as well as their standard deviations, were calculated for the catchment's rainfall events. Daily discharge data calculated along the calibrated segments of the catchment at Bhunter gauging station was also examined for average monthly, seasonal, and yearly runoff, as well as their coefficient of variation and standard deviations, for the above-mentioned period. Further, this seasonal trend of climatic parameters was compared with the discharge pattern and a trend of discharge was carried out. The relationship between different variables annually and seasonally was checked with parametric as well as non-parametric statistical tests.

Understanding variability and temporal trends concerning temperature, rainfall, and discharge are vital (Apaydin et al. 2006). As a result, for the period 1968–2005, monthly, seasonal, and annual temperature, rainfall, and discharge values were used to determine their temporal distribution and trends. To compute seasonal variability, a year was broadly divided into four seasons: winter (December–March), spring (April–June), summer (July–September), and Autumn (October and November), which is in line with other scholars regarding Himalayas (Kumar and Jain 2010; Babel et al. 2014; Singh and Singh 2020a, b).

Analyses of Trends in Standardized Temperature, Rainfall, and Discharge

Hydrological and climatic responses from the catchment system were identified using the standardized indices (SI) for temperature, rainfall, and discharge data obtained from the Parbati catchment. By subtracting the average from the temperature data series and splitting it by the standard deviation, the standardized temperature indices were determined. The standardized indices for rainfall and discharge were calculated in the same way. To recognize the trends in SI data series, the Mann–Kendall test and simple linear regression models were used. The presence of a statistically significant trend was examined at a 95% confidence level. These two methods are defined in the following sections.

Mann–Kendall (MK) Test

The Mann–Kendall test is a non-parametric trend measure (Mann 1945; Kendall 1948). It has been commonly employed in hydrological investigations to determine trends (Zhang et al. 2005; Deka et al. 2013; Singh et al. 2020). The test does not include the presumption of normality, and it only shows the course of important patterns, not their magnitude (Helsel and Hirsch 1992). This test has the advantages of (1) being able to deal with non-normality, censoring, or data that is reported as "less than," missing values, or seasonality, and (2) having greater asymptotic reliability (Fu et al. 2009). Every data value in the time series was compared to all subsequent values in this test. The test statistics (S) were set to zero at the start, and S was increased by one if a statistics value in a successive period was greater than a statistics value in the former period, and vice versa. The Mann–Kendall test is a rank-based test that takes outliers into account, which is distribution-free, and has greater robustness than other tests when autocorrelation is not significant (Duhan and Pandey 2013). To remove the influence of autocorrelation in the data sequence, the Mann–Kendall test was used after pre-whitening and at various seasons (Partal and Kahya 2006; Mohammad and Jha 2014). The final value of S was determined by the outcome of all of these procedures. The Mann–Kendall test (S) statistics are as follows:

$$S = \sum_{k=1}^{n-1} \sum_{j=k+1}^{n} \operatorname{sgn}(x_j - x_k) \tag{2.1}$$

$$\operatorname{sgn}(x_j - x_k) = \begin{cases} +1, & \text{if } x_j - x_k > 0 \\ 0, & \text{if } x_j - x_k = 0 \\ -1, & \text{if } x_j - x_k < 0 \end{cases} \tag{2.2}$$

where x refers to a particular statistics point, x_j and x_k are statistics values at time j and k ($j > k$), separately, n is time series length, and *sign* is the signum function. S

values that are positive (negative) suggest an increasing (declining) trend. This test assumes that the dataset contains a small number of tied values. The variance of S was determined as:

$$\mathrm{Var}(S) = \frac{\left[n(n-1)(2n+5) - \sum_{r} r(r-1)(2r+5)\right]}{18} \quad (2.3)$$

where the terms r and $\sum_r r$ apply to the level of any given tie and the sum of all ties, separately. A tie occurs when a group of sample data has the same value. When the sample size is greater than ten, the typical normal variable Z_s was determined as:

$$Z_s = \begin{cases} \frac{S-1}{\sqrt{\mathrm{Var}(S)}}, & \text{if} S > 0 \\ 0, & \text{if} S = 0 \\ \frac{S+1}{\sqrt{\mathrm{Var}(S)}}, & \text{if} S < 0 \end{cases} \quad (2.4)$$

where $Z_s > 0$ denotes upward trends and $Z_s < 0$ denotes downward trends. Trends were tested at a particular α significance stage, and the null hypothesis was dismissed when $Z_s > Z_{1\text{-}\alpha/2}$.

Simple Linear Regression Method

A basic and extensively used trend identification approach is simple linear regression analysis. It is common to fit a linear equation to observed data to identify a relationship between two variables, which indicates the average periodic variance of the measured parameter (independent and dependent). Here, temperature, rainfall, and discharge were used as dependent variables, whereas time is a separate variable, which is stated as $Y = (\alpha + \beta X + \mu)$. Where Y = temperature or rainfall or discharge, X = time period and μ = error/random term. Positive and negative gradients show increasing and decreasing trends, respectively.

Results and Discussion

Trend Analysis of Temperature Time Series

Natural transition and anthropogenic disruption are two ways that climate change impacts water supplies. The natural transition is a mechanism in which climatic influences change components of the hydrological cycle and geographical conditions, influencing the consistency and temporal-spatial supply of water resources. In a mountainous area, climate change would almost certainly affect the long-term pattern of hydrological processes. Air temperature is one of the most important metrics used to estimate climate change. Figure 2.2 depicts a temperature series in the Parbati catchment from 1968 to 2005. These plots clearly show the increasing temperature in the catchment. The obtained results confirm with other studies (Jaswal and Rao 2010; Mushtaq and Lala 2017) in the Himalayan region of Jammu and Kashmir. From 1968 to 2005, the temperature increased slightly during the winter, summer, and autumn seasons. The increase in the magnitude of temperature for the winter season was observed to be 1.35 °C/100 years, while it was found to be 0.81 °C/100 years for the summer and autumn seasons. Throughout the analysis time, annual temperature also showed an upward trend, with an annual rate of 1.04 °C/100 years. The increasing

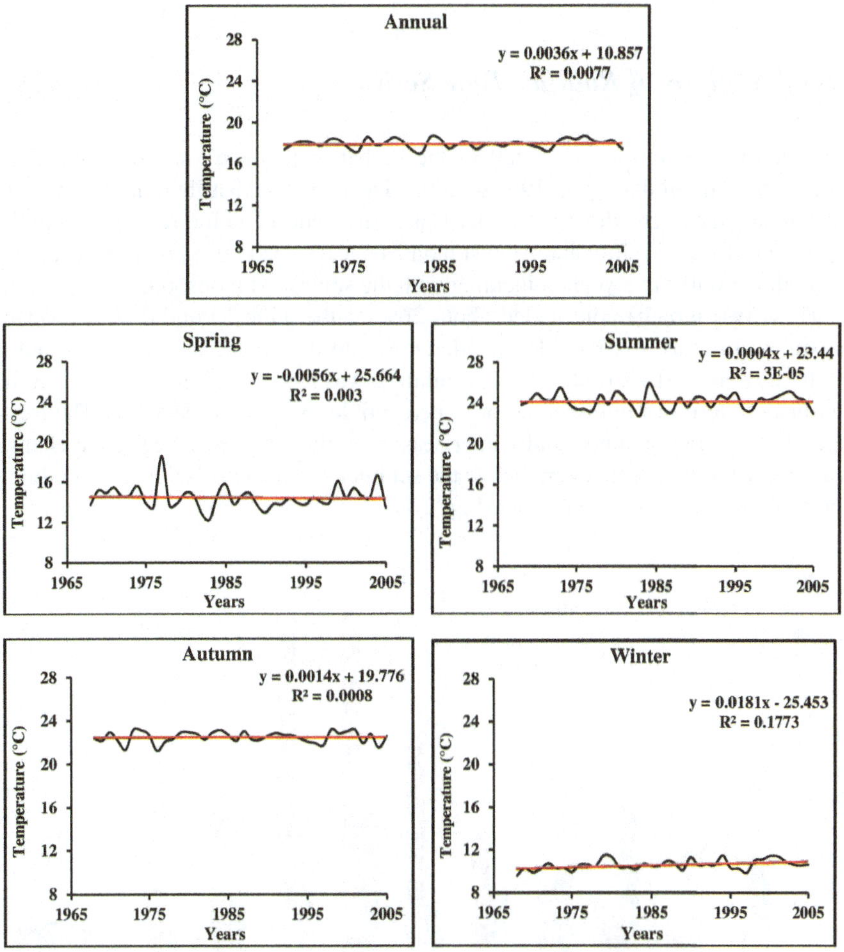

Fig. 2.2 Temporal variation and linear trends in temperature during annual, spring, summer, autumn, and winter in the Parbati catchment

trends in annual temperature were noticed from the year 1995. In addition, the spring season showed no signs of temperature increase in the catchment.

A localized response to climate change and global warming may explain the rise in temperature in the Parbati catchment. Climate change has disastrous consequences for natural habitats. One of the most significant effects of temperature increase in the catchment will change hydrological cycle and runoff variability affecting various other facets of human life (Xu 2000). Additionally, increased urbanization (settlements, transportation, tourism, hydropower activities, air pollution, etc.) with local relief, vulnerability, and atmospheric flow may affect the nature and magnitude of temperature trends in different ways in the catchment.

Trend Analysis of Rainfall Time Series

The seasonal alteration in rainfall in the Parbati catchment was investigated by employing rainfall data from 1968 to 2005. The temporal distribution of rainfall in the Parbati catchment showed a triple-peaked distribution with three maxima in the catchment (Fig. 2.3). July and August months experienced relatively higher rainfall than other months in a year, coinciding with the southwest monsoon.

These two months contributed about 26% of the annual rainfall. A secondary maximum rainfall was recorded in March due to disturbances in the west and it contributed about 16% of the annual rainfall. It was found from the analysis that the seasonal pattern of rainfall showed no perceptible increase or decrease. The trend line of the spring, autumn, and winter season rainfall showed a negligible change over the years in the catchment, while the summer season along with annual rainfall showed a slight rise over the years (Fig. 2.4).

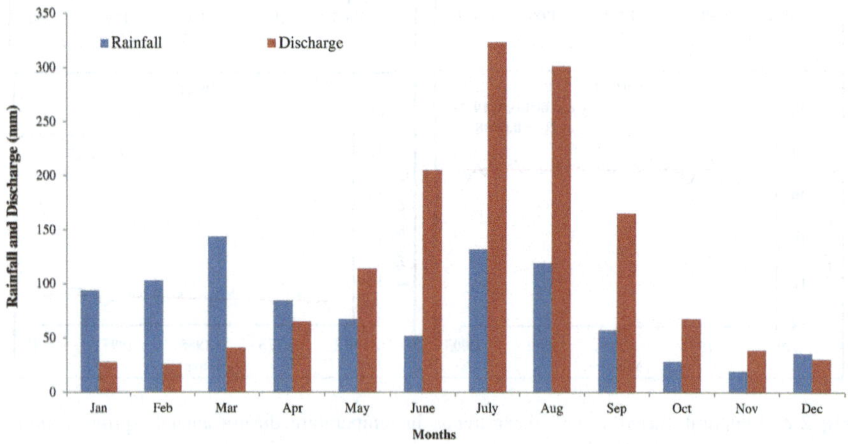

Fig. 2.3 Mean monthly rainfall and discharge in the Parbati catchment during 1968–2005

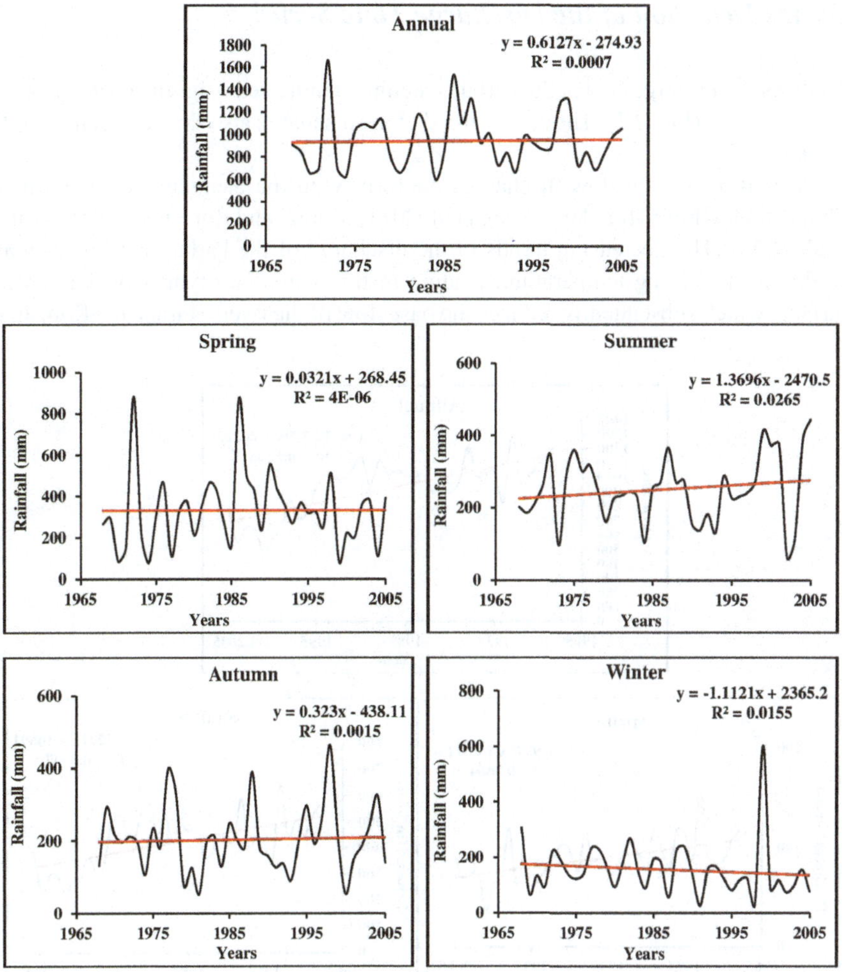

Fig. 2.4 Temporal variation and linear trends in rainfall during annual, spring, summer, autumn, and winter in the Parbati catchment

From 1985 to 1995, there was a discernible rise in summer rainfall, with an abrupt rise beginning in 1995. Similarly, an increase in the annual rainfall was observed from the year 1985. However, these trends of rainfall increase in the catchment were found non-significant during the period of study. These results are not in conformity with a study carried out in the Wular Lake catchment of Kashmir Himalayas (Mushtaq and Lala 2017).

Trends Detection of the Discharge Time Series

Analysis of discharge data of the Parbati catchment indicated a significant decrease in all the seasons (Fig. 2.5). The annual trend of the discharge also showed a remarkable decline.

Decreasing stream flow discharges conform with the Nenjiang River basin of China and Kashmir Himalaya (Feng et al. 2011; Marazi and Romshoo 2018; Rashid et al. 2020). The decreasing trends in the discharge of the Parbati catchment may be due to increasing temperature, leading to the increased dryness of the earth's surface, which subsequently reduces the base flow of the river. Marazi and Romshoo

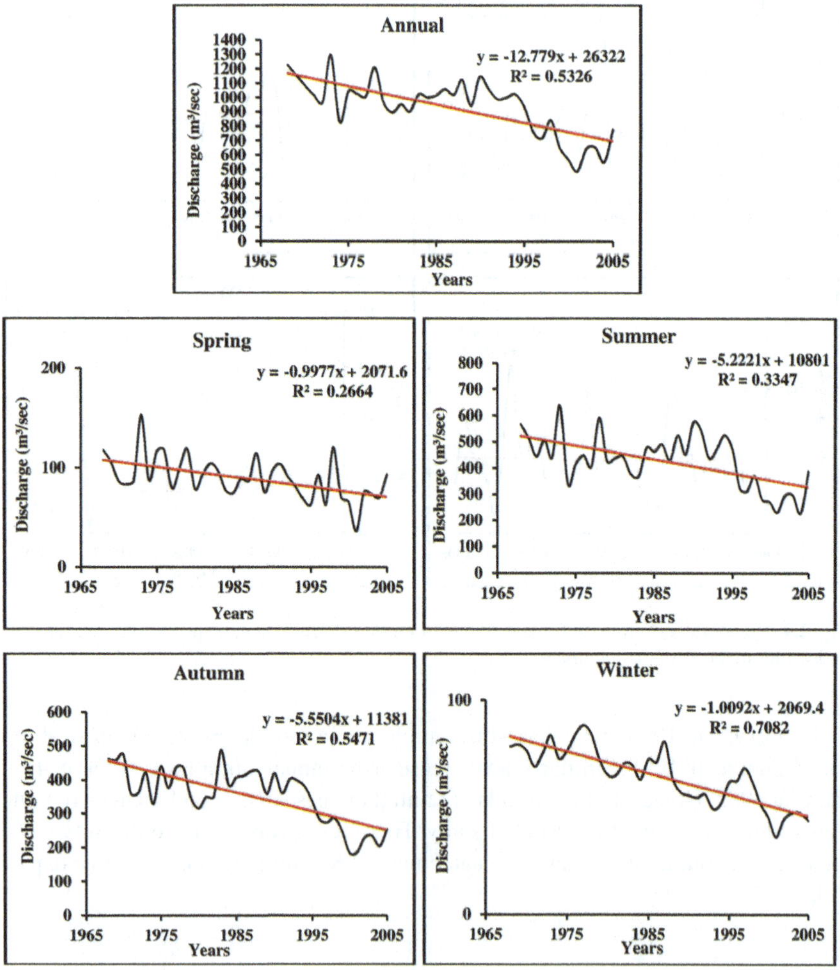

Fig. 2.5 Temporal variation and linear trends in discharge during annual, spring, summer, autumn, and winter in the Parbati catchment

(2018) revealed that decreasing stream flows recorded under the rising precipitation is principally due to the significant reduction in glacial coverage and mass in the Himalayas in the past 5–7 decades. According to recent studies, glacier-fed streams see an increase in discharge for some time due to faster glaciers melting and then exhibit a drop as the ice mass decreases (Rees and Collins 2006; Thayyen and Gergan 2010). Energy production, agriculture, tourism, and even local water availability may be affected as a result of stream flows decline. Meanwhile, the lowest amount of discharge occurs in the winter, and it is largely due to base flow from subglacial melting and groundwater storage. Autumn, summer, and winter season discharge trends also showed a remarkable decline in the flow from 1968 through 2005. Even in the winter, regardless of high air temperatures and average rainfall, discharge is decreasing, confirming that the lower and middle reaches of the catchment have less snow cover. However, a decline in the spring season discharge was observed to be comparatively lower than the other three seasons. Understanding the magnitude of such discharge variability will aid in the operation of hydropower plants and the expansion of water resource management strategies in the Himalayan headwater areas.

The analysis also revealed a decline in the discharge to be maximum after the year 1987. This decrease in discharge may be attributed to anthropogenic accelerated development activities in the catchment, such as deforestation, road construction, and hydropower project construction. The studied catchment in the current study is a keystream of the upper Beas River system in the Kullu valley. In the last four decades of the last century, researchers detected a dramatic reduction in average annual and monsoon discharge in the Beas River of the western Himalayan region (Bhutiyani et al. 2008; Rana et al. 2014). The decrease in discharge has been attributed to changes in rainfall and temperature, as well as the presence of smaller glaciers in the basin, which contribute little to snow and glacier melt and are primarily monsoon fed. In the analysis of long-term trends in river mean discharge as potential measures of climate change, similar findings have been observed in different parts of the world, including the Himalayas (Singh and Jain 2002; Jasper et al. 2004; Cigizoglu et al. 2005; Yang et al. 2006; Thayyen and Gergan 2010). Furthermore, the negative effects of human activities may be the primary cause of the decreasing trend in discharge. Specifically speaking, more and more intensified constructions of hydropower projects are accountable for the declining trend of discharge, which will limit the fiscal progress and safety of human life. Certainly, the decreasing trends of discharge will severely jeopardize the ecological security of the region. A declining trend in stream flow has a direct impact on hydropower production since the decreased stream flows result in lower hydropower output. If the Parbati River's diminishing discharge trend continues, it will have a direct impact on Mullana and other hydroelectric accomplishments. Since several other hydropower plants are proposed in the basin and the hydrology of the basin is altering, therefore their design dimensions need to be reviewed, as in the remote future enough water may not be at hand for power production.

Trend Analysis of Standardized Temperature, Rainfall, and Discharge Indices

Table 2.1 displays the findings of trend analyses of standardized temperature, rainfall, and discharge in the spring, summer, autumn, and winter seasons along with annual data series for the Parbati catchment. These results revealed that the discharge of the Parbati catchment decreased significantly in all the seasons with winter discharge decline being the sharpest.

The variation in rainfall during this time tends to influence the average annual discharge in the Parbati catchment to a large extent (Bhutiyani et al. 2010). Furthermore, decreasing trends in the Parbati catchment's discharge can be due to shifting rainfall patterns, less snow cover in the lower and middle reaches, and the thinning of small glaciers and ice patches over time (Table 2.2).

Glaciers may have become shorter, narrower, and thinner under the effect of atmospheric warming. The Parbati glacier has a total area of 27.92 km^2 in 1976 which has reduced to 5.78 km^2 with a total area of around 22.14 km^2 in 2013 (Rai et al. 2017). Similarly, Ahmad et al. (2017) revealed an average glacier area loss of about 17% from 1962 to 2003 in the Parbati basin with a greater change from 8 to 100% for the sole glacier. If current climate change trends continue, greater and faster glacial melting is projected in the future. As a result of this, long-term in situ

Table 2.1 Results of trend analyses of SRI, STI, and SDI in the Parbati catchment

Data availability	Index	Season	Trend analysis	
			Kendall's non-parametric test	Linear regression coefficient b
1968–2005	Standardized temperature index (STI)	Spring	(–0.007)	(–0.005)
		Summer	(+0.087)	(+0.009)
		Autumn	(+0.090)	(+0.019)
		Winter	(+0.335)*	(+0.039)*
		Annual	(+0.163)	(+0.019)
1968–2005	Standardized rainfall index (SRI)	Spring	(+0.056)	(+0.001)
		Summer	(+0.089)	(+0.010)
		Autumn	(–0.027)	(+0.001)
		Winter	(–0.073)	(+0.001)
		Annual	(+0.059)	(+0.004)
1968–2005	Standardized discharge index (SDI)	Spring	(–0.366)*	(–0.047)*
		Summer	(–0.351)*	(–0.053)*
		Autumn	(–0.516)*	(–0.067)*
		Winter	(–0.664)*	(–0.077)*
		Annual	(–0.502)*	(–0.067)*

* Significant at 95% confidence level. (+) increasing; (–) decreasing

Table 2.2 Glacier fluctuations in the Parbati catchment

Glacier	Total advance/retreat (m)	Total advance/retreat (m)	Rate of retreat/advance (m/yr.)	Rate of retreat/advance (m/yr.)	Areal extent (km^2)	Areal extent (km^2)	Change (km^2)
	1989–2002	2002–2010	1989–2002	2002–2010	1989	2010	1989–2010
Parvati	21.85	116.63	1.7	14.6	22.36	21.71	0.65
Sara Umga	522.95	128.27	40.2	16	50.61	50.15	0.46
Dibika 1	118.92	42.05	9.1	5.3	22.02	21.89	0.13
Dibika 2	126.93	111.16	9.8	13.9	6.43	6.42	0.01
Dibika 3	212.4	102.24	16.3	12.8	17.02	16.68	0.34
Tichu	273.44	84.78	21	10.6	20.55	20.23	0.33
Futiruni	219.8	1.9	16.9	0.2	15.6	15.59	0.01
Girvo Kotli	126.96	42.02	9.8	5.3	19.61	18.88	0.73
Tons 1	132.52	1.69	10.2	0.2	9.57	9.86	–0.29
Tons 2	140.25	68.51	10.8	8.6	10.99	10.8	0.18
Bakar Bihar	174.03	228.27	13.4	28.5	18.85	18.8	0.05

Source Calculation based on TM image (1989), ETM + (2002), and LISS III (2010) image of Parbati catchment, Himachal Pradesh measurements are required to analyze the complex interactions between climate and glaciers to better understand the present processes and give data for trustworthy glacier evolution estimates in the future. It is, therefore, necessary that the monitoring of glaciers in the face of climate change should be treated seriously.

The classification of potential climate change indicators can be done by looking at long-term patterns in river mean stream flow (Singh and Jain 2002; Bhutiyani et al. 2008). Snowmelt runoff, glacier melt, and runoff due to monsoon rainfall are three components of discharge in a glacierized basin. The lowest flows from the Parbati catchment occur from November–March, with the highest run-offs occurring in July and August months. The maximum temperatures, glacier melt, and monsoon rains occur during these two months, making it difficult to estimate the variability in water contributions to Himalayan head water rivers from glaciers and snow cover. This is because changes in monsoon, snow, and glacier regimes all have an effect on river runoff in the lower reaches. During years of strong snowfall, the water yield of non-glacierized catchment regions matches that of the area that has been glacierized, while in years of low snowfall, a non-glacierized region the water yield decreases to half that of the glacierized area (Thayyen et al. 2007). Furthermore, Singh and Bengtsson (2005) detected that yearly melt was declined by about 18% for the snow-fed basin studied in a T + 2 °C scenario, whereas it improved by about 33% for the basin fed by glaciers studied in the Himalayan zone. Interestingly, it is impossible to determine the contribution of each component of the discharge accurately due to the lack of reliable data. However, increased glacier melting may have little impact on river runoff, which is determined by winter snowfall and monsoon precipitation.

The Parbati catchment is representative of rain, snow, and glacier-fed in its different reaches. As a result, changes in precipitation characteristics, especially the extent and duration of winter snow cover, may significantly reduce headwater river flow, while the glacier portion can sustain the low flow. The changing climate of the area is illustrated by the linear negative trends in snow cover extent over Eurasia over the past three decades, as well as the rise in winter temperature over the western Himalayas (Pant et al. 2003). As the buffering capacity of melting glaciers decreases further, downstream of the Himalayan headwater rivers will see greater yearly flow variability in the future (Thayyen et al. 2007). While seasonal snow cover is a significant parameter in a glacierized catchment's hydrological regime, it only plays a role in the spring season, when snowmelt and base flow are two components of the discharge.

Conclusions

The results of this study show that seasonal rainfall, energy budget, and runoff are all interconnected in the Parbati catchment. Trend tests, both parametric and non-parametric, revealed a much lower average annual and seasonal discharge at a 95% of confidence level. Warmer temperatures, not precipitation abundance, determine the discharge regime, as seen by significant decreases in discharge over the basin. The major limitation of this study is that the results are based on the available recorded data of a single station for a limited number of years and it is understood that long-term data at various locations would provide a better understanding of such variations with time and space. In a changing climate, uncertainties in precipitation characteristics, especially winter snowfall, have a greater impact on headwater river runoff variability than runoff variations caused by a melting glacier. Conversely, glaciers are critical for maintaining river flows during years of low summer runoff. Furthermore, time-series data from the Parbati catchment showed substantial upward trends in winter temperature. Increased winter temperature may cause a shift in the amount of rainfall that falls as rain near the freezing point, but this may have little impact at higher elevations where the most accumulation occurs. As a result, runoff from foothills and snow-fed areas of the catchment may be greatly affected, whereas runoff from glacier-fed areas of catchments may be minimally affected. Finally, improvements in the Parbati catchment's snow and glacier regimes would have the greatest impact on the catchment's lower reaches, away from the ice caps, where many hydropower projects are whether expected or already in use.

References

Ahmad S, Khan K, Singh N, Ansari Z, Mulhim M (2017) Temporal change of glacier area and geomorphometric parameters in Parbati Valley Himachal Pradesh India. Environ Earth Sci 76:407

Apaydin H, Erpul G, Bayramin I, Gabriels D (2006) Evaluation of indices for characterizing the distribution and concentration of precipitation: a case for the region of Southeastern Anatolia Project, Turkey. J Hydrol 328:726–732

Babel MS, Bhusal SP, Wahid SM, Agarwal A (2014) Climate change and water resources in the Bagmati River Basin Nepal. Theor Appl Climatol 115:639–654

Barnett TP, Adam JC, Lettenmaier DP (2005) Potential impacts of a warming climate on water availability in snow-dominated regions. Nature 438:303–309

Beniston M (2003) Climatic change in mountainous regions: a review of possible impacts. Clim Change 59:5–31

Bhutiyani MR, Kale VS, Pawar NJ (2008) Changing stream flow patterns in the rivers of north western Himalaya: implications of global warming in the 20th century. Curr Sci 95:618–626

Bhutiyani MR, Kale VS, Pawar NJ (2010) Climate change and the precipitation variations in the north western Himalaya: 1866–2006. Int J Climatol 30:535–548

Borgaonkar HP, Pant GB (2001) Long-term climate variability over Monsoon Asia as revealed by some proxy sources. Mausam 52:9–22

Cigizoglu HK, Bayazit M, Önöz B (2005) Trends in the maximum, mean and low flows of Turkish rivers. J Hydrometeorol 6:281–290

Crowley TJ (2000) Causes of climate change over the past 1000 years. Science 289:270–276

Deka RL, Mahanta C, Pathak H, Nath KK, Das S (2013) Trends and fluctuations of rainfall regime in the Brahmaputra and Barak basins of Assam, India. Theor Appl Climatol 114:61–71

Duhan D, Pandey A (2013) Statistical analysis of long term spatial and temporal trends of precipitation during 1901–2002 at Madhya Pradesh, India. Atmos Res 122:136–149

Feng X, Zhang G, Yin X (2011) Hydrological responses to climate change in Nenjiang River basin, Northeastern China. Water Res Manage 25:677–689

Fu GB, Charles SP, Yu JJ, Liu CM (2009) Decadal climatic variability, trends and future scenarios for the north China plain. J Clim 22:2111–2123

Haijun D, Yaning C, Yang L (2019) Glacier and snow variations and their impacts on regional water resources in mountains. J Geog Sci 29:84–100
Helsel DR, Hirsch RM (1992) Statistical methods in water resources. Elsevier, Amsterdam
Huai B, Wang Y, Li Z, Sun W, Wang X (2018) Glacier changes and its effect on water resources in Urumqi River basin, Tianshan mountains, China, from 1964 to 2014. Arab J Geosci 11:716
Immerzeel WW, Droogers P, De Jong SM, Bierkens MFP (2009) Large-scale monitoring of snow cover and runoff simulation in Himalayan river basins using remote sensing. Remote Sens Environ 113:40–49
IPCC (2001) Climate change 2001: the scientific basis contribution of working group I to the third assessment report of the Intergovernmental Panel on Climate Change (IPCC). Cambridge University Press, London
Jasper K, Calanca P, Gyalistraus D, Fuhrer J (2004) Differential impacts of climate change on the hydrology of two alpine river basins. Climate Res 26:113–129
Jaswal AK, Rao GSP (2010) Recent trends in meteorological parameters over Jammu and Kashmir. Mausam 61:369–382
Kamga FM (2001) Impact of greenhouses gas induced climate change on the runoff of the upper Benue River (Cameroon). J Hydrol 252:145–156
Kendall MG (1948) Rank correlation methods. Econometrica 28:181–183
Kumar V, Jain SK (2010) Trends in seasonal and annual rainfall and rainy days in Kashmir valley in the last century. Quatern Int 212:64–69
Larson RP, Byrne JM, Johnson DL et al (2011) Modelling climate change impacts on spring runoff for the Rocky Mountains of Montana and Alberta I: model development, calibration and historical analysis. Can Water Res J 36:17–34
Latief SU, Rashid SM, Singh R (2016) Impact analysis of climate change on Kolahoi glacier in Liddar valley, north-western Himalayas. Arab J Geosci 9:705
Mann HB (1945) Non parametric tests against trend. Econometrica 13:245–259
Marazi A, Romshoo S (2018) Streamflow response to shrinking glaciers under changing climate in the Liddar Valley, Kashmir Himalaya. J Mt Sci 15:1241–1253
Minaei M, Irannezhad M (2018) Spatio-temporal trend analysis of precipitation, temperature and river discharge in the northeast of Iran in recent decades. Theor Appl Climatol 131:167–179
Mohammad S, Jha MK (2014) Seasonal and annual precipitation time series trend analysis in North Carolina, United States. Atmos Res 137:183–194
Murtaza KO, Romshoo SA (2017) Recent glacier changes in the Kashmir Alpine Himalayas, India. Geocarto Int 32:188–205
Mushtaq F, Lala MGN (2017) Assessment of climatic variability in the catchments of Himalyan Lake, Jammu and Kashmir. Geocarto Int 32:1090–1104
Normatov I, Normatov P (2020) Climate change impact on hydrological characteristics and water availability of the Pamir rivers. Proc IAHS 383:31–41
Oo HT, Zin WW, Thin Kyi CC (2019) Assessment of future climate change projections using multiple global climate models. Civ Eng J 5:2152–2166
Pant GB, Borgaonkar HP, Kumar R (2003) Climate variability over the western Himalaya since the little ice age: dendroclimatic implications. Jalvigyan Sameeksha 18:111–120
Partal T, Kahya E (2006) Trend analysis in Turkish precipitation data. Hydrol Proc 20:2011–2026
Qin DH, Ding YH, Su JL (2005) The change of climate and environment in China. Science Press, Beijing
Rai PK, Mishra VN, Singh S, Prasad R, Nathawat MS (2017) Remote sensing based study for evaluating the changes in glacial area: a case study from Himachal Pradesh, India. Earth Sys Environ 1:1
Rana RS, Singh S, Chander N, Sood R, Sharma R, Aditya, (2014) Impacts of changes in climate on mountain water resources of Himachal Pradesh. Mausam 65:153–160
Rashid I, Majeed U, Aneaus S, Pelto M (2020) Linking the recent glacier retreat and depleting stream flow patterns with land system changes in Kashmir Himalaya, India. Water 12:1168

Rees HG, Collins DN (2006) Regional differences in response of flow in glacier-fed Himalayan rivers to climatic warming. Hydrol Proc 20:2157–2169
Singh J, Singh O (2020a) Exploring seasonality and erosivity of rainfall over a lower Himachal Himalayan Catchment, India. Arab J Geosci 13:700
Singh J, Singh O (2020b) Assessing rainfall erosivity and erosivity density over a Western Himalayan Catchment, India. J Earth Syst Sci 129:97
Singh O, Kasana A, Singh KP, Sarangi A (2020) Analysis of drivers of trends in groundwater levels under rice-wheat ecosystem in Haryana, India. Nat Res Res 29:1101–1126
Singh P, Bengtsson L (2005) Impact of warmer climate on melt and evaporation for the rain fed, snow fed and glacier fed basins in the Himalayan region. J Hydrol 300:140–154
Singh P, Jain SK (2002) Snow and glacier melt in the Sutlej River at Bhakra dam in the Himalayan region. Hydrol Sci J 47:93–106
Singh V, Jain SK, Shukla S (2021) Glacier change and glacier runoff variation in the Himalayan Baspa River basin. J Hydrol 593:125918
Thapa S, Li H, Li B, Fu D, Shi X, Yabo S, Lu L, Qi H, Zhang W (2021) Impact of climate change on snowmelt runoff in a Himalayan basin, Nepal. Environ Monit Assess 193:393
Thayyen RJ, Gergan JT (2010) Role of glaciers in catchment hydrology: a preliminary study of a Himalayan catchment. Cryosphere 4:115–128
Thayyen RJ, Gergan JT, Dobhal DP (2007) Role of glaciers and snow cover on headwater river hydrology in monsoon regime-micro-scale study of Din Gad catchment, Garhwal Himalaya, India. Curr Sci 92:376–382
Xu CY (2000) Modelling the effects of climate change on water resources in central Sweden. Water Res Manage 14:177–189
Yang SL, Gao A, Helenmary HM, Zhu J, Dai SB, Li M (2006) Trends in annual discharge from the Yangtze River to the sea (1865–2004). Hydrol Sci J 50:825–834
Zhang Q, Jiang T, Gemmer M, Becker S (2005) Precipitation, temperature and runoff analysis from 1950 to 2002 in the Yangtze basin, China. Hydrol Sci J 50:65–79

Chapter 3
Tree-Ring-Based Hydrological Records Reconstructions of the Himalayan Rivers: Challenges and Opportunities

Mayank Shekhar, Parminder Singh Ranhotra, Amalava Bhattacharyya, Ayushi Singh, Rupesh Dhyani, and Shaktiman Singh

Abstract Himalayan rivers play a significant role in the survival of the vast population of the Indian subcontinent. Information on the long-term spatio-temporal hydrological variability is required for hydropower generation, irrigation, and flood management. Therefore, there is an urgent need to understand the long-term perspective of the flow of these rivers (also considered as streamflow) which could only be achieved based on proxy data analysis in the absence of observed data. Tree-rings provide an excellent proxy of streamflow to understand the hydrologic variations for a long past beyond the existing instrumental records. The logic behind developing the relationship between tree growth and streamflow is the common climatic factors, mainly precipitation and evapotranspiration controlling the tree growth and streamflow. Here, this chapter presents a review of the available information of tree-ring-based streamflow records and flood events from some of the Himalayan rivers along with fundamental principles followed for the analysis of discharge reconstructions.

Keywords Tree-ring study · Streamflow reconstruction · Climate change · Flood events · Himalayan conifers

Introduction

The Himalayan rivers (e.g., Indus, Ganga, and Brahmaputra) are the primary source of water for nurturing a vast population of the Indian subcontinent. The stable economic activities and the sustainable development of this subcontinent mainly depend on the water availability in all seasons in these rivers. Recent hydrologic changes of

M. Shekhar (✉) · P. S. Ranhotra (✉) · A. Bhattacharyya · A. Singh
Birbal Sahni Institute of Palaeosciences, 53-University Road, Lucknow, Uttar Pradesh, India

R. Dhyani
Amos Campus, Forest Research Institute (IRF), Université du Québec en Abitibi-Témiscamingue (UQAT), Rouyn-Noranda, Canada

S. Singh
School of Geosciences, University of Aberdeen, Aberdeen AB24 3UE, UK

S. Rani and R. Kumar (eds.), *Climate Change*, Springer Climate,
https://doi.org/10.1007/978-3-030-92782-0_3

the Himalayan region could be analyzed based on the observed trends in streamflow (also considered as runoff or discharge) with respect to changes in climate and anthropogenic activities (Gosain et al. 2006; Rees and Collins 2006; Immerzeel et al. 2010; Intergovernmental Panel on Climate Change (IPCC 2013; Singh et al. 2016). As observed by many researchers, the atmospheric circulations significantly influence the temporal variations of hydrologic events (Parthasarathy et al. 1988; Krishna Kumar et al. 1999; Gadgil 2003; Gadgil et al. 2004; Ali and Mishra 2017). Due to rapid climate change, most of the Himalayan rivers are facing a water crisis, which may become acute in the coming years due to global warming (IPCC 2013; Sevellec and Drijfhout 2018).

The increasing dependency on the available surface water resources in the Himalayan region is likely to accelerate the occurrence of hydrological droughts, and it becomes conspicuous in signing the treaty not only between countries of strategic significance but also even across the provincial political boundaries within India. Long-term streamflow records from the rivers of a region could provide spatio-temporal variability of the hydrological fluctuations, which is required for water resource planning and management. The long-term records of river discharge beyond the instrumental data are therefore needed based on the proxy data analysis. In the catalog of biological proxy by Bradley (1999) and Cronin (2009), tree-ring data has a prominent role in streamflow reconstructions. Over the past few decades, tree-ring data have been increasingly applied in streamflow reconstruction as a planning and research tool in water resource studies (Stockton 1975; Cook and Jacoby 1983; Woodhouse 2001; Woodhouse and Meko 2002; Gou et al. 2007; Axelson et al. 2009; Cook et al. 2010, 2013; Liu et al. 2010; Urrutia et al. 2011; Meko and Woodhouse 2011; Bekker et al. 2014; Woodhouse et al. 2016; Ewane and Lee 2017). In addition, through tree-ring analysis, unusual high discharge or floods could also be dated (Alestalo 1971; Stoffel et al. 2010; Stoffel and Corona 2014).

Moreover, there is a high potential for extracting information about past flood events which are indicators of periodic high discharge through tree-ring records. Dating of flood scars in tree-ring is a valuable tool because of the ubiquitous presence of trees with distinct annually grown rings in the temperate as well as subtropical regions. However, after the pioneering work, e.g., Sigafoos (1961, 1964), Shroder (1980), the tree-rings have not been given much importance as another proxy of palaeo flood evidence as the slack water deposits been taken into much consideration (Kochel and Baker 1982; Baker 2008). In the last few years, tree-ring dating of flood-scared trees has also been given impetus in flood or periodical high runoff reconstructions in the Himalayan region due to annually resolved dating (Bhattacharyya et al. 2017; Ballesteros-Cánovas et al. 2017). Fundamentally, trees growing close to the stream channel might be influenced by the unusually high or low water discharge or streamflow; hence, they are useful in understanding the streamflow extremes.

Before the discussion on the status and progress of tree-ring-based streamflow reconstructions from the Himalayan region, the present study added a brief outline of methodologies generally followed for the establishment of tree growth and streamflow relationship keeping in mind its applicability in the Himalayan region. This includes a brief account on the selection of suitable trees and sites for the collection of samples, the procedure of dating of tree-rings, preparation of tree-ring chronology, and correlations of a common period of tree-ring series and recorded water discharge data. Observed relationships would help to find out which months' streamflow is mostly correlated with the corresponding months/years tree-ring data so that the streamflow of those months could be retrieved beyond the observational records from long tree-ring sequences. Overall, this study provides a review of recent trends and developments of dendrohydrological research in India with (i) methodologies used, (ii) hypothesis of the relationship between tree growth and streamflow, and development of the model, and (iii) limitation and challenges of dendrohydrology and future roadmap.

Early History of Tree Growth and Streamflow Relationship

Since long back, tree-ring data have been recognized as a proxy for the streamflow reconstructions. In early studies during the 1900s, there was the estimation of wet and dry periods for pre-gauged streamflow years based on comparisons of tree-ring records with corresponding year's streamflow (Hardman and Reil 1936; Hawley 1937; Schulman 1945, 1951). Systematic dendrohydrological analyses started in the mid-1930s and have been efficiently applied to reconstruct the high-resolution hydrologic variables worldwide (Hardman and Reil 1936; Hawley 1937; Schulman 1945; Helley and LaMarche 1968; Stockton and Jacoby 1976; Meko et al. 2001; Pederson et al. 2001; Polacek et al. 2006; Woodhouse and Meko 2002; Chalise et al. 2003). In Asia, streamflow reconstruction was initiated much later with more magnitude in China (Wu 1992; Hughes et al. 1994; Yuan et al. 2007). The tree-ring-based streamflow reconstructions in the Himalayan region are in progress during the last decade. Initial tree-ring studies in the Himalaya region were toward the selection of sampling sites and evaluation of suitable tree species for the climactic analysis (Bhattacharyya et al. 1988). Tree-ring-based climate reconstructions got momentum from 2000 C.E. onward from the Western Himalayas (Yadav et al. 2004; Yadav and Bhutiyani 2013, Yadav et al. 2013, Yadav et al. 2015, 2017; Bhattacharyya and Shah 2009; Shekhar et al. 2017, 2018) and later from the Eastern Himalayas (Bhattacharyya and Chaudhary 2003; Yadav et al. 2014). Most of the analyses were restricted to the reconstruction of temperature and precipitation (Yadav and Park 2000; Singh and Yadav 2005; Singh et al. 2006; Yadav et al. 2014) and a lesser degree on the reconstruction of streamflow (Shah et al. 2013, 2014; Singh and Yadav 2013; Cook et al. 2013; Misra et al. 2015; Shekhar and Bhattacharyya 2015; Rao et al. 2018).

Hypothesis Regarding the Relationship Between Tree Growth and Streamflow

The trees growing on the slope, close to the river channel are usually considered suitable for streamflow reconstructions. Dated tree-ring data of these trees seem to be a good proxy for streamflow reconstructions because of the similar climate factors, primarily precipitation, evapotranspiration (the loss of water from plant and soil), and atmospheric circulation of moisture, thus affecting both the growth of trees and the amount of water reaching the streams (Fig. 3.1). However, the impacts of climate on the discharge vary from basin to basin depending on the location and orientation of the basin where precipitation received is high or low; population inhabited; closeness to the glacier, and other factors.

Fig. 3.1 Tree and streamflow relationship. **a** Trees growing along the riverside. **b** Surface runoff reaches to river flow. **c** Subsurface flow in the river channels. [Harsil, Uttarakhand Himalaya]

Suitability of Tree and Site Selection

For the tree-ring-based streamflow studies, it is necessary to know the distribution of trees growing close to a river channel of a drainage basin. This information could help in the selection of sites and sensitive trees for the dendrohydrological analyses. Trees growing mostly in the temperate and subalpine forests and also a few trees in subtropical forests of Himalayas have distinct annual rings and are found to be sensitive to climate change (Bhattacharyya and Shah 2009). In the Western Himalayas, the most suitable trees are birch (*Betula utilis*), fir (*Abies spectabilis*), blue pine (*Pinus wallichiana*), spruce (*Picea smithiana*), juniper (*Juniperus*), chilgoza pine (*Pinus gerardina*), and deodar (*Cedrus deodara*) (Bhattacharyya and Shah 2009). In the Eastern Himalayas, *Abies densa*, *Juniperus recurva*, and *Betula utilis* seem to be potential for climate reconstructions (Shekhar 2014; Shekhar and Bhattacharyya 2015; Bhattacharyya and Shekhar 2018). A list of conifers with their distributions, probable age, and their potentiality in tree-ring analysis in different perspectives from the Himalayan region is given in Table 3.1, which could help in tracing the suitable tree or a group of trees for the long record of streamflow.

Observational Data of Streamflow and Climate

In India, there are several places for the depository of measured recent streamflow data, like Central Water Commission (CWC, Ministry of Jal Shakti, Government of India), Bhakra Beas Management Board (BBMB, Ministry of Power: Government of India), and regional hydropower generation units of streamflow gauge stations. Moreover, the compilation of river database is also available through River Discharge Database, Version 1.0 (Riv DIS v1.0); Vörösmarty et al. (1996) of several Himalayan rivers but these data are of very short time and with high missing years. In general, streamflow records for the Himalayan region are available for the limited number of streams or rivers due to the lack of a sufficient number of observatories or measuring stations. Even meteorological data, especially for temperature and precipitation used for validation and reconstruction, are mostly restricted to lower elevations. Despite ~500 meteorological stations of the India Meteorological Department (IMD) in the country, the stations having data for more than a century are very few in the Himalayan region (Attri and Tyagi 2010).

Moreover, tree-ring sampling sites are often located far from the location of meteorological stations. To overcome this problem, climate data are synthesized by using the data from Climate Research Unit (CRU), (Harris et al. 2014) (Fig. 3.2a, b). Other useful gridded temperature products are from the University of Delaware (http://www.esrl.noaa.gov), and precipitation from Global Precipitation Climatology Center (GPCC) (http://gpcc.dwd.de), Tropical Rainfall Measuring Mission (TRMM) (Huffman and Bolvin 2013), and reanalyzed data of the National Center for Atmospheric Research (NCAR) and the National Center for Environmental Prediction

Table 3.1 Distribution and probable age of various Himalayan conifers and their utility in the reconstruction of climate and environment parameters

Sr. No	Name of the species	Distribution	Probable age (Years)	Applicability of tree-ring data in different domains of sciences
1	Taxus wallichiana Zucc	All along the Himalaya from Kashmir Himalaya through Khasi Hills, Naga Hills close to Himalaya to Arunachal Pradesh	Probably over 1000	Ecology Chaudhary et al. (1999), Yadav and Singh (2002)
2	Pinus merkusii Jungh. and de Vriese	Lohit district of Arunachal Pradesh at around 2500 m	350	Climate Buckley et al. (2005), Ecology Shah and Bhattacharyya (2012)
3	Pinus wallichiana A.B. Jacks	All along the Himalayas from Kashmir to Arunachal Pradesh, 1800–3700 m. Rare in Sikkim and in a considerable portion of Kumaun	400–500	Ecology Yadava et al. (2017) Climate, (Bhattacharyya and Yadav 1996; Yadav and Park (2000),, Buckley et al. (2005), Yadav (2009), Shah and Bhattacharyya (2009), Chaudhary et al. (2013), Natural Hazards Bhattacharyya et al. (2017), Flood Ballesteros Cánovas et al. (2017), Glacier Singh and Yadav (2000), Glacier Shekhar et al. (2018)
4	Pinus gerardiana Wallich ex D. don	Bashar westward to Kashmir, Chitral, North Baluchistan and Afghanistan, Kinnaur; 1800–3000 m	1000	Climate Yadav (1991), Singh and Yadav (2007), Singh et al. (2009),Yadav et al.(2013, 2017)
5	Pinus roxburghii Sargent	Entire Himalayas from Pakistan to Arunachal Pradesh at 450–2300 m, except Kashmir Valley proper	327	Climate Pant and Borgaonkar(1984), Singh and Yadav (2014), Shekhar et al. (2017), Natural Hazards Bhattacharyya et al. (2017), Ballesteros Cánovas et al. (2017), Ecology Shah et al. (2013), Forest fire Brown et al. (2011)

(continued)

Table 3.1 (continued)

Sr. No	Name of the species	Distribution	Probable age (Years)	Applicability of tree-ring data in different domains of sciences
6	Pinus kesiya Royle ex Gordon	Khasi and Naga Hills close to Himalaya	350	Ecology Chaudhary and Bhattacharyya (2002), Shah and Bhattacharyya (2012)
7	Tsuga dumosa (D. Don) Eichler [syn.T.brunoniana]	E. Kumaon to Arunachal Pradesh	600	Climate Borgaonkar et al. (2018)
8	Picea smithiana (Wall.) Boiss	In the W. Himalaya, from Afghanistan to Kumaon at 2150–3300 m	331	Climate Borgaonkar et al. (2011), Thapa et al. (2015), Ecology Thapa et al. (2017)
9	Abies pindrow (Royle ex D. Don) Royle	Kashmir Himalaya to Central Himalaya Nepal, 2300–3350 m	360	Climate Chaudhary et al. (2013), Ram and Borgaonkar (2014), Glacier history Bhattacharyya et al. (2001), Shekhar et al. (2018)
10	Abies spectabilis (D. Don) Mirbel	All along the Himalayas from Pakistan to Arunachal Pradesh, Tibet; 2800–4300 m; common between 3300–4100 m	370	Glacier history Bhattacharyya et al. (2001), Shekhar et al. (2018) Ecology Yadav et al. (2004), Shrestha et al. (2017), Singh et al. (2019) Natural Hazards Bhattacharyya et al. (2017)
11	Abies densa Griffith	Toward east of Nepal, Darjeeling to Arunachal Pradesh, at 2750–3950 m	400	Climate Bhattacharyya and Chaudhary (2003) Glacier history Shekhar(2014), Ecology Shah and Bhattacharyya (2012), Streamflow Shekhar and Bhattacharyya (2015), Palaeoseismic Dating Bhattacharyya et al. (2008)
12	Cedrus deodara (Roxburgh ex D. Don) G. Don	Kashmir Himalaya, Garhwal, Kurnauli Valley (West Nepal); 1200–3300 m	1200	Climate Bhattacharyya and Yadav (1989), Singh and Yadav (2005), Yadav (2009), Yadav et al. (2015, 2017), Shah et al. (2018), Snowfall (Yadav and Bhutiyani(2013), Glacier history Borgaonkar et al. (2009), Ecology and Streamflow Shah et al. (2013), Misra et al. (2015), Natural Hazards Laxton and Smith (2009)

(continued)

Table 3.1 (continued)

Sr. No	Name of the species	Distribution	Probable age (Years)	Applicability of tree-ring data in different domains of sciences
13	Larix griffithii Hooker f	E. Nepal to Arunachal Pradesh, Chumbi Valley (Tibet), NE Upper Myanmar, 2400–3700 m; common from 2900 to 3300 m	250	Climate Chaudhary and Bhattacharyya (2000), Bhattacharyya and Chaudhary (2003), Shekhar (2014), Yadava et al. (2015), Streamflow Shah et al. (2014)
14	Juniperus polycarpos K. Koch	Afghanistan, Baluchistan, Cagan Valley, Kashmir Lahul to over Kumaon, W Tibet, 250O–4300 m	Probably over 2000	Ecology Yadav et al. (2006)

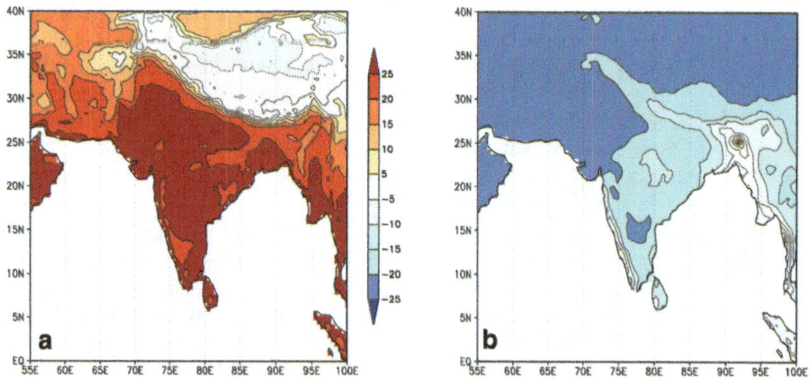

Fig. 3.2 Plot of **a** mean annual temperature (MAT) and **b** mean annual precipitation (MAP) variability over the Himalaya and Indian subcontinent based on CRU TS 4.01, gridded data (http://climexp.knmi.nl)

(NCEP). ERA-Interim reanalysis (ERA stands for ECMWF, i.e., European Center for Medium-Range Weather Forecasts (Kalnay et al. 1996) and MERRA (MERRA is a NASA reanalysis for the satellite data), etc. nearest to the study site (Dee et al. 2011).

Tree-Ring Chronology Development

From the river basin, a good number of tree-ring cores collected from suitable trees (usually 20–40) of the same species are required to prepare a robust tree-ring chronology and to maximize the common climate signal through replication of tree-ring data (Fritts 1976). Each tree-ring or growth ring was "cross-dated" and assigned a precise calendar year. The measured ring widths from multiple trees were then averaged into one site "chronology." Tree-ring widths of each dated core were measured on a hundredth of a millimeter accuracy using the increment measuring stage attached to a microcomputer. To check the dating accuracy, a computer program COFECHA Holmes (1983) was used. To assign an exact calendar date to each ring, the skeleton plot technique of cross-dating was used (Stokes and Smiley 1968; Fritts 1976).

Tree-Ring Data Standardization

In tree-ring studies, it is a common practice to standardize tree-ring series and to create average values to produce composite chronologies. Standardization of the individual tree-ring series minimizes the uncommon variability (Fritts 1976) that is not related to climate (age trends and local disturbances). Several standardization trials were carried out using the Auto-Regressive Standardization (ARSTAN) computer program (Cook 1985; Cook and Krusic 2005). In this program, ring-width series are fitted with the straight line or negative exponential function or cubic smoothing spline, etc. depending on the slope of the series. The detrend data from individual tree cores were then averaged using a bi-weight robust mean to develop the standard and residual chronologies (Cook and Kairiukstis 1990).

Further, the quality of tree-ring chronologies was evaluated estimating the mean correlation between all series (RBAR) and expressed population signal (EPS). The RBAR is a measure of the common variance between all series in a chronology (Wigley et al. 1984), whereas the EPS is a measure of the similarity between a chronology and a hypothetical chronology that has been infinitely replicated (Briffa 1995). The EPS statistic is particularly sensitive to the chronology replication and points out the minimum number of samples in a chronology required to maintain a robust common signal. The quantitative and qualitative characteristics of tree-ring chronologies are assessed based on these characteristics: (i) data homogeneity, (ii) sample replication, (iii) growth coherence, (iv) chronology development, and (v) climate signal including the correlation with instrumental data (Esper et al. 2016). Shekhar and Bhattacharyya (2015) used the double-detrending approach, where a first negative exponential curve (Fritts 1976), then the series were detrended a second time with 30 years smoothing spline curve to reduce the impact of biotic factors such as competition and defoliation on the radial growth (Fig. 3.3a, b).

a

b

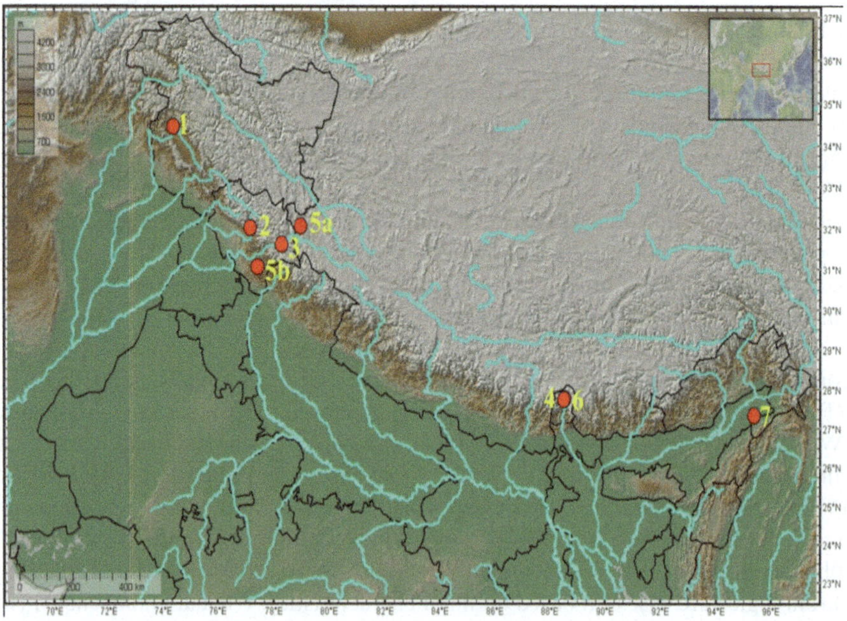

Fig. 3.3 **a** Stem injuries induced by hydrogeomorphic processes were modified after Bhattacharyya et al. (2017). **b** Map showing the tree-ring-based streamflow reconstruction sites (filled red circles) from the Himalayan region. (1) Upper Indus Basin Cook et al. (2013), (2) Beas River Shah et al. (2013), (3) Satluj River Singh and Yadav (2013), (4) Lachen River Shah et al. (2014), (5a,b) Satluj River Misra et al. (2015), (6) Lachen River Shekhar and Bhattachryya (2015), and (7) Lohi River Shah et al. (2019)

In some cases, negative exponential curve fitting has been used for reconstruction (Singh and Yadav 2013; Misra et al. 2015). Radial growth trends show an exponential decay with time after the juvenile period of high radial increment. Negative exponential and cubic spline curve fitting are therefore efficient methods to remove the growth trends related to age and size in the ring-width series of trees growing as open stands; as to preserve long-term fluctuations in the series, the negative exponential curve has been used to detrend the ring-width measurement series (Fritts 1976; Cook et al. 1990). In few cases, the cubic spline method was used where suppression and release in growth were noted due to competition. Shah et al. (2014) used the cubic spline approach that retains the maximum common climate signal and removes the non-climatic trend in the tree-ring standardization (Fritts 1976).

Models for Streamflow Reconstructions

Extension of Gauged Records

In most cases, gauged streamflow records in India are of short temporal length. The skillful management of water resources requires comprehensive knowledge of the natural variability over a longer time in the records of streamflow (Woodhouse and Lukas 2006). To establish the relationships between tree growth and streamflow data, tree-ring data are compared with the available streamflow data of the corresponding year (Fritts 1976). There are several statistical models to extend the streamflow data, viz., Bayesian Regression (BR) used for Indus River, Upper Indus Basin (Rao et al. 2018), and linear regression for Zemuchuu River, Sikkim Eastern Himalaya (Shekhar and Bhattacharyya 2015), etc. Generally, among the several statistical methods, multiple linear regressions are commonly used to determine the subset of tree-ring chronologies that estimate best the gauged streamflow, resulting in a regression equation. The skill of the model was evaluated using independent data or on subsets of the calibration data. The model was then applied to the full length of the tree-ring records for generating the streamflow data that extend back hundreds of years depending on the length of tree-ring chronology. The ultimate goal of streamflow reconstructions is to provide the basic information for the understanding of cause and effect in the climate system, and thus, despite clear restrictions and uncertainties, high-resolution reconstructions represent an attempt to extend the instrumental record back in time (Ammann and Wahl 2007). Depending on what questions to be addressed, one can exploit to enhance the signals of the proxy information by diluting or deleting the noise with different techniques. In the Indian context, Shah et al. (2013) performed tree growth and climate (temperature and precipitation) relationship based on bootstrap correlation using R package bootRes (Zang and Biondi 2013). For the reconstruction of the most significant monthly or seasonal streamflow variables, we carried out a correlation analysis between the residual chronology and monthly streamflow data. A simple linear regression-based transfer function model was then used for reconstructing the strongest seasonal streamflow signals. For other discharge reconstructions from both Western and Eastern Himalayas, a linear regression model was used (Singh and Yadav 2013; Shah et al. 2014; Shekhar and Bhattacharyya 2015; Misra et al. 2015).

Reconstruction of Floods from Flood Scared Trees

Fundamentally, trees growing close to the stream channel and or relatively away might be influenced by the unusually high or low streamflow, thus useful to understand the streamflow extremes. Reconstruction of past flood events implies the identification and precise dating of the anatomical growth responses (Fig. 3.3a). In the flood study, event years associated with flood occurrence were identified through the dating of the following growth disturbances (GD): (i) onset of compression wood (Timell 1986), (ii) the first year with abrupt growth suppression Kogelnig (Mayer et al. 2013), (iii) release (Mundo et al. 2007), (iv) onset of callus tissue formation (Stoffel and Klinkmüller 2013), and (v) formation of tangential rows of traumatic resin ducts (TRD) (Stoffel and Beniston 2006; Stoffel 2008; Schneuwly et al. 2009; Stoffel et al. 2010). In the case of TRD formed in consecutive years, only the first TRD year has been assigned as an event year.

Method for Flood Reconstruction

The tree-ring-based natural hazards study is based on the process-event-response concept, (Shroder 1978, 1980). Here, we use a yearly index value defined as the ratio between trees showing growth responses and all sampled trees being alive in that year. In the "conventional" approach, an index value I_t was calculated for each year t, based on the following formula:

$$I_t = \left(\left(\sum_i^n \mathrm{Rt}\right) / \left(\sum_i^n \mathrm{At}\right)\right) * 100$$

where R is the number of trees showing a GD as a response to flood events in year t, and A is the total number of sampled trees alive in year t (Ballesteros-Cánovas et al. 2017).

Validation of Reconstruction Models

There are several methods to determine the validity of the regression models, but the commonly used one is the cross-validation technique. It is useful for overcoming the problem of over-fitting (Firtts 1976). There are different methods to evaluate the performance of reconstruction models, viz., data split (cross-validation), leave-one-out cross-validation (LOOCV), Jackknife and bootstrap, K-fold cross-validation, etc. The most common one is the data split method where the data are divided into two sets. First, we build the model based on a calibration data set and then apply the model to another data set (e.g., tree-ring) to make predictions; subsequently, the validation set is used to test the model by estimating the prediction error. For the streamflow reconstructions, several statistical parameters, viz., correlation coefficient (r), reduction of error (RE), sign test (ST) and product mean test (PMT), coefficient of determination (*R* Square), $F = F$ test; (RE) reduction of error; (CE) coefficient of efficiency, have been used (Shah et al. 2013; Shekhar 2014; Singh and Yadav 2014; Shekhar and Bhattacharyya 2015; Misra et al. 2015). Generally, the sample split calibration verification method has been used in Indian streamflow papers (Shekhar 2014; Singh and Yadav 2014; Shekhar and Bhattacharyya 2015; Misra et al. 2015). In addition, Shah et al. (2013) used the leave-one-out cross-validation test when the data were short and not enough to be divided into two equal parts (Fritts 1976).

Results and Discussion

Streamflow Reconstructions

Streamflow reconstructions available to date are meager in comparison to the vast number of rivers in the Himalayan region. So far, a total of six rivers from both the Eastern and Western parts of the Himalayas analyzed for the streamflow analyses (Fig. 3.3b). Streamflow data from the central part of the Himalayas are yet to come.

The Western Himalaya

From this region tree-ring data of two conifers, *Cedrus deodara* (Deodar) and *Pinus gerardiana* are of much potential for the long record of river flow, though several other conifers growing at this region are also recorded potential for the dendroclimatic analysis (Bhattaharyya et al. 2012). These trees may also be suitable for the streamflow. Shah et al. (2013) reconstructed the flow of the Beas River using the tree-ring data of *Cedrus deodara* (Deodar) from the Kullu valley, Himachal Pradesh (India). Their study showed positive relationships with the early summer months (March–April) based on the correlation between tree-ring chronology of deodar trees and discharge data of the Beas River basin from several gauge stations. Linear regression transfer function model going back to 1834 C.E. was used for the streamflow reconstruction of early summer months. The reconstructed streamflow data from 1834 to 1984 C.E. showed five high and low discharge phases (Table. 3.2).

In the Himalayan region, the longest streamflow data (1295–2005 C.E) were developed from Satluj River, Kinnaur, Western Himalaya (Singh and Yadav 2013).

Table 3.2 Tree-ring-based streamflow reconstructions (high and low flow years) in and around adjoining areas of the Himalaya

Serial No	Site name	Species	Time span of the reconstruction (C.E.)	High flow year (C.E.)	Low flow year (C.E.)	Reference
1	Beas River Kullu (Western Himalaya)	Cedrus deodara (Roxburgh ex D. Don) G. Don	1834–1998	1840–1843, 1893–1897, 1922–1926, 1935–1940 and 1942–1945	1847–1851, 1863–1866, 1927–1929, 1960–1962 and 1972–1974	Shah et al. (2013)
2	Kinnaur, Himachal Pradesh (Western Himalaya)	Pinus gerardiana Wallich ex D. don and Cedrus deodara (Roxburgh ex D. Don) G. Don	1295–2005	1817–1846, 1954–1983, 1953–2002	1673–1722 and 1450s–1510s, 1540s–1560s, 1610s–1710s, and 1770s–1820s	Singh and Yadav (2013)
3	Kinnaur, Himachal Pradesh (Western Himalaya)	Cedrus deodara (Roxburgh ex D. Don) G. Don	1660–2004	1751–1760, 1922–1931, 1730–1739 and 1822–1831	1779–1788, 1812–1821, 1740–1749 and 1907–1916	Misra et al. (2015)
4	Upper Indus Basin (Hunza Valley region of northern Pakistan)	Cedrus deodara (Roxburgh ex D. Don) G. Don Pinus gerardiana Wallich ex D. don Pinus wallichiana A.B. Jacks. Picea smithiana (Wall.) Boiss, etc	1452–2008	1988–2008	1572–1683, 1637–1663	Cook et al. (2013)
5	Sikkim, the (Eastern Himalaya)	Larix griffithii Hooker f		1823–1835, 1879–1890, 1926–1946 and 1980–1989	1791–1805, 1813–1822 and 1914–1925	Shah et al. (2014)

(continued)

Table 3.2 (continued)

Serial No	Site name	Species	Time span of the reconstruction (C.E.)	High flow year (C.E.)	Low flow year (C.E.)	Reference
6	Sikkim, (Eastern Himalaya)	Abies densa Griffith	1775–1996	1778, 1786, 1791,1792, 1802, 1810, 1821, 1823, 1830, 1833, 1835, 1842, 1851, 1852, 1853, 1870, 1873, 1900, 1915, 1930, 1931, 1946, and 1975	1782, 1783, 1799, 1814, 1815, 1825, 1837,1839, 1840, 1844, 1848, 1860, 1864, 1866, 1905, 1921, 1932, 1940, 1952, 1967 and 1975	Shekhar and Bhattacharyya (2015)

This reconstruction is based on a combined ring-width chronology of *Cedrus deodara* from three sites and *Pinus gerardiana* from one site. The reconstruction revealed 50 years of low and high river discharge respectively in the eighteenth and nineteenth centuries. A consistency was observed between the decreasing trends both in river discharge and winter precipitation of the region since the 1990s.

Recently, Misra et al. (2015) from the same river, developed October (previous year) to June (current year) streamflow reconstruction based on a network of deodar ring-width chronologies of seven sites in the Kinnaur district of Himachal Pradesh. The reconstruction extends back to 1660 C.E. and captured 40% of the variance in the calibration period (C.E. 1923–1952). Since the 1990s, a consistency has been reported between the flow of this river and winter precipitation, showing a decreasing trend. The reconstruction showed ten years of high and 20 years of low streamflow records in the twentieth century. From the adjacent Karakorum Mountain, Cook et al. (2013) reconstructed streamflow of May–September for the Upper Indus Basin. They reported that the flow level in the 1990s was higher during last 1452–2008 C.E. Further, the streamflow reconstruction was extended by Rao et al. (2018) covering the period from 1430 to 2008 C.E. In their reconstruction, dry periods were reported during the mid-1600 s, late 1700s, and early 1900s C.E. but unprecedented increasing streamflow trend during late 1990s C.E. The increasing trend of streamflow in the recent years may indicate its linkage with the increasing temperature and global warming; however further in the long term, the trend might decrease (Singh et al. 2016).

The Eastern Himalaya

The rivers of the Eastern Himalayas receive a high amount of rainfall during the rainy season through the southwest monsoon, which often causes floods during

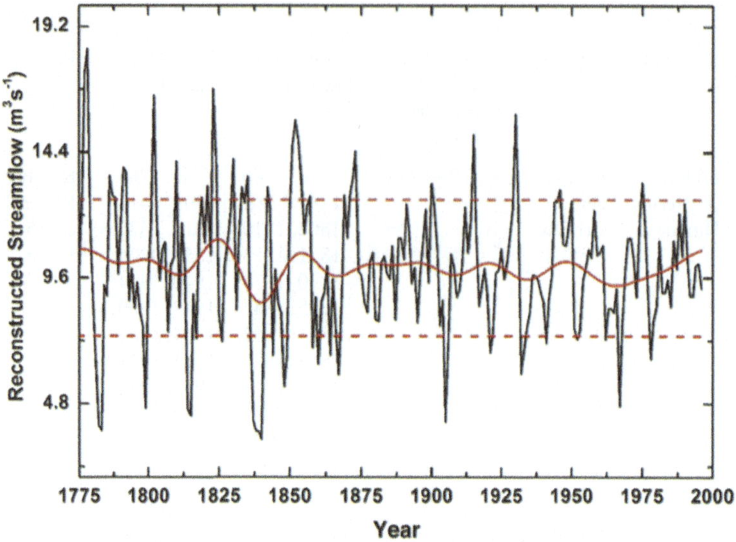

Fig. 3.4 January–April (J–A) streamflow reconstruction of "Zemuchuu," Eastern Himalaya (thick black line). Smooth red line indicating ten-year Fourier Transform (FTT) filter of reconstructed streamflow. Red dotted upper and lower horizontal lines represent, respectively, the wet years (mean + 1 standard deviation) and drought years (mean + 1 Standard deviation)

that period and at the same time acute shortage of water during the pre-monsoon period. This pre-monsoon reconstruction of river flow would be of great significance in knowing the acute water scarcity events and causes, but the long reconstructed discharge data are available only from the upper tributary of the Teesta River. Shekhar and Bhattacharyya (2015) reconstructed January–April (J-A) mean discharge of the Zemuchuu and Teesta River at Lachen, North Sikkim, since 1775. They reported 23 high discharge and 21 extremely low discharge years over the past 1775–1996 C.E. (Fig. 3.4).

Further, we recorded a significant correlation between the J–A streamflow of these rivers and the precipitation of the same months, which validates the main role of precipitation in streamflow (Table 3.2; Fig. 3.5). Based on earlywood width (EWW) and latewood width (LWW), chronologies of *Larix griffithiana* (Shah et al. 2014) built the relationship of streamflow records of Lachen River, Lachen, North Sikkim, Eastern Himalaya. The chronologies significantly correlate with the observed discharge of the Lachen River, where 61.2% variance of streamflow was explained by the EWW chronology. Discharge of the Lachen River, extending back to 1790 C.E., was reconstructed for the period of the previous year March to the current year February using EWW chronology. In the smoothed reconstructed data, the three and four phases of extreme low and high streamflow were observed, respectively (Table. 3.2). The streamflow was also found to be lower than the average during the monsoon failure of 1792–1796 C.E and 1876–1878 C.E. The suppressed tree

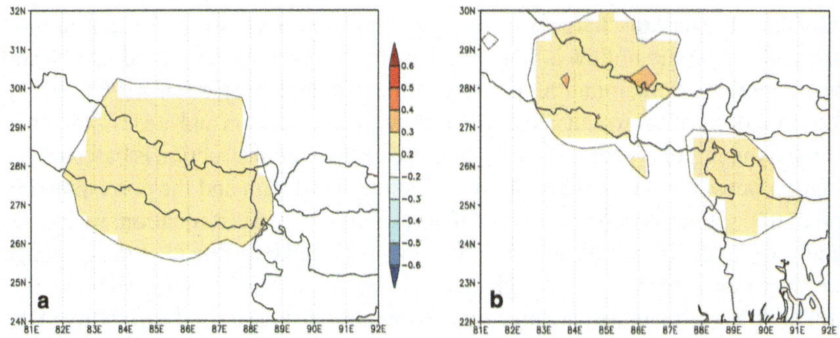

Fig. 3.5 Spatial correlation patterns of January–April reconstructed streamflow **a** with January–April gridded precipitation data of CRU TS.322 and **b** drought indices of the Standardized Precipitation-Evapotranspiration Index (SPEI) for January–April months

growth during 1816–1822 C.E. According to Shah et al. (2014), such reduction in streamflow could be related to the Tambora volcanic eruption of Indonesia.

Flood Reconstruction from the Himalaya

Flood is an indication of high discharge for a short period. Recently, tree-ring dating has also been given much impetus in flood reconstruction from the Himalayan region. Fundamentally, trees closer to the stream channel and those further away might be influenced by unusually high or low streamflow. These trees with datable tree-ring sequences are potential to analyze the streamflow extremes. Under the Indo-Swiss collaborative research project, we explored the dating of flash floods from Beas valley of the Kullu Himalaya (Bhattacharyya et al. 2017) based on tree-ring dating of flood scar in both conifers and broad-leaved taxa of this region. The eleven flooding events in the Sainj River, five in the Beas River, nine in the Thiertan River, and eight in the Parvati River have been reconstructed (Ballesteros-Cánovas et al. 2017).

Limitations, Challenges of Dendrohydrology, and Future Roadmap for Water Resource Management

For water resource management, the database of past high and low flow events for a long period is necessary to understand the natural variability in a river system. As discussed earlier, hydrological reconstructions using tree-ring chronologies in principle are based on their relationships with precipitation. Thus, in principle, tree-rings seem to be an indirect proxy of the river flow reconstruction, as precipitation that directly affects tree growth also influences the river flow. In the mountainous region to understand the past discharge behavior, the information regarding the contribution of snowmelt in the streamflow is vital when most of the glaciers are receding fast and predicted increased precipitation in future climate change projections (IPCC 2013). The contributions of snowmelt in the streamflow are not uniform and are less known in the Himalayan region. As discussed earlier, streamflow reconstruction is based on the correlation between the available short period observed data and the corresponding period of tree growth records. Based on that resultant regression equation, we extend streamflow in past. However, uncertainty in the observed data due to climate change and accelerated melting of snow/glaciers, the relationship becomes site-specific. Precipitation and streamflow are projected to increase in amount and variability in the future (IPCC 2013). It has been recorded that especially in the High Asian region; the glaciers contribute substantially to water resources forming the headwaters of many of the continent's largest rivers (Hock 2005). Immerzeel et al. (2010) found that the glacier melt portion of streamflow in five major Southeast Asian rivers is significant. There are several studies on the contributions of glacier melt in many rivers. Those are by Singh et al. (1997) from Akhnoor for the Chenab River, 49.1% (1982–1992); Singh and Jain (2002) from Bhakra Dam for the Satluj River, 59% (1986–1996) and Kumar et al. (2007) from Pandoh Dam for the Beas River 37.4% (1998–2004). However, with increased warming, our estimates of glacier contribution fractions may increase as mass loss accelerates but ultimately would be more canceled by the loss of glacier areas. In the coming years, the eco-physiological basis of using tree growth as a measure of hydro-meteorological conditions is expected to be properly established. Physiological understanding of water and tree growth relationship based on oxygen isotopes data of both tree-ring and water would be a vital question need to be addressed. It would also be possible to quantify glacier/snowmelt contribution in the main river system and their long record reconstructions. The water flow is also often affected by human activities, and this may reduce the climate signal. Human-induced possible disturbances could have resulted in a larger noise of the chronologies. In addition, the chronologies could be more reliable by using more samples and older trees.

Conclusions

Short-term observational data of streamflow in the Himalayan region restrict meaningful and robust hydrological modeling. In the past few decades, tree-rings as the proxy have been progressively applied as a tool in the research and planning of water resources. A review of tree-ring-based streamflow and flood reconstructions of the Himalayan region discussed in this study reveals that the data are available only for a few rivers of both the Eastern and Western parts of the Himalayan region. However, concerning a large number of rivers in the Himalayan region, data on the reconstruction of river flow are in the initial stage and still not adequate to understand its natural variability spatio-temporally. A network of climatically sensitive trees growing in the contrasting climatic zones of both northwestern and northeastern parts would provide a large number of trees for preparing longer tree-ring chronologies for the long records of streamflow. Due to anthropogenic pressure, old trees are becoming endangered throughout the Himalayas. Extensive efforts are now required in the collection of tree-ring samples from new geographical areas and ancient wood used for various purposes. Collecting samples from leftover stumps of logging in forest sites, old timbers used in the construction of houses, bridges, etc., and subfossil wood may provide materials for extending the tree-ring chronologies in a longer time frame. Long-term analysis of the detailed climatic dynamics vis-à-vis river flow of the Indian subcontinent would provide information regarding the impact of climate on river flow or streamflow during the Medieval Warm Period, Little Ice Age, and recent global warming. For this, it is felt necessary to explore more trees under both conifers and broad-leaved segments. A significant part of the Himalayas is also covered with broad-leaved taxa, and possibilities of their radial growth seem to be limited by the vagaries of the monsoon. Most of the tree-ring analyses are based on ring width, while other tree-ring parameters such as density, isotopes, cell size, and vessel area are not considered seriously for understanding the past climatic change vis-à-vis streamflow behavior. The stable isotopic ratio in tree-ring could also be taken into consideration to evaluate the impact of the source of natural streamflow regimes. Although tree-ring $d^{18}O$ and precipitation relationship have been studied in the Himalaya (Sano et al. 2010, 2013), the growth model/streamflow and the isotopic relationship have not been explored yet in the Himalayan region.

Acknowledgements The authors express their gratitude to Director, Birbal Sahni Institute of Palaeosciences (BSIP), Lucknow, Uttar Pradesh, India, for encouragement and permission to publish this work (vide: BSIP/RDCC/Publication No.32/2018–19) under the in-house 5.8 project. One of the authors (MS) also acknowledges the Birbal Sahni Research Associate grant from BSIP, Lucknow, Uttar Pradesh, India, and Department of Science and Technology (DST), New Delhi, India, startup grant for Young Scientist under Grant No. SR/FTP/ES–127/2014 sss for the work.

References

Alestalo J (1971) Dendrochronological interpretation of geomorphic processes. Fennia 105:1–140

Ali H, Mishra V (2017) Contrasting response of rainfall extremes to increase in surface air and dew point temperatures at urban locations in India. Sci Rep 7(1):1228. https://doi.org/10.1038/s41598-017-01306-1

Ammann CM, Wahl ER (2007) The importance of the geophysical context in statistical evaluations of climate reconstruction procedures. Clim Chang 85(1–2):71–88. https://doi.org/10.1007/s10584-007-9276-x

Attri SD, Tyagi A (2010) Climate profile of India. Environ Monit Res Cent India Meteorol Dep: New Delhi India

Axelson JN, Sauchyn DJ, Barichivich J (2009) New reconstructions of streamflow variability in the South Saskatchewan River Basin from a network of tree ring chronologies, Alberta, Canada. Water Res Res 45(9):09422. https://doi.org/10.1029/2008WR007639

Baker VR (2008) Paleoflood hydrology: origin, progress, prospects. Geomorphology 101:p1-13. https://doi.org/10.1016/j.geomorph.2008.05.016
Ballesteros-Cánovas JA, Stoffel M, St George S et al (2015) A review of flood records from tree rings. ProgPhys Geogr 39(6):794–816. https://doi.org/10.1177/0309133315608758
Bekker MF, Justin DeRose R, Buckley BM et al (2014) A 576-year weber river streamflow reconstruction from tree rings for water resource risk assessment in the wasatch front, Utah. J Am Water Resource Assoc 50(5):1338–1348. https://doi.org/10.1111/jawr.12191
Bhattacharyya A, Chaudhary V (2003) Late-summer temperature reconstruction of the eastern Himalayan region based on tree-ring data of Abiesdensa. Arct Antarc Alp Res 35(2):196–202. https://doi.org/10.1657/1523-0430(2003)035[0196:LTROTE]2.0.CO;2
Bhattacharyya A, Chaudhary V, Gergan JT (2001) Tree ring analysis of Abiespindrow around DokrianiBamak (Glacier), western Himalayas, in relation to climate and glacial behaviour: preliminary results. Palaeobotanist 50:71–75
Bhattacharyya A, LaMarche Jr VC, Telewski FW (1988) Dendrochronological reconnaissance of the conifers of northwest India. Tree-Ring Bull
Bhattacharyya A, Shah SK (2009) Tree-ring studies in India past appraisal, present status and future prospects. IAWA J 30(4):361–370
Bhattacharyya A, Shah SK, Chaudhary V (2008) Feasibility of tree-ring data in palaeoseismic dating in northeast Himalaya. J Geol Soc India 71(3):419–424
Bhattacharyya A, Shekhar M (2018) Scenario of tree ring study of the Himalayas—present status and future outlook. In: Das AD, Bera S (eds) Plant diversity in the Himalaya Hotspot Region. Bishen Singh Mahendra Pal Singh, Dehra Dun (India), pp 668–676
Bhattacharyya A, Shekhar M, Shah SK (2012) Role of tree-ring study in forest management: prospects in Indian context. In: Panda S, Ghosh C (eds) Diversity and conservation of plants and traditional knowledge. Bishen Singh Mahendra Pal Singh, Dehradun, pp 287–298
Bhattacharyya A, Stoffel M, Shekhar M et al (2017) Dendrogeomorphic potential of the Himalaya–case studies of process dating of natural hazards in Kullu valley, Himachal Pradesh. Curr Sci pp 2317–2324. https://doi.org/10.18520/cs%2Fv113%2Fi12%2F2317-2324
Bhattacharyya A, Yadav RR (1989) Growth and climate relationship in Cedrusdeodara from Joshimath, Uttar Pradesh. Palaeobotanist 38:411–414
Bhattacharyya A, Yadav RR (1996) Dendrochronological reconnaissance of Pinuswallichiana to study glacial behaviour in the western Himalaya. Curr Sci 70:739–744
Borgaonkar HP, Gandhi N, Ram S et al (2018) Tree-ring reconstruction of late summer temperatures in northern Sikkim (eastern Himalayas). Palaeogeogr Paleoclimatol Palaeoecol 504:125–135. https://doi.org/10.1016/j.palaeo.2018.05.018
Borgaonkar HP, Pant GB, Kumar KR (1999) Tree-ring chronologies from western Himalaya and their dendroclimatic potential. IAWA J 20(3):295–309
Borgaonkar HP, Ram S, Sikder AB (2009) Assessment of tree-ring analysis of high-elevation Cedrusdeodara D. Don from Western Himalaya (India) in relation to climate and glacier fluctuations. Dendrochronologia 27:59–69. https://doi.org/10.1016/j.dendro.2008.09.002
Borgaonkar HP, Sikder AB, Ram S (2011) High altitude forest sensitivity to the recent warming: a tree-ring analysis of conifers from western Himalaya, India. Quat Int 236:158–166. https://doi.org/10.1016/j.quaint.2010.01.016
Bradley RS (1999) Paleoclimatology: reconstructing climates of the quaternary. Elsevier
Briffa KR (1995) Statistical aspects of high-resolution proxy climate data: the example of dendroclimatology. In: Von Storch H, Navarra A (eds) Analysis of climate variability: Application of statistical techniques Springer Berlin, pp 77–94
Brown PM, Bhattacharyya A, Shah SK (2011) Potential for developing fire histories in Chir Pine (Pinusroxburghii) Forests in the Himalayan Foothills. Tree-Ring Res 67(1):57–62
Buckley BM, Cook BI, Bhattacharyya A et al (2005) Global surface temperature signals in pine ring-width chronologies from southern monsoon Asia. Geophys Res Lett 32(20). https://doi.org/10.1029/2005GL023745

Cánovas JB, Trappmann D, Shekhar M et al (2017) Regional flood-frequency reconstruction for Kullu district, Western Indian Himalayas. J Hydrol 546:140–149. https://doi.org/10.1016/j.jhydrol.2016.12.059

Chalise SR, Kansakar SR, Rees G et al (2003) Management of water resources and low flow estimation for the Himalayan basins of Nepal. J Hydrol 282(1–4):25–35. https://doi.org/10.1016/S0022-1694(03)00250-6

Chaudhary V, Bhattacharyya A (2000) Tree ring analysis of Larixgriffithiana from the Eastern Himalayas in the reconstruction of past temperature. Curr Sci 79:1712–1716

Chaudhary V, Bhattacharyya A (2002) Suitability of Pinuskesiya in Shillong Meghalaya for tree-ring analyses. Curr Sci 83(8):1010–1015

Chaudhary V, Bhattacharyya A, Guior J (2013) Reconstruction of August-September temperature, in North-Western Himalaya since AD 1773, based on tree-ring data of Pinuswallichiana and Abiespindrow. In: Kotlia BS (ed) Holocene: perspectives, environmental dynamics and impact events. Nova Science Publishers, pp 145–156

Chaudhary V, Bhattacharyya A, Yadav RR (1999) Tree-ring studies in the eastern Himalayan region: prospects and problems. IAWA J 20(3):317–324

Cook ER (1985) A time series analysis approach to tree-ring standardization (Dendrochronology, Forestry, Dendroclimatology, Autoregressive Process).

Cook ER, Anchukaitis KJ, Buckley BM et al (2010) Asian monsoon failure and megadrought during the last millennium. Science 328(5977):486–489. https://doi.org/10.1126/science.1185188

Cook ER, Briffa KR, Shiyatov S et al (1990) Tree-ring standardization and growth-trend estimation. Methods of dendrochronology: Appl Environ Sci pp104–123

Cook ER, Jacoby GC (1983) Potomac river streamflow since 1730 as reconstructed by tree rings. J Appl Meteorol Climatol 22:1659–1672. https://doi.org/10.1175/1520-0450(1983)022%3c1659:PRSSAR%3e2.0.CO;2

Cook ER, Kairiukstis LA (1990) Methods of dendrochronology. In: Applications in the environmental sciences, vol 394. Kluwer Academic Publishers, Dordrecht. http://dx.doi.org/https://doi.org/10.1007/978-94-015-7879-0

Cook ER, Kairiukstis LA (2013) Methods of dendrochronology: Appl Environ Sci Springer Sci Bus Media

Cook ER, Krusic PJ (2005) Program ARSTAN: a tree-ring standardization program based on detrending and autoregressive time series modeling, with interactive graphics. Columbia University, Palisades, NY, Lamont-Doherty Earth Observatory

Cook ER, Palmer JG, Ahmed M et al (2013) Five centuries of upper Indus River flow from tree rings. J Hydrol 486:365–375. https://doi.org/10.1016/j.jhydrol.2013.02.004

Cronin TM (2009) Paleoclimates: understanding climate change past and present, vol 448. Columbia University Press

Dee DP, Uppala SM, Simmons AJ et al (2011) The ERA-Interim reanalysis: configuration and performance of the data assimilation system. Q J R Meteorol Soc 137(656):553–597. https://doi.org/10.1002/qj.828

Esper J, Krusic PJ, Ljungqvist FC et al (2016) Ranking of tree-ring based temperature reconstructions of the past millennium. Quat Sci Rev 145:134–151. https://doi.org/10.1016/j.quascirev.2016.05.009

Ewane BE, Lee HH (2017) Tree-ring reconstruction of streamflow for Palgong Mountain forested watershed in southeastern South Korea. J Mt Sci 14(1):60–76. https://doi.org/10.1007/s11629-016-3860-3

Fritts HC (1976) Tree rings and climate. Press London 1991

Gadgil S (2003) The Indian monsoon and its variability. Annu Rev Earth Planet Sci 31:429–467

Gadgil S, Vinayachandran PN, Francis PA et al (2004) Extremes of the Indian summer monsoon rainfall, ENSO and equatorial Indian Ocean oscillation. Geophys Res Lett 31(12) https://doi.org/10.1029/2004GL019733

Global Precipitation Climatology Centre (GPCC). National Center for Atmospheric Research, USA. http://gpcc.dwd.de

Gosain AK, Rao S, Basurya D (2006) Climate change impact assessment on hydrology of Indian River Basins, climate change and India. Curr Sci 90:346–353

Gou X, Chen F, Cook E et al (2007). Streamflow variations of the Yellow River over the past 593 years in western China reconstructed from tree rings. Water Resour Res 43(6). https://doi.org/10.1029/2006WR005705

Hardman G, Reil OE (1936) relationship between tree-growth and stream runoff in the Truckee River Basin. California-Nevada. https://doi.org/10.1029/TR017i002p00491

Harris IPDJ, Jones PD, Osborn TJ et al (2014) Updated high-resolution grids of monthly climatic observations–the CRU TS3. 10 Dataset. Int J Climatol 34(3):623–642. https://doi.org/10.1002/joc.3711

Hawley FM (1937) Relationship of southern cedar growth to precipitation and run off. Ecol 18(3):398–405

Helley EJ, LaMarche VC (1968) December 1964, a 400-year flood in Northern California. US GeolSurv Prof Pap 600:34–37

Hock R (2005) Glacier melt: a review of processes and their modelling. Prog Phys Geogr 29(3):362–391. https://doi.org/10.1191/0309133305pp453ra

Holmes RL (1983) Computer-assisted quality control in tree-ring dating and measurement. Tree-ring bull

Huffman GJ, Bolvin DT (2013) TRMM and other data precipitation data set documentation. ftp://meso-a.gsfc.nasa.gov/pub/trmmdocs/3B42_3B43_doc.pdf

Hughes MK (1992) Dendroclimatic evidence from Western Himalaya. In Bradley RS, Jones PD (eds) Climate since AD 1500 Routledge, London, pp 415–431

Hughes MK, Xiangding W, Xuemei S et al (1994) A preliminary reconstruction of rainfall in north-central China since AD 1600 from tree-ring density and width. Quat Res 42(1):88–99

Immerzeel WW, Van Beek LP, Bierkens MF (2010) Climate change will affect the Asian water towers. Science 328(5984):1382–1385. https://doi.org/10.1126/science.1183188

IPCC (2013) In: Stocker TF, Qin D, Plattner GK et al (eds) Climate change 2013: the physical science basis. contribution of working group I to the fifth assessment report of the intergovernmental panel on climate change. Cambridge University Press, Cambridge and New York

Kalnay E, Kanamitsu M, Kistler R et al (1996) The NCEP/NCAR 40-year reanalysis project. Bull Am Meteorol Soc 77(3):437–472. https://doi.org/10.1175/1520-0477(1996)077<0437:TNYRP>2.0.CO;2

Kharal DK, Meilby H, Rayamajhi S et al (2014) Tree ring variability and climate response of Abiesspectabilis along an elevation gradient in Mustang, Nepal. Banko Janakari 24(1):3–13

Kochel RC, Baker VR (1982) Paleoflood hydrology. Science 215:353–361. https://doi.org/10.1126/science.215.4531.353

Kogelnig-Mayer B, Stoffel M, Schneuwly-Bollschweiler M (2013) Four-dimensional growth response of mature Larix decidua to stem burial under natural conditions. Trees 27(5):1217–1223. https://doi.org/10.1007/s00468-013-0870-4

Krishna Kumar K, Rajagopalan B, Cane MA (1999) On the weakening relationship between the Indian Monsoon and ENSO. Science 284:5423. https://doi.org/10.1126/science.284.5423.2156

Kumar V, Singh P, Singh V (2007) Snow and glacier melt contribution in the Beas river at Pandoh dam, Himachal Pradesh, India. Hydrol Sci J 52(2):376–388. https://doi.org/10.1623/hysj.52.2.376

Laxton SC, Smith DJ (2009) Dendrochronological reconstruction of snow avalanche activity in the Lahul Himalaya, Northern India. Nat Hazards 49(3):459–467. https://doi.org/10.1007/s11069-008-9288-5

Liu Y, Sun J, Song H et al (2010) Tree-ring hydrologic reconstructions for the Heihe River watershed, western China since AD 1430. Water Res 44(9):2781–2792. https://doi.org/10.1016/j.watres.2010.02.013

Meko D, Woodhouse CA (2011) Application of streamflow reconstruction to water resources management. Dendroclimatology. Springer, Dordrecht, pp 231–261

Meko DM, Therrell MD, Baisan CH et al (2001) Sacramento River flow reconstructed to AD 869 from tree rings. J Am Water Res Assoc 37(4):1029–1039. https://doi.org/10.1111/j.1752-1688.2001.tb05530.x

Michaelsen J, Haston L, Davis FW (1987) 400 years of central california precipitation variability reconstructed from tree-rings. J Am Water Resour Assoc 23(5):809–818

Misra KG, Yadav RR, Misra S (2015) Satlujriver flow variations since AD 1660 based on tree-ring network of Himalayan cedar from western Himalaya, India. Quat Int 371:135–143. https://doi.org/10.1016/j.quaint.2015.01.015

Mundo IA, Barrera MD, Roig FA (2007) Testing the utility of Nothofagus pumilio for dating a snow avalanche in Tierra del Fuego, Argentina. Dendrochronologia 25:19–28

Pant GB, Borgaonkar HP (1984) Growth rate of Chir pine (Pinusroxburghii) trees in Kumaun area in relation to regional climatology. Himalayan Res Dev 3(2):1–5

Parthasarathy B, Diaz HF, Eischeid JK (1988) Prediction of all-India summer monsoon rainfall with regional and large-scale parameters. J Geophys Res Atmos 93(D5):5341–5350. https://doi.org/10.1029/JD093iD05p05341

Pederson N, Jacoby GC, D'Arrigo RD et al (2001) Hydrometeorological reconstructions for northeastern Mongolia derived from tree rings: 1651–1995. J Clim 14(5):872–881. https://doi.org/10.1175/1520-0442(2001)014%3c0872:HRFNMD%3e2.0.CO;2

Polacek D, Kofler W, Oberhuber W (2006) Radial growth of Pinussylvestris growing on alluvial terraces is sensitive to water-level fluctuations. New Phytol 169(2):299–308

Ram S, Borgaonkar HP (2014) Climatic response of various tree ring parameters of fir (Abiespindrow) from Chandanwadi in Jammu and Kashmir, western Himalaya, India. Curr Sci 106(11):1568–1576

Rao MP, Cook ER, Cook BI et al (2018) Six centuries of Upper Indus Basin streamflow variability and its climatic drivers. Water Resource Res 54(8):5687–5701. https://doi.org/10.1029/2018WR023080

Rees HG, Collins DN (2006) Regional differences in response of flow in glacier-fed Himalayan rivers to climatic warming. Hydrol Process 20:2157–2169. https://doi.org/10.1002/hyp.6209

Sano M, Sheshshayee MS, Managave S et al (2010) Climatic potential of δ18O of Abies spectabilis from the Nepal Himalaya. Dendrochronologia 28(2):93–98

Sano M, Tshering P, Komori J et al (2013) May–September precipitation in the Bhutan Himalaya since 1743 as reconstructed from tree ring cellulose d18O. J Geophys Res Atmos 118(15):8399–8410

Schneuwly DM, Stoffel M, Bollschweiler M (2009) Formation and spread of callus tissue and tangential rows of resin ducts in Larix decidua and Piceaabies following rockfall impacts. Tree Physiol 29(2):281–289

Schulman E (1945) Root growth-rings and chronology. Tree-Ring Bull Boston 1024–1029

Schulman E (1951) Tree-ring indies of rainfall, temperature, and river flow. Compendium of meteorology. American Meteorological Department Society, Boston, pp p1024-1029

Sévellec F, Drijfhout SS (2018) A novel probabilistic forecast system predicting anomalously warm 2018–2022 reinforcing the long-term global warming trend. Nat Commun 9(1):3024. https://doi.org/10.1038/s41467-018-05442-8

Shah SK, Bhattacharyya A (2009) Tree-ring analysis of sub-fossil woods of Pinuswallichiana from Ziro Valley, Arunachal Pradesh, Northeast Himalaya. J Geol Soc India 74:503–508

Shah SK, Bhattacharyya A (2012) Spatio-temporal growth variability of three Pinus species of Northeast Himalaya with relation to climate. Dendrochronologia 30(4):266–278

Shah SK, Bhattacharyya A, Chaudhary V (2014) Streamflow reconstruction of Eastern Himalaya River, Lachen 'Chhu', North Sikkim, based on tree-ring data of Larixgriffithiana from Zemu Glacier basin. Dendrochronologia 32(2):97–106. https://doi.org/10.1016/j.dendro.2014.01.005

Shah SK, Bhattacharyya A, Shekhar M (2013) Reconstructing discharge of Beas river basin, Kullu valley, western Himalaya, based on tree-ring data. Quat Int 286:138–147

Shah SK, Pandey U, Mehrotra N (2018) Precipitation reconstruction for the Lidder Valley, Kashmir Himalaya using tree-rings of Cedrusdeodara. Int J Climatol 38:758–773. https://doi.org/10.1002/joc.5405

Shekhar M (2014) Application of Multi-proxy tree-ring parameters in the reconsutcion of climate vis-à-vis glacial fluctuation from the Eastern Himalaya. Ph.D thesis. Department of Botnay, University of Lucknow, Lucknow, Uttar Pradesh, India http://shodhganga.inflibnet.ac.in/bitstream/10603/43041/1/mayank_shekhar_ph_d_botany_2013.pdf

Shekhar M, Bhardwaj A, Singh S et al (2017) Himalayan glaciers experienced significant mass loss during later phases of little ice age. Sci Rep 7(1):1–14. https://doi.org/10.1038/s41598-017-09212-2

Shekhar M, Bhattacharyya A (2015) Reconstruction of January–April discharge of ZemuChuu–a first stage of Teesta River North Sikkim Eastern Himalaya based on tree-ring data of fir. J Hydrol: Reg Stud 4:776–786. https://doi.org/10.1016/j.ejrh.2015.06.019

Shekhar M, Pal AK, Bhattacharyya A et al (2018) Tree-ring based reconstruction of winter drought since 1767 CE from Uttarkashi, Western Himalaya. Quat Int 479:58–69. https://doi.org/10.1016/j.quaint.2017.08.029

Shrestha KB, Chhetri PK, Bista R (2017) Growth responses of Abiesspectabilis to climate variations along an elevational gradient in Langtang National Park in the central Himalaya, Nepal. J Res 22(5):274–281. https://doi.org/10.1080/13416979.2017.1351508

Shroder JF Jr (1978) Dendrogeomorphological analysis of mass movement on Table Cliffs Plateau, Utah. Quat Res 9:168–185. https://doi.org/10.1016/0033-5894(78)90065-0

Shroder JF Jr (1980) Dendrogeomorphology: review and new techniques of tree-ring dating. Prog Phys Geogr 4(2):161–188. https://doi.org/10.1177/030913338000400202

Sigafoos RS (1961) Vegetation in relation to flood frequency near Washington, DC. US. Geol Surv Prof Pap 424:248–249

Sigafoos RS (1964) Botanical evidence of floods and flood-plain deposition, vol 485. US Government Printing Office

Singh J, Park WK, Yadav RR (2006) Tree-ring-based hydrological records for western Himalaya, India, since AD 1560. Clim Dyn 26:295–303. https://doi.org/10.1007/s00382-005-0089-1

Singh J, Singh N, Chauhan P et al (2019) Tree-ring δ18O records of abating June–July monsoon rainfall over the Himalayan region in the last 273 years. Quat Int 532:48–56

Singh J, Yadav RR (2000) Tree-ring indications of recent glacier fluctuations in Gangotri, Western Himalaya, India. Curr Sci 79(11):1598–1601

Singh J, Yadav RR (2005) Spring precipitation variations over the western Himalaya, India, since AD 1731 as deduced from tree rings. J Geophys Res Atmos 110:D01110. https://doi.org/10.1029/2004JD004855

Singh J, Yadav RR (2007) Dendroclimatic potential of millennium-long ring-width chronology of Pinusgerardiana from Himachal Pradesh, India. Curr Sci 833–836

Singh J, Yadav RR (2013) Tree-ring-based seven century long flow records of Satluj River, western Himalaya, India. Quat Int 304:156–162. https://doi.org/10.1016/j.quaint.2013.03.024

Singh J, Yadav RR (2014) Chir pine ring-width thermometry in western Himalaya, India. Curr Sci 735–738 https://doi.org/10.18520/cs%2Fv106%2Fi5%2F735-738

Singh J, Yadav RR, Wilmking M (2009) A 694-year tree-ring based rainfall reconstruction from Himachal Pradesh, India. Clim Dyn 33(7–8):1149. https://doi.org/10.1007/s00382-009-0528-5

Singh P, Jain SK (2002) Snow and glacier melt in the Satluj River at Bhakra Dam in the western Himalayan region. Hydrol Sci J 47(1):93–106. https://doi.org/10.1080/02626667.2018.1506128

Singh P, Jain SK, Kumar N (1997) Estimation of snow and glacier contribution to the Chenab River Western Himalaya. Mt Res Dev 17(1):49–56

Singh S, Kumar R, Bhardwaj A et al (2016) Changing climate and glacio-hydrology in Indian Himalayan Region: a review. Wiley Interdiscip Rev Clim Chang 7(3):393–410. https://doi.org/10.1002/wcc.393

Stockton CW (1975) Long-term streamflow records reconstructed from tree-rings. University of Arizona, Tucson, Arizona, p 111

Stockton CW, Jacoby GC (1976) Long-term surface water supply and streamflow trends in the Upper Colorado River Basin, Lake Powell Research Project Bulletin No. 18. Nation Science Foundation, Arlington, VA
Stoffel M (2008) Dating past geomorphic processes with tangential rows of traumatic resin ducts. Dendrochronologia 26(1):53–60. https://doi.org/10.1016/j.dendro.2007.06.002
Stoffel M, Beniston M (2006) On the incidence of debris flows from the early Little Ice Age to a future greenhouse climate: a case study from the Swiss Alps. Geophys Res Lett 33:L16404. https://doi.org/10.1029/2006GL026805
Stoffel M, Bollschweiler M, Butler DR et al (2010) Tree rings and natural hazards: an introduction. Tree Ring Nat Hazards. Springer Dordrecht, pp 3–23
Stoffel M, Corona C (2014) Dendroecological dating of geomorphic disturbance in trees. Tree-Ring Res 70(1):3–20
Stoffel M, Klinkmüller M (2013) 3D analysis of wound reactions in European conifers after mechanical stem wounding: qualitative insights from X-ray computed tomography. Trees—Struct Function 27:1805–1811. https://doi.org/10.1007/s00468-013-0900-2
Stokes MA, Smiley TL (1968) An introduction to tree-ring dating. University of Chicago Press, Chicago
Thapa UK, Shah SK, Gaire NP et al (2015) Spring temperatures in the far-western Nepal Himalaya since AD 1640 reconstructed from Piceasmithiana tree-ring widths. Climdyn 45(7):2069–2081. https://doi.org/10.1007/s00382-014-2457-1
Thapa UK, St George S, Kharal DK et al (2017) Tree growth across the Nepal Himalaya during the last four centuries. Prog Phys Geogr 41(4):478–495. https://doi.org/10.1177/030913331771 4247
Timell TE (1986) Compression wood in gymnosperms, vol 3. Springer-Verlag, Berlin
Urrutia RB, Lara A, Villalba R et al (2011) Multicentury tree ring reconstruction of annual streamflow for the Maule River watershed in south central Chile. Water Resour Res 47(6) https://doi.org/10.1029/2010WR009562
Vörösmarty CJ, Fekete B, Tucker BA (1996) River Discharge Database, Version 1.0 (RivDIS v1.0), Volumes 0 through 6. A contribution to IHP-V Theme 1. Technical Documents in Hydrology Series
Wigley T, Briffa K, Jones P (1984) On the average value of correlated time series, with applications in dendroclimatology and hydrometeorology. J Appl Meteorol Climatol 23:201–213
Woodhouse C, Meko DM (2002) Introduction to tree-ring-based streamflow reconstructions. Southwest Geol pp. 14–15
Woodhouse CA (2001) A tree-ring reconstruction of streamflow for the Colorado Front Range. J Am Water Resour Assoc 37(3):561–569. https://doi.org/10.1111/j.1752-1688.2001.tb05493.x
Woodhouse CA, Lukas JJ (2006) Drought, tree rings and water resource management in Colorado. Can Water Resour J 31(4):297–310. https://doi.org/10.4296/cwrj3104297
Woodhouse CA, Pederson GT, Morino K et al (2016) Increasing influence of air temperature on upper Colorado River streamflow. Geophys Res Lett 43(5):2174–2181. https://doi.org/10.1002/2015GL067613
Wu XD (1992) Studies in China. Climate since A.D. 1500. In: Bradley RS, Jones PD (eds) Chapman and Hall pp 437–443
Yadav RR (1991) Tree ring research in India: an overview. Palaeobotanist 40:394–398
Yadav RR (2009) Tree ring imprints of long-term changes in climate in western Himalaya. India. J Biosci 34(5):699
Yadav RR (2013) Tree ring–based seven century drought records for the Western Himalaya, India. J Geophys Res Atmos 118(10):4318–4325. https://doi.org/10.1002/jgrd.50265
Yadav RR, Bhutiyani MR (2013) Tree-ring based snowfall record for cold arid western Himalaya, India since AD 1460. J Geophys Res Atmos 118(14):7516–7522. https://doi.org/10.1002/jgrd.50583
Yadav RR, Gupta AK, Kotlia BS et al (2017) Recent wetting and glacier expansion in the northwest Himalaya and Karakoram. Sci Rep 7(1):1–8. https://doi.org/10.1038/s41598-017-06388-5

Yadav RR, Misra KG, Kotlia BS et al (2014) Premonsoon precipitation variability in Kumaon Himalaya, India over a perspective of 300 years. Quat Int 325:213–219. https://doi.org/10.1016/j.quaint.2013.09.005
Yadav RR, Misra KG, Yadava AK et al (2015) Tree-ring footprints of drought variability in last~300 years over Kumaun Himalaya, India and its relationship with crop productivity. Quat Sci Rev 117:113–123. https://doi.org/10.1016/j.quascirev.2015.04.003
Yadava AK, Yadav RR, Misra KG et al (2015) Tree ring evidence of late summer warming in Sikkim, northeast India. Quat Int 371:175–180. https://doi.org/10.1016/j.quaint.2014.12.067
Yadav RR, Park WK (2000) Precipitation reconstruction using ring-width chronology of Himalayan cedar from western Himalaya: preliminary results. J Earth Syst Sci 109(3):339–345
Yadav RR, Singh J (2002) Tree-ring analysis of Taxusbaccata from the western Himalaya, India, and its dendroclimatic potential. Tree-Ring Res 58(1/2):23–29
Yadav RR, Singh J, Dubey B et al (2004) Varying strength of relationship between temperature and growth of high-level fir at marginal ecosystems in western Himalaya, India. Currsci 86:1152–1156
Yadav RR, Singh J, Dubey B, Misra KG (2006) A 1584-year ring width chronology of juniper from Lahul, Himachal Pradesh: prospects of developing millennia-long climate records. Curr Sci 1122–1126
Yadava AK, Braeuning A, Singh J et al (2016) Boreal spring precipitation variability in the cold arid western Himalaya during the last millennium, regional linkages, and socio-economic implications. Quat Sci Rev 144:28–43. https://doi.org/10.1016/j.quascirev.2016.05.008
Yadava AK, Sharma YK, Dubey B et al (2017) Altitudinal treeline dynamics of Himalayan pine in western Himalaya, India. Quat Int 444:44–52. https://doi.org/10.1016/j.quaint.2016.07.032
Yuan Y, Shao X, Wei W et al (2007) The potential to reconstruct Manasi River streamflow in the northern Tien Shan Mountains (NW China). Tree-Ring Res 63(2):81–93
Zang C, Biondi F (2013) Dendroclimatic calibration in R: the bootRes package for response and correlation function analysis. Dendrochronologia 31(1):68–74

Chapter 4
Hydroclimatic and Glacial Variabilities in the Himalayan and Tibetan Regions Since Last Glacial Maxima: A Synthesis

Ipsita Roy, Nidhi Tomar, Ayushi Singh, Mayank Shekhar, Parminder Singh Ranhotra, Amalava Bhattacharyya, and Yogesh K. Sharma

Abstract Hydroclimatic conditions of the Himalayan–Tibetan region have been studied to understand their impacts on the vegetation and glacial dynamics since the Last Glacial Maxima by referring to published datasets pan India and China. Though the signatures of major climatic events are reported globally, their intensity and behavior vary across different precipitation regimes. Here, the emphasis has been laid on the past climate records from Indian Summer Monsoon, Western Disturbances, and East Asian Summer Monsoon dominant regions of the Himalayas and Tibet, captured by the sedimentary and speleothem proxy archives. Review showed a cold-arid Last Glacial Maxima (~20 ka BP) which ameliorated ca. 15–12.5 ka BP (Bølling–Allerød) with intermittent dry events. The Pleistocene/Holocene transition (ca. 11.7 ka BP) recorded a decline in precipitation with a conspicuous increase at ca. 11.5 ka BP as reported from the Tibetan Plateau. Short-lived climate events were found less pronounced with asynchrony observed regionally in the time-periods of Medieval Climatic Anomaly and Little Ice Age. Further comparison of biotic and abiotic proxies suggested their differential behavior in different precipitation regimes. Finer-resolution pollen-spore frequency datasets would be helpful to delineate the abrupt and short-lived climatic phases following the annually resolved speleothem records.

Keywords Indian summer monsoon · East Asian summer monsoon · Western disturbances · Hydroclimatic variability · Pollen · Speleothem

---

I. Roy · N. Tomar · A. Singh · M. Shekhar · P. S. Ranhotra (✉) · A. Bhattacharyya
Birbal Sahni Institute of Palaeosciences, 53 University Road, Lucknow, Uttar Pradesh, India

Y. K. Sharma
Department of Botany, University of Lucknow, 53 University Road, Lucknow, Uttar Pradesh, India

S. Rani and R. Kumar (eds.), *Climate Change*, Springer Climate,
https://doi.org/10.1007/978-3-030-92782-0_4

Introduction

The Himalayan–Tibetan highlands play an important role in the climate dynamics of the Asian region by controlling the wind circulation patterns, mainly the Indian Summer Monsoon (ISM) and the Western Disturbances (WD), critical in managing the South Asian hydrology, floristic, civilizations, and agro-economy. The latitudinal shifts of the Inter-Tropical Convergence Zone (ITCZ) are controlled by solar irradiance (Chao and Chen 2001; Fleitmann et al. 2007; Wünnemann et al. 2010; Xu et al. 2019) that governs the variations in monsoonal precipitation. The years with weak solar insolation reduces the northward shift of the ITCZ resulting in a weak monsoon year and vice-versa (Saha and Saha 1980; Neff et al. 2001; Agnihotri et al. 2002; Fleitmann et al. 2007; Tierney and Russell 2007; Gray et al. 2010).

The diverse precipitation systems make the Himalayan and Tibetan regions unique for studying their climate and inferring the monsoonal and glacial dynamics (Bookhagen et al. 2005). Recent studies on the Himalayan glaciers (Raina 2009; Scherler et al. 2011; Bolch et al. 2012; Shekhar et al. 2017) show an alarming reduction rate in most of their mass balance in view of unprecedented global warming (IPCC 2001, 2007, 2013; Masson-Delmotte et al. 2018), putting up a major concern for future hydrological changes in the South Asian region as well as vegetation change and tree-line shifts to the higher altitudes. Hence, it becomes necessary to understand the long-term past monsoonal variability and its impacts on the vegetation and glacier dynamics for ascertaining their behavior in the future. The meteorological data from the Himalayas are temporally too short, i.e., not beyond 1850 C.E. (Parthasarathy et al. 1994; Kotlia et al. 2012), making it difficult to decipher the climate change and underlying driving mechanisms on a long-term basis. Also, the measured instrumental climate data beyond a century (if available) are from very limited stations and, therefore, could not represent the entire Himalayan climate. To get a comprehensive scenario of the Himalayan and Tibetan climate systems and glacial vis-à-vis vegetation dynamics, the up-scaling of climate data is required at fine temporal and spatial scale, which is possible only through different available biotic and abiotic proxies.

The proxies tend to preserve the climate signals over time. The signals of vegetation dynamics through ages are best preserved as pollen-spore in the lacustrine and glacio-fluvial deposits within the valleys providing the base for the past climate reconstructions on a decadal to centennial-scale. Speleothem and tree-ring archives preserve the climatic signals at an annual to decadal scale, but it is difficult to find the preserved older samples of such archives extending much into the past. The available studies from sedimentary archives inferred the climatic changes qualitatively over coarser time resolution, thus putting an information gap for the high-resolution datasets of a longer period, that is, entire Holocene or beyond. This information gap comes as a barrier for more precise reconstructions of the climatic events and their teleconnections on a global scale.

In this review, the paleoclimatic studies considering the biotic (pollen-spore) and abiotic ($\delta^{13}C$ and $\delta^{18}O$) proxies through sedimentary and speleothem archives have been compared to understand the role of ISM, WD, and East Asian Summer Monsoon (EASM) in deciding the climatic fluctuations over the Himalayan and Tibetan highlands. The records from the Himalayas have been grouped under different precipitation regimes, namely WD dominated, ISM-WD influenced, and core ISM zones. The studies from the Tibetan Plateau and surrounding landmass have been considered accordingly under the influence of ISM, WD, and EASM. A detailed description of the study sites is provided in Table 4.1.

Study Area

The Himalaya, also taken as the third pole and water tower of Asia (Immerzeel et al. 2010), is home to around 35,000 glaciers (https://www.isro.gov.in/earth-observation/snow-and-glacier) which are sources of many major Himalayan and Trans-Himalayan rivers. The entire Himalayas can be divided into three precipitation regimes where the eastern Himalayan region receives precipitation entirely through ISM; the central to western Himalayan regions are affected by both, the winter precipitation through WD and summer rains through ISM; and the north-west part of Himalayas is enriched mainly by WD (Benn and Owen 1998). Also, the range encompasses vegetation varying from tropical in the foothills to grasslands in the alpine vegetation zones (Champion and Seth 1968). Within a short latitudinal shift, the flora varies drastically as a result of the different climatic regimes. On a regional scale, there is a noticeable variation among the flora of eastern and western Himalayas, where the flora of eastern Himalayas is mainly evergreen type as compared to deciduous vegetation in the western Himalayas (Singh and Singh 1987). Eastern Himalayas are more diverse with the species of *Larix griffithiana, Picea spinulosa, Cephalotaxus griffithii, Gnetum montanum, Cycas pectinata,* etc. and tree-ferns, that otherwise are absent in the western part. While *Picea smithiana, Pinus gerardiana, Cedrus deodara, Juniperus polycarpos*, and several spp. of *Ephedra* are found in the western Himalayas but are absent in the eastern Himalayan part (Hajra and Rao 1990).

Tibetan Plateau, the highest plateau in the world with an average altitude of ~4500 m above mean sea level (AMSL), has Himalayas to the south and the Kunlun Mountains in the north (Cai et al. 2012), and about 57,000 km^2 of glaciated area (Thompson et al. 1995). The Tibetan Plateau receives precipitation from different sources, namely the Indian Ocean (ISM), the Pacific (South-east Asian Monsoon), the Arctic (northerlies), and the Atlantic (westerlies) (Owen and Benn 2005; Bershaw et al. 2012). The precipitation decreases from south-east to north-west which is also reflected in the vegetation pattern. The warm-wet margins of the plateau is characterized by the growth of mixed conifers (*Pinus, Picea, Abies, Cedrus, Tsuga*) and broadleaved vegetation (*Quercus, Betula, Rhododendron*) etc., up to 3,000 m AMSL. This gets occupied by moist-loving sub-alpine shrubs (*Potentilla* and *Salix*) on the wetter slopes while temperate steppe (Amaranthaceae, *Artemisia*, and Poaceae) on the drier side of the plateau (Yu et al. 2001; Herzschuh et al. 2006). Variations in the precipitation systems along with topographical control across the mountain

Table 4.1 Description of study sites from the Himalayan–Tibetan region

S. No.	Study site	Latitude (N)	Longitude (E)	Proxy	Time Period (ka BP)	Reference
S1	Lamayuru, Ladakh	34°	76°	PSp	35-22	Ranhotra et al. (2007)
S2	Tsokar Lake, Ladakh	33°10’	78°	PSp	30-9	Bhattacharyya (1989)
S3	Tso kar Lake, Ladakh	33°10’	78°	PSp	15.2-1.3	Demske et al. (2009)
S4	Tso Moriri Lake, Ladakh	32°54’	78°19’	PSp	12-Recent	Leipe et al. (2014)
S5	Chandra valley, Lahaul , Himachal Pradesh	32°28’	77°36’	PSp, $\delta^{13}C$	12.88-0.35	Rawat et al. (2015)
S6	Rukti valley, Kinnaur, Himachal Himalaya	31°25’	77°45’	PSp	16.6-3.5	Ranhotra et al. (2018)
S7	Sangla, Kinnaur, Himachal Pradesh	31°25’	77°45’	PSp, $\delta^{13}C$	10.450-1	Chakraborty et al. (2006)
S8	Baspa Valley, Kinnaur, Himachal Pradesh	31°	77°45’	PSp	11.79-Recent	Ranhotra and Bhattacharyya (2010)
S9	Naradu Glacier Valley, Kinnaur, Himachal Pradesh	31°	77°45’	PSp	12.72 -10.19	Bhattacharyya et al. (2006)
S10	Dokriani Glacier, Uttarakhand	30°48’	78°40’	PSp	12.41-5.41	Bhattacharyya et al. (2011a)
S11	Dokriani Glacier, Uttarakhand	30°50’	78°47’	PSp	7.8-Recent	Phadtare (2000)
S12	Chorabari Glacier, Uttarakhand	30°43’	79°04’	PSp, $\delta^{13}C$	8-Recent	Srivastava et al. (2017)
S13	Bhujbas, Gangotri Glacier, Uttarakhand	30°43’22"	79°16’34"	PSp	2-Recent	Kar et al. (2002)

(continued)

Table 4.1 (continued)

S. No.	Study site	Latitude (N)	Longitude (E)	Proxy	Time Period (ka BP)	Reference
S14	Pindari Glacier, Uttarakhand	30° 15' 36"	79° 59' 52"	PSp	7-Recent	Bali et al. (2015)
S15	Phulara palaeolake, Uttarakhand	29°20'	80°08'	PSp, $\delta^{13}C$	21.5-3.5	Kotlia et al. (2010)
S16	Darjeeling, West Bengal	27°01'45.24"	88° 19'18.71"	PSp, $\delta^{13}C$	24-Recent	Ghosh et al. (2018)
S17	Darjeeling, West Bengal	26°53'30"	88°46'23"	PSp, $\delta^{13}C$	46.4-3.5	Ghosh et al. (2015)
S18	Chopta Valley, Sikkim	27° 54' 16.28"	88° 31'31.32"	$\delta^{13}C$	12.7-Recent	Ali et al. (2018)
S19	Kamrup District, Assam	26°02' 10.5"	91°25'20.7"	PSp	14.9-Recent	Dixit and Bera (2013)
S20	Paradise Lake, Arunachal Pradesh	27°30.324'	92°06.269'	PSp, $\delta^{13}C$	1.8-Recent	Bhattacharyya et al. (2007)
S21	Ziro Lake, Arunachal Pradesh	27° 32'1.22"	93° 49' 49.24"	PSp	19-3.8	Ghosh et al. (2014)
S22	Srinagar, South-Tripura	22°59'57.76"	91°34'26.95"	PSp	6.8-Recent	Bhattacharyya et al. (2011b)
S23	Koucha Lake, Tibetan Plateau	34°	97.2°	PSp	16.4-Recent	Herzschuh et al. (2009)
S24	Qinghai Lake, Tibetan Plateau	36°32'	99°36'	PSp, $\delta^{13}C$	18.3-Recent	Ji et al. (2005)
S25	Daotang Pond, Tibetan Plateau	36°	100°	PSp	1.2-Recent	Müller (2017)
S26	Gonghai Lake, North China	38°54'	112°14'	PSp	14.7-Recent	Chen et al. (2015)
S27	Naleng Lake, Tibetan Plateau	31.10°	99.75°	PSp	17.7-8.5	Kramer et al. (2010)
S28	Zoige Plateau, Tibetan Plateau	31°51'	101°31'	PSp, $\delta^{13}C$	30-Recent	Yan et al. (1999)

(continued)

Table 4.1 (continued)

S. No.	Study site	Latitude (N)	Longitude (E)	Proxy	Time Period (ka BP)	Reference
S29	Wuxu Lake, Tibetan Plateau	29°9'11.48"	101°24'21.6"	PSp	12.3-recent	Zhang et al. (2016)
C1	Bittoo Cave, Uttarakhand	30°47'25"	77°46'35"	$\delta^{18}O$	280-Recent	Kathayat et al. (2016)
C2	Sainji Cave, Uttarakhand	30°16'07"	79°18'14"	$\delta^{18}O$	4-Recent	Kotlia et al. (2015)
C3	Chulerasim Cave, Uttarakhand	29°53'08"	79°21'06"	$\delta^{18}O$	0.36-Recent	Kotlia et al. (2012)
C4	Panigarh Cave, Uttarakhand	29°33'10"	80°07'03"	$\delta^{18}O$	0.7-Recent	Liang et al. (2015)
C5	Dharamjali Cave, Uttarakhand	29°31'27.8"	80°12'40.3"	$\delta^{18}O$	1.8-Recent	Sanwal et al. (2013)
C6	Timta Cave, Uttarakhand	29°50'17"	80°02'01"	$\delta^{18}O$	15.2-11.7	Sinha et al. (2005)
C7	Mawmluh Cave, Meghalaya	25°15'44"	91°52'54"	$\delta^{18}O$	33.8-5.5	Dutt et al. (2015)
C8	Mawmluh Cave, Meghalaya	25°15'32"	91°42'45"	$\delta^{18}O$	4.44-3.78	Kathayat et al. (2018)
C9	Shenqi cave, Tibetan Plateau	28°56'	103°06'	$\delta^{18}O$	6.6-Present	Tan et al. (2018)
C10	Furong Cave, China	29°13'44"	107°54'13"	$\delta^{18}O$	2-Recent	Li et al. (2011)
C11	Dongge Cave, China	25°17'	108°5'	$\delta^{18}O$	16-Recent	Dykoski et al. (2005)
C12	Dongge Cave, China	25°17'	108°5'	$\delta^{18}O$	9-Recent	Wang et al. (2005)

'S' stands for Sedimentary deposits and 'C' stands for Cave deposits
PSp= Pollen-spore; $\delta^{13}C$= carbon isotope; $\delta^{18}O$= oxygen isotope

range result in changes in the extent of glaciations throughout the Quaternary time (Derbyshire 1981; Benn and Owen 1998; Owen and Benn 2005; Owen et al. 2008).

Material and Methods

For this synthesis, significant studies from the Himalayas and the Tibetan Plateau have been considered (Fig. 4.1; Table 4.1) covering the later part of the late Quaternary period i.e. since the Last Glacial Maxima (LGM) time. The climatic interpretations based on pollen, carbon isotope ($\delta^{13}C$), and oxygen isotope of speleothem ($\delta^{18}O$ speleothem) have been discussed. The sub-surface sedimentary archives cover a longer timescale but provide climate reconstruction of decadal to centennial resolution. Speleothem though describe the climatic fluctuations at annual to decadal timescale but only a few of its records extend to deeper geologic time. Due to the absence of literature on cave deposits from the Eastern Himalayas, the surrounding areas have also been included. Similarly, there are only a few speleothem records from the Tibetan Plateau because of which studies from other parts of China have also been interpreted.

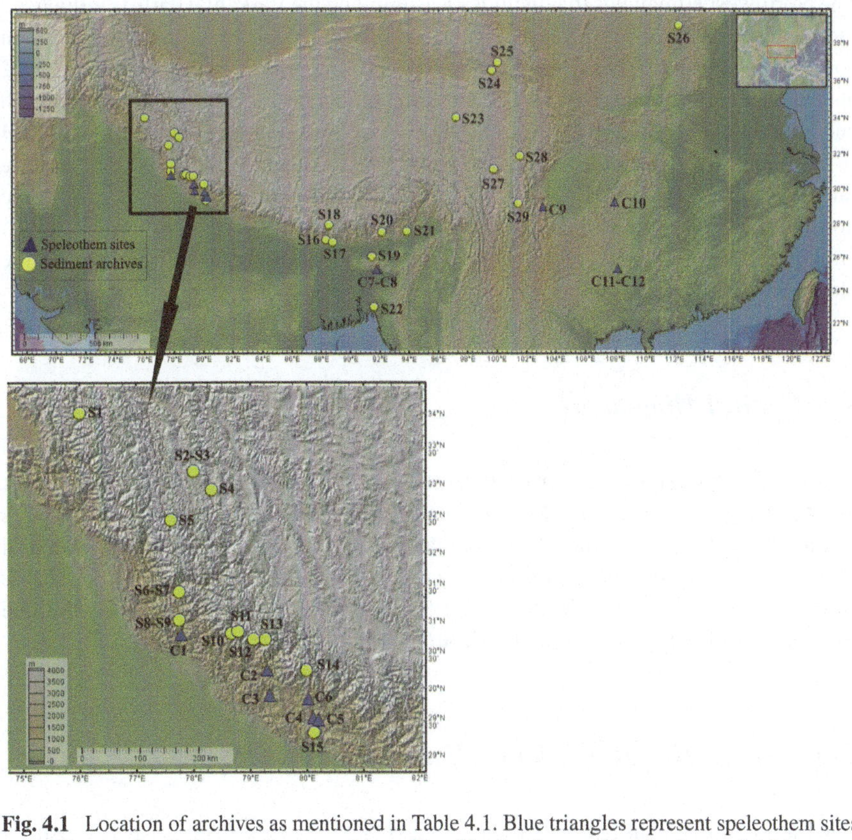

Fig. 4.1 Location of archives as mentioned in Table 4.1. Blue triangles represent speleothem sites and yellow circles indicate sediment archives. *Map Plotting Source* GeoMapApp v. 3.6.14

The Trans-Himalaya

The Trans-Himalayan region lies in the rain shadow zone and receives negligible precipitation through ISM. But winter precipitation through WD is in the form of snowfall. The mean annual temperature (MAT) and mean annual precipitation (MAP) remain respectively around 0.2 °C–5.5 °C and 54–115 mm year^{-1} (Demske et al. 2009). Due to cold and dry conditions, the region is devoid of arboreal vegetation, except shrubby and non-arboreal taxa at the localized moist areas. This region though less explored but provides long climatic records much beyond the Holocene. The climatic fluctuations as recorded by different proxies have been enlisted in Table 4.2.

The Northwest and Western Himalayas

The Northwest Himalayas that includes Kashmir and the Pir Panjal parts is influenced mainly by the WD (485 mm year^{-1}) in the form of snow and the summer precipitation contributes ~29% (196 mm year^{-1}) of MAP. The temperature ranges between 1.6 °C and 24.3 °C (Jeelani et al. 2017). The western Himalayas includes Himachal Pradesh and Uttarakhand has a MAP range of ~1200–2000 mm year^{-1}, affected by both ISM (>75%) and WD (Kotlia et al. 2010; Bajpai and Kar 2018). Here, the temperature ranges from −20 °C to 31 °C (Kotlia et al. 2010; Ranhotra et al. 2018). From the northwest region, past climate studies are limited, whereas a good number of past climate reconstructions are available from the western part covering beyond the Holocene time frame (Table 4.3).

The Central Himalayas

The Central Himalayas, covering the Indian part and Nepal Himalayas, receives good precipitation through ISM. The intensity of WD winter precipitation reduces toward the eastern part of the Central Himalayan region. The studies from the Central Himalayas of the Indian region (Juyal et al. 2004, 2009; Bali et al. 2013, 2015) and Nepal region (Vishnu-Mittre 1984; Fujii and Sakai 2002) also cover the late Pleistocene time. Significant studies from the region are listed in Table 4.4.

The Eastern Himalayas and Adjoining Area

The Eastern Himalayas has been relatively less explored for paleoclimate reconstruction in a longer time frame. The temperature in this part of the Himalayas can vary between 0 °C (winter) and 32 °C (summer) (Rao 2006). The region is much moist and

Table 4.2 Paleoclimatic studies from the Trans-Himalaya

Geological time	Time intervals (ka BP)[a]	Inferred Climate	References	Proxy
Late Pleistocene to Pleistocene–Holocene Boundary	~22	Arid (LGM)	Ranhotra et al. (2007)	Pollen-spore
	21–18	Amelioration	"	"
	18–11.8	Amelioration	Bhattacharyya (1989)	"
	15.2–14	Dry/ cold	Demske et al. (2009)	"
	12.9–12.5	Improved moisture	"	"
	12.2–11.8	Weak monsoon (YD)	"	"
Early Holocene	10.9–9.2	Maximum monsoon	"	"
	10.45–4.31	Warm/ moist	Ranhotra and Bhattacharyya (2010)	"
	10 till recent	Amelioration	Bhattacharyya (1989)	"
	9.2–4.8	Weak monsoon	Demske et al. (2009)	"
	~8	Increased precipitation	Demske et al. (2009)	"
Late Holocene	< 4.8	Arid	Demske et al. (2009)	"
	2.8–1.3	Weakening of monsoon	Demske et al. (2009)	"
	2–1	Cool/dry	Mazari et al. (1995)	"
	1–0.4	Warm/moist	Mazari et al. (1995)	"
	~0.4 till recent	Cold/dry	Mazari et al. (1995)	"

[a]All dates in "ka BP" are based on 1950 C.E. *LGM* last glacial maxima, *YD* younger dryas

ISM influenced with the MAP ranging from 3000 to 4000 mm year^{-1} at the foothills with Arunachal Pradesh receiving 1400–6000 mm year^{-1} rainfall and Cherrapunji in Meghalaya province receives an average annual rainfall of 11,000 mm year^{-1} (Rao 2006). Gupta (1969), Bhattacharya, and Chanda (1987, 1992) have defined vegetation history, paleoecology, and biostratigraphy with respect to the procured pollen taxa. However, due to age constraints, the climate cannot be inferred much at a finer scale. Recently, new data have been added to the existing timescale, and also attempts

Table 4.3 Paleoclimatic studies from the North-West and Western Himalaya

Geological time period	Time intervals (ka BP)*	Inferred Climate	References	Proxy
Late Pleistocene to Pleistocene – Holocene Boundary	20.2 – 18.5	Cold (LGM)	Kotlia et al. (2010)	PSp, $\delta^{13}C$
	18.5 – 14.3	Warm/wet	"	"
	~ 16.6	Warm/moist	Ranhotra et al. (2018)	PSp
	15.2 – 11.7	Enhanced ISM	Sinha et al. (2005)	$\delta^{18}O$
	14.3 – 13.8	Arid; Weak ISM (OD)	Kotlia et al. (2010)	PSp, $\delta^{13}C$
	14.1- 13.2	Decline in ISM	Sinha et al. (2005)	$\delta^{18}O$
	13.8 – 13.5	Warm (BA)	Kotlia et al. (2010)	PSp, $\delta^{13}C$
	13.5 – 12	Decline in ISM; Cool (YD)	"	"
	~ 13.3	Cool	Ranhotra et al. (2018)	PSp
	~ 12.88 – 11.64	Decline in ISM (YD)	Rawat et al. (2015)	PSp, $\delta^{13}C$
	~ 12.5	Decrease in ISM (YD)	Sinha et al. (2005)	$\delta^{18}O$
	12.4 – 10.6	Cool/dry	Bhattacharyya et al. (2011).a	PSp
	~12.7	Less moist	Bhattacharyya et al. (2006)	"
	12 – 8.3	Intensification of monsoon	Kotlia et al. (2010)	PSp, $\delta^{13}C$
Early Holocene	~ 11.64 – 8.81	Increase in ISM	Rawat et al. (2015)	"
	~ 11.5	Enhanced ISM	Ranhotra et al. (2018)	PSp
	11 – 9.6	Increased moisture	Leipe et al. (2014)	"
	~ 10.63	Warm/ moist	Bhattacharyya et al. (2011).a	"
	~10.45 – 4.31	Warm/ moist; enhanced ISM	Chakraborty et al. (2006)	PSp, $\delta^{13}C$
	~ 10.40 – 9.78	Decrease in ISM	Rawat et al. (2015)	"
	~10.2	Less moist	Bhattacharyya et al. (2006)	PSp

(continued)

Table 4.3 (continued)

Geological time period	Time intervals (ka BP)*	Inferred Climate	References	Proxy
	9.67 – 9	Cool/ dry	Bhattacharyya et al. (2011).a	”
	~ 8.7 – 7.8	Cool (8.2 ka BP)	Ranhotra et al. (2018)	”
	~ 8.81 – 8.12	Cold/ dry (8.2 ka cold event)	Rawat et al. (2015)	PSp, $\delta^{13}C$
	~ 8.81 – 6.73	Weak summer monsoon	”	”
	8.3 – 6	Cold (8.2 ka event)	Kotlia et al. (2010)	PSp, $\delta^{13}C$
Mid Holocene	~ 8	Cool/ moist	Bhattacharyya (1988)	PSp
	~ 7.8	Cold/ wet; moderate monsoon precipitation	Phadtare (2000)	”
	7.8 – 5	Enhanced monsoon	”	”
	~ 7.1	Cool/ dry	Bhattacharyya et al. (2011).a	PSp
	~ 6.73 – 3.34	Warm/ wet	Rawat et al. (2015)	PSp, $\delta^{13}C$
	~ 6.5 – 5.4	Warm/ wet	Srivastava et al. (2017)	PSp
	6 – 4.5	Warm/ humid (mid Holocene climate optimum)	Phadtare (2000)	”
	~ 6	Warm/ moist	Bhattacharyya (1988)	”
	~ 5.5	Warmer	Srivastava et al. (2017)	”
	5.4 – 3.8	Variable climate (overall drier/ colder with warm/ wet intervals)	”	”
	5.2 – 3	Dry	Leipe et al. (2014)	”
	5 – 4	Arid	Kotlia et al. (2010)	PSp, $\delta^{13}C$
	~4.81 – 4.33	Cold/ dry	Rawat et al. (2015)	”
	~ 4.8 – 3.2	Warm/ humid	Srivastava et al. (2017)	PSp

(continued)

Table 4.3 (continued)

Geological time period	Time intervals (ka BP)*	Inferred Climate	References	Proxy
	4.5 – 4.3	Increased aridity	Leipe et al. (2014)	”
	~4.31 – 1.8	Dry; reduced ISM	Chakraborty et al. (2006)	PSp, $\delta^{13}C$
Late Holocene	4 – 3.6	Increased aridity	Leipe et al. (2014)	PSp
	4 – 3.5	Cooling / Reduced monsoon	Phadtare (2000)	”
	4 – 3	Decrease in precipitation	Kotlia et al. (2015)	$\delta^{18}O$
	3.8 – 3	Enhanced ISM	Srivastava et al. (2017)	PSp
	~ 3.5	Cool/ moist	Bhattacharyya (1988)	”
	~ 3.34 – 2.03	Warm/ wet	Rawat et al. (2015)	PSp, $\delta^{13}C$
	~ 3.2	Dry	Kotlia et al. (2015)	$\delta^{18}O$
	~ 3.2	Increased aridity	Leipe et al. (2014)	PSp
	3 – 2	Increased precipitation	Kotlia et al. (2015)	$\delta^{18}O$
	~ 3 – 1.4	Dry/ cool	Srivastava et al. (2017)	PSp
	~ 3	Strengthening of monsoon	Phadtare (2000)	”
	2.5 – 2.3	Warm/ wet	Srivastava et al. (2017)	”
	~ 2.03 – 1.16	Weakening of ISM	Rawat et al. (2015)	PSp, $\delta^{13}C$
	2 – 1.8	Cool/ dry	Bhattacharyya (1988)	PSp
	2 – 0.8	Fluctuations in precipitation	Kotlia et al. (2015)	$\delta^{18}O$
	~2	Cool/moist	Kar et al. (2002)	PSp
	~1.8–1	Climate fluctuated between dry and wet phases	Chakraborty et al. (2006)	PSp, $\delta^{13}C$
	1.8 – 1.37	Warm/ moist	Bhattacharyya (1988)	PSp

(continued)

Table 4.3 (continued)

Geological time period	Time intervals (ka BP)*	Inferred Climate	References	Proxy
	~1.7	Increase in precipitation and temperature	Kar et al. (2002)	"
	~ 1.2 – 0.7	Warm/ wet (MWP)	Srivastava et al. (2017)	"
	~ 1.16 – 0.65	Warm/ moist (MWP)	Rawat et al. (2015)	PSp, $\delta^{13}C$
	1 – 0.4	Increased humidity	Leipe et al. (2014)	PSp
	0.85	Drier	Kar et al. (2002)	"
	~ 0.8 till recent	Strengthening of monsoon	Phadtare (2000)	"
	0.8 – 0.5	Warm/ moist	Bhattacharyya (1988)	"
	0.72 – 0.62	Cold/ dry	"	"
	~ 0.65 – 0.34	Decrease in monsoon (LIA)	Rawat et al. (2015)	PSp, $\delta^{13}C$
	0.65 – 0.34 (~1,303 – 1,609 C.E.)	Cold/ dry (LIA)	"	"
	~ 0.6 – 0.25	Cold/low monsoon (LIA)	Srivastava et al. (2017)	PSp
	0.5 – 0.25	High precipitation	Kotlia et al. (2015)	$\delta^{18}O$
	~ 0.50 till recent	Cool/dry	Bhattacharyya (1988)	PSp
	0.34 – 0.09	Strengthening of ISM	Rawat et al. (2015)	PSp, $\delta^{13}C$
	Recent	Warm/ moist	Kar et al. (2002)	PSp

* All dates in "ka BP" are based on 1950 C.E. PSp—Pollen-spores; $\delta^{13}C$—Carbon isotope; $\delta^{18}O$—Oxygen isotope of speleothem; OD—Older Dryas; BA—Bølling–Allerød; YD—Younger Dryas; MWP—Medieval Warm Period; LIA—Little Ice Age

have been made to study the changes quantitatively (Ghosh et al. 2014, 2015, 2018; Ali et al. 2018). The available literatures detailing the climatic history of the seven sister states comprising of the Himalayas and the tropical belts have been illustrated in Table 4.5.

Table 4.4 Paleoclimatic studies from the Central Himalaya

Geological time period	Time intervals (cal ka BP)*	Inferred Climate	References	Proxy
Late Pleistocene to Pleistocene – Holocene Boundary	17.4	Cold/ dry	Vishnu-Mittre and Sharma (1984)	Pollen-spore
Mid Holocene	~ 7	Cold/ dry	Bali et al. (2015)	"
	7 – 4.9	Warm/ moist	"	"
Late Holocene	4.9 – 1.75	Cold/ dry	"	"
	1.75 - 0.9	Warm/ moist (MWP)	"	"
	0.9 – 0.2	Cold/ dry (LIA)	"	"
	0.3	Warm/ moist	"	"

* All dates in "ka BP" are based on 1950 C.E. MWP—Medieval warm Period; LIA—Little Ice Age

The Tibetan Plateau and Surrounding Areas

Several paleoclimatic studies have been carried out from the Tibetan Plateau covering the maximum part of the Quaternary period. Tibet as a whole is influenced by the ISM, EASM, and WD. So it becomes essential to compare the different precipitation regimes both on temporal and spatial scales. The monsoonal influence is highest on eastern Tibet with pronounced summer precipitation which falls as snow on the higher altitudes (Owen and Benn 2005). In contrast, western Tibet receives heavy snowfall during winters from the mid-latitudinal westerlies with negligible contribution from summer monsoon (Owen and Dortch 2014). Based on the precipitation sources and the topography, Tibet has three types of glaciers: 1) continental interior glacier in the western and central Tibet; 2) maritime monsoonal type in the southeastern Tibet, and 3) continental monsoonal type glacier in eastern and northeastern Tibet (Derbyshire 1981). The studies describing the past climate of the different precipitation regimes have been included in Table 4.6. The primary focus has been given to the Tibetan Plateau, but also the other parts of China have been included due to the lack of adequate speleothem records in Tibet.

Table 4.5 Paleoclimatic studies from in and around the Eastern Himalaya

Geological time period	Time intervals (cal ka BP)*	Climate	References	Proxy
Late Pleistocene to Pleistocene – Holocene Boundary	~18.3 – 15.6	Increased monsoon	Ghosh et al. (2015)	PSp, $\delta^{13}C$
	17 – 15	Very dry	Dutt et al. (2015)	$\delta^{18}O$
	15 – 12.9	Increased monsoon (BA)	”	”
	14.80 – 12.45	Humid/ fluvial environment	Dixit and Bera (2013)	PSp
	12.7 – 11.6	Weaker monsoon	Ali et al. (2018)	PSp, $\delta^{13}C$
	~12.4	Abrupt increase in precipitation	”	”
	12.45 – 10.81	Cool/dry (YD)	Dixit and Bera (2013)	PSp
Early Holocene	~11.7 – 11.4	Decline in monsoon	Ali et al. (2018)	PSp, $\delta^{13}C$
	10.81 – 7.68	Cool/dry	Dixit and Bera (2013)	PSp
	~10.6 – 8	Increased precipitation	Ali et al. (2018)	PSp, $\delta^{13}C$
	10 – 6.5	Increased monsoon	Dutt et al. (2015)	$\delta^{18}O$
Mid Holocene	~8 – 3	Decrease in ISM	Ali et al. (2018)	PSp, $\delta^{13}C$
	7.68 – 6.78	Humid; high fluvial activity	Dixit and Bera (2013)	PSp
	~ 6.8	Drier	Bhattacharyya et al. (2011b)	PSp
	6.78 – 1.95	High precipitation	Dixit and Bera (2013)	”
	6.5 – 5.5	Decrease in monsoon	Dutt et al. (2015)	$\delta^{18}O$
	5.4 – 4.3	Increased precipitation	Ghosh et al. (2015)	PSp, $\delta^{13}C$
Late Holocene	4.3 – 3.5	Decreased monsoon	”	”
	~ 3.8 till recent	Dry	Bhattacharyya et al. (2011b)	PSp
	2.8	Enhanced ISM	Ali et al. (2018)	PSp, $\delta^{13}C$
	2.36 – 1.82 (364 BCE – 131 C.E.)	Humid	Ghosh et al. (2018)	”
	2.1	Enhanced ISM	Ali et al. (2018)	”

(continued)

Table 4.5 (continued)

Geological time period	Time intervals (cal ka BP)*	Climate	References	Proxy
	1.95 – 0.99	Warm/ Humid (MWP)	Dixit and Bera (2013)	PSp
	1.8 – 1.3 (131 – 624 C.E.)	Dry	Ghosh et al. (2018)	PSp, $\delta^{13}C$
	1.8	Warm/ humid (similar to present day climate)	Bhattacharyya et al. (2007)	PSp
	1.3 – 0.4	Enhanced ISM	Ali et al. (2018)	PSp, $\delta^{13}C$
	1.1	Warmer	Bhattacharyya et al. (2007)	"
	0.99 till recent	Warm and relatively dry	Dixit and Bera (2013)	"
	0.83 (1,118 C.E.)	Humid	Ghosh et al. (2018)	PSp, $\delta^{13}C$
	0.58 – 0.15 (1367 – 1802 C.E.)	Increased precipitation	"	"
	0.55	Cooler/ less moist	Bhattacharyya et al. (2007)	PSp

* All dates in "ka BP" are based on 1950 C.E. PSp—Pollen-spores; $\delta^{13}C$—Carbon isotope; $\delta^{18}O$—Oxygen isotope; BA—Bølling–Allerød; YD—Younger Dryas; MWP—Medieval Warm Period

Discussion

Hydroclimate and Glacier Dynamics Since the Last Glacial Maxima in the Himalayas and Tibet

The Last Glacial Maxima ca. 23–18 ka (Gasse 2000; Yokoyama et al. 2000; Clark et al. 2009) has been referred to as the cold phase of Pleistocene when the glacial ice masses in the Northern Hemisphere expanded to their maxima with high vertical ice thickness (Sharma and Owen 1996; Ranhotra 2007; Mehta et al. 2012, 2014; Ali et al. 2013; Owen and Dortch 2014; Shukla et al. 2018; Kumar et al. 2020). However, due to the lack of dating the glacial successions, it gets ambiguous to understand the extent of glaciations in the region (Owen 2009). There are a few sedimentary and speleothem records from the Himalayas and Tibet which extend up to the LGM time frame and have altogether inferred cold-dry conditions during this phase (Yan et al. 1999; Ranhotra 2007; Kotlia et al. 2010; Dutt et al. 2015; Kasper et al. 2015). Also, Owen et al. (2002) showed that most of the Himalayan glaciations during the LGM occurred due to lowering of temperature at the high altitudes. The cold-dry phase continued till the Older Dryas ca. 18–13.6 ka (Ji et al. 2005; Sinha et al. 2005; Kramer et al. 2010) interrupted by an intermittent warm-moist Bølling–Allerød ca.

Table 4.6 Paleoclimatic studies from the Tibetan Plateau and surrounding areas

Geological time period	Time intervals (cal ka BP)*	Inferred Climate	References	Proxy
Late Pleistocene to Pleistocene – Holocene Boundary	18 – 16.9	Very cold/dry	Ji et al. (2005)	PSp, $\delta^{13}C$
	17.7 – 14.8	Low effective moisture	Kramer et al. (2010)	PSp
	16.7 – 14.6	Cold/dry	Herzschuh et al. (2009)	"
	16.7 – 8.9	Arid climate	Tao et al. (2010)	"
	14.8 – 12.5	Warmer/wetter (BA)	Kramer et al. (2010)	"
	14.6 – 6.6	Warm/wet	Herzschuh et al. (2009)	"
	~14.7 – 12.7	Warm/wet (BA)	"	"
	14.7 – 7.0	Increase in monsoon	Chen et al. (2015)	"
	14.1 – 6.5	Warm/wet	Ji et al. (2005)	PSp, $\delta^{13}C$
	14.2 – 10	Fluctuation between cooling and warming	Yan et al. (1999)	PSp
	13 – 12.5	Warming	"	"
	12.8 – 11.6	Cooling	"	"
	12.5 – 11.7	Cooling/reduced moisture (YD) and upward retreat of tree-line	Kramer et al. (2010)	"
	12.3 – 11.3	Cold winters and dry summers (YD)	Zhang et al. (2016)	"
	~12	Cold/dry	Herzschuh et al. (2009)	"
Early Holocene	11.7 – 10.7	Increase of summer monsoon; onset of Holocene	Kramer et al. (2010)	"
	~11.6	Sudden drop in temperature	Yan et al. (1999)	"
	11.3 – 10.4	Cold winters and increased summer precipitation	Zhang et al. (2016)	"
	11.2	Increased monsoon intensity	Dykoski et al. (2005)	$\delta^{18}O$
	10.8	Increased monsoon intensity	"	"

(continued)

Table 4.6 (continued)

Geological time period	Time intervals (cal ka BP)*	Inferred Climate	References	Proxy
	10.4 – 4.9	Warm winters; High summer precipitation (Holocene climatic optimum)	Zhang et al. (2016)	PSp
	9.2	Increased monsoon intensity	Dykoski et al. (2005)	$\delta^{18}O$
	9 – 8.8	Decrease in humidity	Yan et al. (1999)	PSp
	9 – 7	Increase in Asian monsoon	Wang et al. (2005)	$\delta^{18}O$
	8.9 – 7.9	Amelioration	Tao et al. (2010)	PSp
	8.7 – 8.3	Cold; low humidity	Herzschuh et al. (2006)	"
	8.3	Decrease in Asian monsoon	Wang et al. (2005)	$\delta^{18}O$
Mid Holocene	8.2	Increased monsoon intensity	Dykoski et al. (2005)	"
	7.9 – 4.3	Increase in effective moisture	Tao et al. (2010)	PSp
	7.8 – 5.3	Maximum monsoon	Chen et al. (2015)	"
	~7.4	Cold; low humidity	Herzschuh et al. (2006)	"
	7.3 – 4.4	Wet and warm	"	"
	6.7 – 5.2	Decrease in humidity	Yan et al. (1999)	"
	4.9 – 2.6	Warm winters; reduced summer precipitation	Zhang et al. (2016)	"
	4.4	Decrease in Asian monsoon	Wang et al. (2005)	$\delta^{18}O$
	4.3 – 3.8	Extremely arid	Tao et al. (2010)	PSp
Late Holocene	4.1 – 3.5	Cold	Herzschuh et al. (2006)	"
	4.5	Colder/drier	Ji et al. (2005)	PSp, $\delta^{13}C$
	4.4	Decrease in Asian monsoon	Wang et al. (2005)	$\delta^{18}O$
	3.8 – 0.53	Climate was optimum	Tao et al. (2010)	PSp
	3.5	Decreased monsoon intensity	Dykoski et al. (2005)	$\delta^{18}O$
	3.3 till recent	Rapid decline in precipitation	Chen et al. (2015)	PSp

(continued)

Table 4.6 (continued)

Geological time period	Time intervals (cal ka BP)*	Inferred Climate	References	Proxy
	2.6 to recent	Humid	Zhang et al. (2016)	"
	2 – 1.7	Wet	Li et al. (2011)	$\delta^{18}O$
	1.9 – 1.7	Wet	Tan et al. (2018)	"
	1.8 – 0.6	Increase in humidity	Yan et al. (1999)	PSp
	1.7 – 0.8	Dry	Li et al. (2011)	$\delta^{18}O$
	1.6 – 1.4	Wet	Tan et al. (2018)	"
	1.6	Decrease in Asian monsoon	Wang et al. (2005)	"
	1.4 – 1.1	Cold	Herzschuh et al. (2006)	PSp
	0.8 – 0.9	Decrease in precipitation	Müller (2017)	"
	0.8 – 0.5	Wet	Li et al. (2011)	$\delta^{18}O$
	0.7 – 0.55	Decrease in precipitation	Müller (2017)	PSp
	0.6	Decrease in precipitation	Yan et al. (1999)	"
	0.53 till recent	Climate deteriorated	Tao et al. (2010)	"
	0.5 – 1.1	Warm/moist	Müller (2017)	"
	0.5 till recent	Decrease in precipitation	"	"
	0.5 – 0.35	Dry	Li et al. (2011)	$\delta^{18}O$
	0.5	Decrease in Asian monsoon	Wang et al. (2005)	"
	0.3 till recent	Wet	Li et al. (2011)	"
	0.2 till recent	Drought	Tan et al. (2018)	"

* All dates in "ka BP" are based on 1950 C.E. PSp—Pollen-spores; $\delta^{13}C$—Carbon isotope; $\delta^{18}O$—Oxygen isotope; BA—Bølling–Allerød; YD—Younger Dryas

14.5–12.9 ka with enhanced ISM (Ji et al. 2005; Demske et al. 2009; Herzschuh et al. 2009; Kramer et al. 2010; Dutt et al. 2015).

The Pleistocene–Holocene transition was demarcated by a cold-dry phase of Younger Dryas ca. 12.8–11.6 ka (Lamb 1977; Alley et al. 1993; Stuiver et al. 1995; Grootes and Stuiver 1997; Bradley 1999; Alley 2000; Morrill et al. 2003; Björck 2007) which was also experienced by the Himalayan and Tibetan highlands with weak monsoons (Sinha et al. 2005; Kramer et al. 2010; Kasper et al. 2015; Rawat et al. 2015; Zhang et al. 2016). Late-glacial glacier advances are apparent throughout the Himalayan–Tibetan regions (Seong et al. 2009), but Younger Dryas advancement is less evident owing to the dating uncertainties (Owen 2009). Comparatively, during the early Holocene, the climate ameliorated in the ISM-affected parts of the Tibetan Plateau while the parts affected by WD and EASM do not show any change; the latter areas experienced ameliorated climate during the mid-Holocene (Herzschuh 2006). Overall, both in the Himalayas and Tibet, the early Holocene was demarcated as moist with notable glacier advancement followed by minor advances during the mid-Holocene (Owen 2009). However, Holocene moraines are better preserved than Pleistocene moraines, providing better potential for them to be dated (Owen 2009). Seong et al. (2009) reviewed the glaciations from west Tibet and surrounding areas to conclude advances during 11.2, 10.2, 8.4, 6.7, 4.2, 3.3, and 1.4 ka BP.

Optimal moisture was recorded during the mid-Holocene (ca. 8–5 ka) (Yan et al. 1999; Herzschuh 2006). A well-defined 8.2 ka cool event is inconspicuous in several studies from the Himalayas, while from the Tibetan Plateau, inferences are made of a decline in humidity between 9 and 8.8 ka BP (Yan et al. 1999) and an increased monsoon intensity at ca. 8.2 ka BP (Dykoski et al. 2005; Wang et al. 2005). Precipitation increased subsequently around the mid-Holocene (Bhattacharyya 1988; Herzschuh et al. 2006; Dixit and Bera 2013) in the areas affected by the ISM. While lesser evidence is found from the Trans-Himalaya, receiving negligible rainfall from ISM (Demske et al. 2009) which have noted higher precipitation during mid-Holocene inferred from an increase in *Artemisia* steppe pollen grains. While Wünnemann et al. (2010) have reported no such evidence during the mid-Holocene from their record. The summer monsoon weakened around 4.2 ka BP with the collapse of the Neolithic culture of Central China (Wenxiang and Tungsheng 2004; Wang et al. 2005), Longshan culture in the Huaihe River Basin of Eastern China (Zhang et al. 2010) and Harappan civilization of Indus valley (Staubwasser et al. 2003) etc. Though an overall weakening of the monsoonal precipitation is evident from the Himalayas and Tibet, yet there is evidence of high summer monsoon precipitation in central and southern China ca. 4.2 ka (Hu et al. 2008; Tan et al. 2018; Zhang et al. 2018), which otherwise was weak in the northern and southwestern China (Wang et al. 2005; Xiao et al. 2014). The weakening of summer monsoon was discussed as a result of reduced Atlantic Meridional Overturning Circulation leading to weak ITCZ.

The late Holocene time period fluctuated between moist and dry with a decrease in moisture in the Tibetan Plateau (Herzschuh 2006; Chen et al. 2015) ca. 3 ka. There are evidence of glacial episodes during this part of the late Holocene though they are less defined and less extensive (Owen 2009). Little Ice Age (LIA) and Medieval Warm Period (MWP) events were also heterogeneous with asynchronous responses in different precipitation regimes (Morrill et al. 2003; Dixit and Tandon 2016). During the last 2 millennia, the most extensive glacier expansion in the south-eastern Tibetan Plateau was ca. 400–600 C.E. when glacier advance was also noted from the central Himalayas at ~1000 C.E. (Yang et al. 2008).

Long-term variability in the WD and ISM frequencies is linked to changes in the Northern Hemisphere insolation. While their annual to decadal-scale variability is attributed to changes in the internal climate system such as changes in the Eurasian snow cover, El-Nino Southern Oscillation (ENSO), tropical sea-surface temperatures, and volcanic eruptions (Sirocko et al. 1991; Prell and Kutzbach 1992). Glacier advances during Holocene in the Himalayan–Tibetan orogen are a consequence of long-term orbital forcing and oceanic circulations (Solomina et al. 2015; Saha et al. 2018). The early Holocene cooling and glacier advances were the resultant of northerly movement of ITCZ and enhanced summer monsoon while that of mid- and late Holocene correspond to the North Atlantic cooling teleconnection to western disturbances (Saha et al. 2018).

Regional Hydroclimatic Variability and Mechanisms

A review study by Morrill et al. (2003) from ISM and EASM influenced areas, discussed the timing and spatial pattern of occurrences of the abrupt climatic events and probable mechanisms behind them. The increase (decrease) in the monsoonal strength at varying geologic periods since the last deglaciation was found nearly synchronous with the warming (cooling) in the North Atlantic. However, heterogeneity has been observed in the nature and timing of the late Holocene events on a regional scale, as also discussed in the earlier sections. Dixit and Tandon (2016) presented a synthesis of the past millennium hydroclimatic variabilities from the Indian sub-continent and discussed globally asynchronous warm (MWP) and cold (LIA) phases during the last two millennia. The ISM was inferred to be strong during MWP and relatively weak during the LIA while WD was weak during MWP and strengthened during LIA. Chen et al. (2019) also reviewed paleoclimatic studies from the westerly and monsoon-dominated regions of Asia and highlighted that the moisture variations between the areas anti-phased on different timescales of the Holocene. Majorly during MWP, the monsoonal regions were deduced wetter and the arid regions were drier, whereas the LIA phase in the arid regions was recorded relatively wetter. Cave deposits from Central China affected by EASM show the opposite relationship with $\delta^{18}O$ stalagmite record of Central India for the last 98 years (Tan et al. 2015), thus inferring $\delta^{18}O$ speleothem records from different precipitation regimes behaving differently.

Low pressure over Indian landmass results in the northward shifting of ITCZ, thus controlling the ISM from the Bay of Bengal (BOB) and the Arabian Sea (AS) branches (Mooley and Parthasarathy 1982; Agnihotri et al. 2002; Fleitmann et al. 2003; Gupta et al. 2003; Bhattacharyya and Narasimha 2005). The Western and Central Himalayas receive precipitation from the AS and BOB branches of the ISM during summer months (JJAS) and from extra-tropical Westerlies during winter months (DJF) in association with an orographic uplifting of monsoonal winds (Dimri and Mohanty 2009; Dimri et al. 2013; Polanski et al. 2014). Moreover, the total precipitation including the summer and winter rainfall over the Himalayas shows an inverse correlation with the core ISM areas of the Indian subcontinent (Kripalani et al. 2003). External forcing, mainly the changes in solar insolation, combined with the internal forcing (oceanic circulations, volcanic eruptions) plays a crucial role in monsoonal variations. The strength of ISM is determined by the variations in the frequency of El Niño and La Niña events over time, controlled by the solar irradiance

(Clement et al. 1999; Chang et al. 2001; Terray and Dominiak 2005). Low (high) irradiance increases the frequency of El Niño (La Niñas) responsible for the weak (strong) ISM (Mann et al. 2005; Borgaonkar et al. 2010; Dimri et al. 2016). The high El Niño conditions during LIA also might have reduced the flow of warm ocean water to higher northern latitudes causing the cooling of the North Atlantic Ocean and Eurasian landmass that resulted in high snow cover over Eurasia and Himalayas due to the strengthened WDs and weakening of ISM and EASM. This inverse relationship between the winter/spring and summer monsoon precipitations (Kripalani et al. 2003) is evident by the spatial distribution pattern of speleothem records from peninsular India and the Himalayan–Tibetan region for the LIA phase (Dimri et al. 2016; Dixit and Tandon 2016; Kumar et al. 2019).

The palynological (pollen-spore) data recorded dry LIA in the Himalayan–Tibetan region when the ISM was weak and WDs were strong. On the contrary, the speleothem records show wetter conditions in the Himalayas during the LIA phase. Hydroclimatic changes primarily affect the growth of vegetation. The proper moisture availability during the growing season of spring to pre-winter months is vital for the growth activities of plants (Kemp 1983; Pangtey et al. 1990; Rawal et al. 1991). Changes in moisture availability could affect plant growth as they depend on various climatic factors (Walker et al. 1995; Bijalwan et al. 2013). The phases of strong WDs coupled with weak ISM might have provided extended days of snow cover over the Himalayan region, thus affecting the growth dynamics of vegetation by prolonging the freezing soil conditions and might have supported the growth of dry taxa. Inversely, the weaker winter precipitation and strong summer monsoons could provide a longer growing period to the vegetation. The amount of snow-melt-water seeping in the caves resulted in the growth of speleothem inside the caves changing the isotopic concentration value of $\delta^{18}O$ based on the isotope composition of winter precipitation. The vegetation thus responds to the weak and strong ISM by the expansion of dry and moist taxa, respectively, and therefore could be taken as an indicator of ISM dynamics, while the speleothem could provide the signatures of winter precipitation dynamics as well.

In the Himalayas, most of the studies concentrated on building the climatic scenario from the Western and Northwestern parts of the Himalayas receiving summer rainfall from ISM and winter snowfall from WDs. The Eastern and Trans-Himalayan regions have limited studies either related to site inaccessibility due to environmental or other factors. The studies covering the longer time frame beyond the LGM, i.e., before ~20 ka BP are less and discontinuous. Also, a rich dataset is available from the northwest, west, and central parts of the Himalayas but on spatial constraints, the studies are much scattered, not covering the different vegetation zones altitudinally. The higher reaches above the tree-line are less explored. The higher altitudes being more vulnerable to temperature changes, the studies from higher altitudes are important in understanding the tree-line and glacial dynamics concerning past climate changes. Despite the availability of studies from different valleys across the Himalayas, most of these either cover a short time duration (Kar et al. 2002; Sinha et al. 2005; Bhattacharyya et al. 2006; Berkelhammer et al. 2012; Kotlia and Joshi 2013; Liang et al. 2015) due to inaccessibility or unavailability of longer time-frame sedimentary sequences or are deficit of inferring short-lived climatic phases (Sharma and Owen 1996; Bookhagen et al. 2005; Berkelhammer et al. 2012; Bhattacharyya et al. 2014; Kathayat et al. 2016) due to either poor temporal resolution per sample or chronological uncertainties. Fine temporal resolution data at decadal to centennial-scale is a must for unfolding the short-lived climatic events, which is only possible by sampling at the finer depth intervals. Though studies suggest maximum glacial advances in the Himalayan–Tibetan region during the last glacial period but due to lack of proper chronologies, evidence of glacier advances cannot be authenticated. The high errors in absolute dates, due to lack of enough organic matter in the sediments from semi-arid and arid zones for radiocarbon dating and poor luminescence signals in the sediments for Optically Stimulated Luminescence (OSL) and Thermoluminescence (TL) dating, could bias the results.

Conclusions

Literature review shows that though the Himalayas and Tibetan Plateau have been explored to generate the datasets on past climatic fluctuations, still more studies are needed to fill in the existing gaps identified under the temporal, spatial, chronological, and sample-resolution facets. A better understanding of the timings and extent of the climatic events both in temporal and spatial scales would provide an insight into the mechanisms of each precipitation type. Identification of proper study sites, good dating materials (peat, organic-rich sediment, and organic remains), and more dates might minimize the dating errors. Studies with well-defined short-lived climatic phases of the ISM and WD-dominated parts are essential to draw the interplay of different monsoonal winds in the Himalayan and Tibetan regions. There is a noted discrepancy in the proxy response as well. Vegetation dynamics captured the LIA as dry phase while speleothem records recorded it as wet. It is inferred that in the Himalayan region, the vegetation growth corresponds to the variability of ISM precipitation while speleothem responds to the WD signals as well. A comparative analysis of different proxies from the study area under the same hydroclimatic regime would be helpful to discriminate the role of different precipitation systems that is ISM and WD in the Himalayan region and additionally EASM in Tibet.

Acknowledgements The authors are indebted to the Director, Birbal Sahni Institute of Palaeosciences (BSIP), Lucknow, Uttar Pradesh, India for permission to carry out this work and for publication (BSIP/RDCC/77/2020–21). The author IR is thankful to the Department of Science and Technology (DST) Innovation in Science Pursuit for Inspired Research (INSPIRE) program (IF150998) of the Ministry of Science and Technology for funding her fellowship during the preparation of this manuscript. PSR is thankful to the Director, BSIP for in-house funding to carry out the work.

References

Agnihotri R, Dutta K, Bhushan R et al (2002) Evidence for solar forcing on the Indian monsoon during the last millennium. Earth Planet Sci Lett 198(3–4):521–527

Ali SN, Biswas RH, Shukla AD et al (2013) Chronology and climatic implications of Late Quaternary glaciations in the Goriganga valley, central Himalaya, India. Quat Sci Rev 73:59–76

Ali SN, Dubey J, Ghosh R et al (2018) High frequency abrupt shifts in the Indian summer monsoon since Younger Dryas in the Himalaya. Sci Rep 8(1):1–8

Alley RB (2000) The Younger Dryas cold interval as viewed from central Greenland. Quat Sci Rev 19(1–5):213–226

Alley RB, Meese DA, Shuman CA et al (1993) Abrupt increase in Greenland snow accumulation at the end of the Younger Dryas event. Nature 362(6420):527–529

Bajpai R, Kar R (2018) Modern pollen deposition in glacial settings in the Himalaya (India): abundance of *Pinus* pollen and its significance. Palynology 42(4):475–482

Bali R, Ali SN, Agarwal KK et al (2013) Chronology of late Quaternary glaciation in the Pindar valley, Alaknanda basin, Central Himalaya (India). J Asian Earth Sci 66:224–233

Bali R, Ali SN, Bera SK et al (2015) Impact of Anthropocene vis-a-vis Holocene climatic changes on central Indian Himalayan glaciers. In: Engineering geology for society and territory. Springer, Cham, pp 467–471

Benn DI, Owen LA (1998) The role of the Indian summer monsoon and the mid-latitude westerlies in Himalayan glaciation: review and speculative discussion. J Geol Soc 155(2):353–363

Berkelhammer M, Sinha A, Stott L et al (2012) An abrupt shift in the Indian monsoon 4000 years ago. Geophys Monogr Ser 198:75–87

Bershaw J, Penny SM, Garzione CN (2012) Stable isotopes of modern water across the Himalaya and eastern Tibetan Plateau: Implications for estimates of paleoelevation and paleoclimate. J Geophys Res Atmos 117(D2)

Bhattacharya K, Chanda S (1987) Late-Quaternary vegetational history, palaeoecology and biostratigraphy of some deposits of Brahmaputra Basin, Upper Assam, India. J Palynol 23:225–237

Bhattacharya K, Chanda S (1992) Late Quaternary vegetational history of Upper Assam. India. Rev Palaeobot Palynol 72(3–4):325–333

Bhattacharyya A (1988) Vegetation and climate during post glacial period in the vicinity of Rohtang Pass. Great Himalayan Range. Pollen Spores 30(3–4):417–427

Bhattacharyya A (1989) Vegetation and climate during the last 30,000 years in Ladakh. Palaeogeogr Palaeoclimatol Palaeoecol 73(1–2):25–38

Bhattacharyya A, Ranhotra PS, Gergan JT (2011a) Vegetation vis-a-vis climate and glacier history during 12,400 to5,400year BP from Dokriani valley, Garhwal Himalaya. India. J GeolSoc India 77(5):401–408

Bhattacharyya A, Mehrotra N, Shah SK (2011b) Holocene vegetation and climate of south Tripura based on palynological analysis. J Geol Soc India 77(6):521–526

Bhattacharyya A, Mehrotra N, Shah SK et al (2014) Analysis of vegetation and climate change during late Pleistocene from Ziro valley, Arunachal Pradesh, eastern Himalaya region. Quat Sci Rev 101:111–123

Bhattacharyya A, Ranhotra PS, Ganjoo RK et al (2006) The early Holocene vegetation and climate in Naradu Glacier Valley of Kinnaur District. Himachal Pradesh. Palaeobotanist 55(1–3):89–96

Bhattacharyya A, Sharma J, Shah SK et al (2007) Climatic changes during the last 1800 yrs BP from paradise lake, Sela Pass, Arunachal Pradesh, Northeast Himalaya. Curr Sci 93:983–987

Bhattacharyya S, Narasimha R (2005) Possible association between Indian monsoon rainfall and solar activity. Geophys Res Lett 32(5)

Bijalwan R, Vats M, Joshi SP (2013) Plant phenological response to microclimatic variations in an alpine zone of Garhwal Himalaya. J Appl Nat Sci 5(1):47–52

Björck S (2007) Younger Dryas oscillation. In: Elias SA (ed) Encyclopedia Quaternary Science. Elsevier, New York, pp 1985–1993

Bolch T, Kulkarni A, Kääb A et al (2012) The state and fate of Himalayan glaciers. Science 336(6079):310–314
Bookhagen B, Thiede RC, Strecker MR (2005) Late Quaternary intensified monsoon phases control landscape evolution in the northwest Himalaya. Geology 33(2):149–152
Borgaonkar HP, Sikder AB, Ram S et al (2010) El Niño and related monsoon drought signals in 523-year-long ring width records of teak (*Tectona grandis* LF) trees from south India. Palaeogeogr Palaeoclimatol Palaeoecol 285(1–2):74–84
Bradley RS (1999) Paleoclimatology: reconstructing climates of the Quaternary. Elsevier
Cai Y, Zhang H, Cheng H et al (2012) The Holocene Indian monsoon variability over the southern Tibetan Plateau and its teleconnections. Earth Planet Sci Lett 335:135–144
Chakraborty S, Bhattacharya SK, Ranhotra PS et al (2006) Palaeoclimatic scenario during Holocene around Sangla valley, Kinnaur northwest Himalaya based on multi proxy records. Curr Sci (00113891):91(6)
Champion SH, Seth SK (1968) A revised survey of the forest types of India. Manager of Publications, Delhi
Chang CP, Harr P, Ju J (2001) Possible roles of Atlantic circulations on the weakening Indian monsoon rainfall–ENSO relationship. J Clim 14(11):2376–2380
Chao WC, Chen B (2001) The origin of monsoons. J Atmos Sci 58(22):3497–3507
Chen F, Chen J, Huang W et al (2019) Westerlies Asia and monsoonal Asia: Spatiotemporal differences in climate change and possible mechanisms on decadal to sub-orbital timescales. Earth-Sci Rev 192:337–354
Chen F, Xu Q, Chen J et al (2015) East Asian summer monsoon precipitation variability since the last deglaciation. Sci Rep 5:11186
Clark PU, Dyke AS, Shakun JD et al (2009) The last glacial maximum. Science 325(5941):710–714
Clement AC, Seager R, Cane MA (1999) Orbital controls on the El Nino/Southern Oscillation and the tropical climate. Paleoceanography 14(4):441–456
Demske D, Tarasov PE, Wünnemann B et al (2009) Late glacial and Holocene vegetation, Indian monsoon and westerly circulation in the Trans-Himalaya recorded in the lacustrine pollen sequence from TsoKar, Ladakh. NW India. Palaeogeogr Palaeoclimatol Palaeoecol 279(3–4):172–185
Derbyshire E (1981) Glacier regime and glacial sediment facies: a hypothetical framework for the Qinghai-Xizang Plateau. In: Proceedings of symposium on Qinghai-Xizang (Tibet) Plateau vol 2, pp 1649–1656
Dimri AP, Mohanty UC (2009) Simulation of mesoscale features associated with intense western disturbances over western Himalayas. Meteorol Appl 16(3):289–308
Dimri AP, Yasunari T, Wiltshire A et al (2013) Application of regional climate models to the Indian winter monsoon over the western Himalayas. Sci Total Environ 468:S36–S47
Dimri AP, Yasunari T, Kotlia BS et al (2016) Indian winter monsoon: present and past. Earth-Sci Rev 163:297–322
Dixit S, Bera SK (2013) Pollen-inferred vegetation vis-à-vis climate dynamics since Late Quaternary from Western Assam, Northeast India: signal of global climatic events. Quat Int 286:56–68
Dixit Y, Tandon SK (2016) Hydroclimatic variability on the Indian subcontinent in the past millennium: review and assessment. Earth-Sci Rev 161:1–15
Dutt S, Gupta AK, Clemens SC et al (2015) Abrupt changes in Indian summer monsoon strength during 33,800 to 5500 years BP. Geophys Res Lett 42(13):5526–5532
Dykoski CA, Edwards RL, Cheng H et al (2005) A high-resolution, absolute-dated Holocene and deglacial Asian monsoon record from Dongge Cave. China. Earth Planet Sci Lett 233(1–2):71–86
Fleitmann D, Burns SJ, Mangini A et al (2007) Holocene ITCZ and Indian monsoon dynamics recorded in stalagmites from Oman and Yemen (Socotra). Quat Sci Rev 26(1–2):170–188
Fleitmann D, Burns SJ, Mudelsee M et al (2003) Holocene forcing of the Indian monsoon recorded in a stalagmite from southern Oman. Science 300(5626):1737–1739
Fujii R, Sakai H (2002) Paleoclimatic changes during the last 2.5 myr recorded in the Kathmandu Basin, Central Nepal Himalayas. J Asian Earth Sci 20(3):255–266

Gasse F (2000) Hydrological changes in the African tropics since the Last Glacial Maximum. Quat Sci Rev 19(1–5):189–211

Ghosh R, Bera S, Sarkar A et al (2015) A ~ 50 ka record of monsoonal variability in the Darjeeling foothill region, eastern Himalayas. Quat Sci Rev 114:100–115

Ghosh R, Biswas O, Paruya DK et al (2018) Hydroclimatic variability and corresponding vegetation response in the Darjeeling Himalaya, India over the past ~2400 years. CATENA 170:84–99

Ghosh R, Paruya DK, Khan MA et al (2014) Late Quaternary climate variability and vegetation response in Ziro Lake Basin, Eastern Himalaya: a multiproxy approach. Quat Int 325:13–29

Gray LJ, Beer J, Geller M et al (2010) Solar influences on climate. Rev Geophys 48(4): RG4001

Grootes PM, Stuiver M (1997) Oxygen 18/16 variability in Greenland snow and ice with 10^{-3} to 105 year time resolution. J Geophys Res: Oceans 102(C12):26455–26470

Gupta AK, Anderson DM, Overpeck JT (2003) Abrupt changes in the Asian southwest monsoon during the Holocene and their links to the North Atlantic Ocean. Nature 421:354–357

Gupta HP (1969) Pollen-analytical investigations of some upper Pleistocene samples from Tockalai, Cinnamara, Assam. Palaeobotanist 18:234–236

Hajra PK, Rao RR (1990) Distribution of vegetation types in northwest Himalaya with brief remarks on phytogeography and floral resource conservation. Proc Plant Sci 100(4):263–277

Herzschuh U (2006) Palaeo-moisture evolution in monsoonal Central Asia during the last 50,000 years.Quat Sci Rev 25(1–2):163–178

Herzschuh U, Kramer A, Mischke S et al (2009) Quantitative climate and vegetation trends since the late glacial on the northeastern Tibetan Plateau deduced from Koucha Lake pollen spectra. Quat Res 71(2):162–171

Herzschuh U, Winter K, Wünnemann B et al (2006) A general cooling trend on the central Tibetan Plateau throughout the Holocene recorded by the Lake Zigetang pollen spectra. Quat Int 154:113–121

Hu C, Henderson GM, Huang J et al (2008) Quantification of Holocene Asian monsoon rainfall from spatially separated cave records. Earth Planet Sci Lett 266(3–4):221–232

Immerzeel WW, Van Beek LP, Bierkens MF (2010) Climate change will affect the Asian water towers. Science 328:1382–1385

Intergovernmental Panel on Climate Change (2013) The physical science basis. Contribution of working group I to the 5th assessment report of the intergovernmental panel on climate change. Cambridge University Press, Cambridge

IPCC CC (2001) The scientific basis. Contribution of working group I to the 3rd assessment report of the intergovernmental panel on climate change. Cambridge University Press, Cambridge, United Kingdom, pp 881

IPCC, Solomon S, Qin D, Manning M et al (2007) The physical science basis. In: Contribution of working group I to the 4th assessment report of the intergovernmental panel on climate change. Cambridge University Press, Cambridge, United Kingdom and New York

Jeelani G, Deshpande RD, Shah RA et al (2017) Influence of southwest monsoons in the Kashmir Valley, western Himalayas. Isot Environ Health Stud 53(4):400–412

Ji S, Xingqi L, Sumin W et al (2005) Palaeoclimatic changes in the Qinghai Lake area during the last 18,000 years. Quat Int 136(1):131–140

Juyal N, Pant RK, Basavaiah N et al (2004) Climate and seismicity in the higher Central Himalaya during 20–10 ka: evidence from the Garbayang basin, Uttaranchal. India. Palaeogeogr Palaeoclimatol Palaeoecol 213(3–4):315–330

Juyal N, Pant RK, Basavaiah N et al (2009) Reconstruction of Last Glacial to early Holocene monsoon variability from relict lake sediments of the Higher Central Himalaya, Uttarakhand. India. J Asian Earth Sci 34(3):437–449

Kar R, Ranhotra PS, Bhattacharyya A et al (2002) Vegetation vis-à-vis climate and glacial fluctuations of the Gangotri Glacier since the last 2000 years. Curr Sci 82(3):347–351

Kasper T, Haberzettl T, Wang J et al (2015) Hydrological variations on the Central Tibetan Plateau since the Last Glacial Maximum and their teleconnection to inter-regional and hemispheric climate variations. J Quat Sci 30(1):70–78

Kathayat G, Cheng H, Sinha A et al (2016) Indian monsoon variability on millennial-orbital timescales. Sci Rep 6:24374

Kathayat G, Cheng H, Sinha A et al (2018) Evaluating the timing and structure of the 4.2 ka event in the Indian summer monsoon domain from an annually resolved speleothem record from Northeast India. Clim Past 14(12):1869–1879

Kemp PR (1983) Phenological patterns of Chihuahuan Desert plants in relation to the timing of water availability. J Ecol 427–436

Kotlia BS, Ahmad SM, Zhao JX et al (2012) Climatic fluctuations during the LIA and post-LIA in the Kumaun Lesser Himalaya, India: evidence from a 400 years old stalagmite record. Quat Int 263:129–138

Kotlia BS, Joshi LM (2013) Late Holocene climatic changes in Garhwal Himalaya. Curr Sci 104(7):911–919

Kotlia BS, Sanwal J, Phartiyal B et al (2010) Late Quaternary climatic changes in the eastern Kumaun Himalaya, India, as deduced from multi-proxy studies. Quat Int 213(1–2):44–55

Kotlia BS, Singh AK, Joshi LM et al (2015) Precipitation variability in the Indian Central Himalaya during last ca. 4,000 years inferred from a speleothem record: Impact of Indian Summer Monsoon (ISM) and Westerlies. Quat Int 371:244–253

Kramer A, Herzschuh U, Mischke S et al (2010) Late glacial vegetation and climate oscillations on the southeastern Tibetan Plateau inferred from the Lake Naleng pollen profile. Quat Res 73(2):324–335

Kripalani RH, Kulkarni A, Sabade SS (2003) Western Himalayan snow cover and Indian monsoon rainfall: A re-examination with INSAT and NCEP/NCAR data. Theor Appl Climatol 74(1–2):1–18

Kumar A, Dutt S, Saraswat R et al (2020) A late Pleistocene sedimentation in the Indus Fan, Arabian Sea, IODP Site U1457. Geol Mag 157(6):920–928

Kumar P, Sanwal J, Dimri AP et al (2019) Contribution of diverse monsoon precipitation over Central and Northern India during mid to Late Holocene. Quat Int 507:217–223

Lamb HH (1977) Climate: past, present and future. Volume 2: climate, history and the future. London: Methuen

Leipe C, Demske D, Tarasov PE et al (2014) A Holocene pollen record from the northwestern Himalayan lake Tso Moriri: implications for palaeoclimatic and archaeological research. Quat Int 348:93–112

Li HC, Lee ZH, Wan NJ (2011) The $\delta^{18}O$ and $\delta^{13}C$ records in an aragonite stalagmite from Furong Cave, Chongqing, China: A-2000-year record of monsoonal climate. J Asian Earth Sci 40(6):1121–1130

Liang F, Brook GA, Kotlia BS et al (2015) Panigarh cave stalagmite evidence of climate change in the Indian Central Himalaya since AD 1256: monsoon breaks and winter southern jet depressions. Quat Sci Rev 124:145–161

Mann ME, Cane MA, Zebiak SE et al (2005) Volcanic and solar forcing of the tropical Pacific over the past 1000 years. J Clim 18(3):447–456

Masson-Delmotte V, Zhai P, Pörtner HO et al (2018) Summary for policymakers. An IPCC special report on the impacts of global warming of 1.5 C above pre-industrial levels and related global greenhouse gas emission pathways, in the context of strengthening the global response to the threat of climate change

Mazari RK, Bagati TN, Chauhan MS et al (1995) Palaeoclimatic record of last 2000 years in trans-Himalayan Lahaul-Spiti region. In: Proceedings of Nagoya IGBP-PAGES/PEP-II Symposium pp 262–269

Mehta M, Dobhal DP, Pratap B et al (2014) Late quaternary glacial advances in the tons river valley, Garhwal Himalaya. India and Regional Synchronicity. the Holocene 24(10):1336–1350

Mehta M, Majeed Z, Dobhal DP et al (2012) Geomorphological evidences of post-LGM glacial advancements in the Himalaya: a study from Chorabari Glacier, Garhwal Himalaya. India. J Earth System Sci 121(1):149–163

Mooley DA, Parthasarathy B (1982) Fluctuations in the deficiency of the summer monsoon over India, and their effect on economy. Arch Meteorol Geophys Bioclimatol Series B 30(4):383–398

Morrill C, Overpeck JT, Cole JE (2003) A synthesis of abrupt changes in the Asian summer monsoon since the last deglaciation. The Holocene 13(4):465–476

Müller C (2017) Bi-decadal climate reconstruction derived from a 1200-year long pollen record from the NE Qinghai-Tibet Plateau and its implications for discussion of the late Holocene environmental dynamics. Quat Int 444:1–10

Neff U, Burns SJ, Mangini A et al (2001) Strong coherence between solar variability and the monsoon in Oman between 9 and 6 kyr ago. Nature 411(6835):290

Owen LA (2009) Latest Pleistocene and Holocene glacier fluctuations in the Himalaya and Tibet. Quat Sci Rev 28(21–22):2150–2164

Owen LA, Benn DI (2005) Equilibrium-line altitudes of the Last Glacial Maximum for the Himalaya and Tibet: an assessment and evaluation of results. Quat Int 138:55–78

Owen LA, Caffee MW, Finkel RC et al (2008) Quaternary glaciation of the Himalayan-Tibetan orogen. J Quat Sci 23(6–7):513–531

Owen LA, Dortch JM (2014) Nature and timing of Quaternary glaciation in the Himalayan-Tibetan orogen. Quat Sci Rev 88:14–54

Owen LA, Finkel RC, Caffee MW (2002) A note on the extent of glaciation throughout the Himalaya during the global Last Glacial Maximum. Quat Sci Rev 21(1–3):147–157

Pangtey YPS, Rawal RS, Bankoti NS et al (1990) Phenology of high-altitude plants of Kumaun in Central Himalaya. India. Int J Biometeorol 34(2):122–127

Parthasarathy B, Munot AA, Kothawale DR (1994) All-India monthly and seasonal rainfall series: 1871–1993. Theor Appl Climatol 49(4):217–224

Phadtare NR (2000) Sharp decrease in summer monsoon strength 4000–3500 cal yr BP in the Central Higher Himalaya of India based on pollen evidence from alpine peat. Quat Res 53(1):122–129

Polanski S, Fallah B, Befort DJ et al (2014) Regional moisture change over India during the past Millennium: a comparison of multi-proxy reconstructions and climate model simulations. Glob Planet Chang 122:176–185

Prell WL, Kutzbach JE (1992) Sensitivity of the Indian monsoon to forcing parameters and implications for its evolution. Nature 360(6405):647–652

Raina VK (2009) Himalayan glaciers: a state-of-art review of glacial studies, glacial retreat and climate change. Kosi Katarmal, Ministry of Environment and Forests. G.B. Pant Institute of Himalayan Environment and Development. (MoEFF Discussion Paper.)

Ranhotra P (2007) Vegetational and climatic changes in the Himalayan and Trans-Himalayan Region since late Pleistocene. University of Lucknow, Lucknow. (Ph.D. thesis)

Ranhotra PS, Bhattacharyya A (2010) Holocene palaeoclimate and glacier fluctuations within Baspa valley, Kinnaur. Himachal Pradesh. J Geol Soc India 75(3):527–532

Ranhotra PS, Bhattacharyya A, Kotlia BS (2007) Vegetation and climatic changes around Lamayuru, Trans-Himalaya during the last 35 kyr BP. Palaeobotanist 56:117–126

Ranhotra PS, Sharma J, Bhattacharyya A et al (2018) Late Pleistocene-Holocene vegetation and climate from the palaeolake sediments, Rukti valley, Kinnaur, Himachal Himalaya. Quat Int 479:79–89

Rao VVK (2006) Hydropower in the Northeast: potential and harnessing analysis. Background paper, 6

Rawal RS, Bankoti NS, Samant SS et al (1991) Phenology of tree layer species from the timber line around Kumaun in Central Himalaya. India. Vegetatio 93(2):108–118

Rawat S, Gupta AK, Sangode SJ et al (2015) Late Pleistocene-Holocene vegetation and Indian summer monsoon record from the Lahaul, northwest Himalaya, India. Quat Sci Rev 114:167–181

Saha S, Owen LA, Orr EN et al (2018) Timing and nature of Holocene glacier advances at the northwestern end of the Himalayan-Tibetan orogen. Quat Sci Rev 187:177–202

Saha S, Saha K (1980) A hypothesis on onset, advance and withdrawal of the Indian summer monsoon. Pure Appl Geophys 118(2):1066–1075

Sanwal J, Kotlia BS, Rajendran C et al (2013) Climatic variability in Central Indian Himalaya during the last ~1800 years: evidence from a high resolution speleothem record. Quat Int 304:183–192
Scherler D, Bookhagen B, Strecker MR (2011) Spatially variable response of Himalayan glaciers to climate change affected by debris cover. Nat Geosci 4(3):156–159
Seong YB, Owen LA, Yi C et al (2009) Quaternary glaciation of Muztag Ata and Kongur Shan: Evidence for glacier response to rapid climate changes throughout the Late Glacial and Holocene in westernmost Tibet. Geol Soc Am Bull 121(3–4):348–365
Sharma MC, Owen LA (1996) Quaternary glacial history of NW Garhwal, central Himalayas. Quat Sci Rev 15(4):335–365
Shekhar M, Bhardwaj A, Singh S et al (2017) Himalayan glaciers experienced significant mass loss during later phases of little ice age. Sci Rep 7(1):1–14
Shukla T, Mehta M, Jaiswal MK et al (2018) Late Quaternary glaciation history of monsoon-dominated Dingad basin, central Himalaya, India. Quat Sci Rev 181:43–64
Singh JS, Singh SP (1987) Forest vegetation of the Himalaya. Bot Rev 53(1):80–192
Sinha A, Cannariato KG, Stott LD et al (2005) Variability of Southwest Indian summer monsoon precipitation during the Bølling-Allerød. Geology 33(10):813–816
Sirocko F, Sarnthein M, Lange H et al (1991) Atmospheric summer circulation and coastal upwelling in the Arabian Sea during the Holocene and the last glaciation. Quat Res 36(1):72–93
Solomina ON, Bradley RS, Hodgson DA et al (2015) Holocene glacier fluctuations. Quat Sci Rev 111:9–34
Srivastava P, Agnihotri R, Sharma D et al (2017) 8000-year monsoonal record from Himalaya revealing reinforcement of tropical and global climate systems since mid-Holocene. Sci Rep 7(1):1–10
Staubwasser M, Sirocko F, Grootes PM et al (2003) Climate change at the 4.2 ka BP termination of the Indus valley civilization and Holocene south Asian monsoon variability. Geophys Res Lett 30(8)
Stuiver M, Grootes PM, Braziunas TF (1995) The GISP2 $\delta^{18}O$ climate record of the past 16,500 years and the role of the sun, ocean, and volcanoes. Quat Res 44(3):341–354
Tan L, Cai Y, Cheng H et al (2015) Climate significance of speleothem $\delta^{18}O$ from central China on decadal timescale. J Asian Earth Sci 106:150–155
Tan L, Cai Y, Cheng H et al (2018) Centennial-to decadal-scale monsoon precipitation variations in the upper Hanjiang River region, China over the past 6650 years. Earth Planet Sci Lett 482:580–590
Tao S, An C, Chen F et al (2010) Pollen-inferred vegetation and environmental changes since 16.7 ka BP at Balikun Lake, Xinjiang. Chinese Sci Bull 55(22):2449–2457
Terray P, Dominiak S (2005) Indian Ocean sea surface temperature and El Niño-Southern Oscillation: A new perspective. J Clim 18(9):1351–1368
Thompson LG, Mosley-Thompson E, Davis ME et al (1995) A 1000 year climate ice-core record from the Guliya ice cap, China: its relationship to global climate variability. Ann Glaciol 21:175–181
Tierney JE, Russell JM (2007) Abrupt climate change in southeast tropical Africa influenced by Indian monsoon variability and ITCZ migration. Geophys Res Lett 34(15)
Vishnu-Mittre SC (1984) Vegetation and climate during the last glaciation in the Kathmandu Valley. Nepal. Pollen Spores 26:69–94
Walker MD, Ingersoll RC, Webber PJ (1995) Effects of interannual climate variation on phenology and growth of two alpine forbs. Ecology 76(4):1067–1083
Wang Y, Cheng H, Edwards RL et al (2005) The Holocene Asian monsoon: links to solar changes and North Atlantic climate. Science 308(5723):854–857
Wenxiang W, Tungsheng L (2004) Possible role of the "Holocene Event 3" on the collapse of Neolithic Cultures around the Central Plain of China. Quat Int 117(1):153–166
Wünnemann B, Demske D, Tarasov P et al (2010) Hydrological evolution during the last 15 kyr in the TsoKar lake basin (Ladakh, India), derived from geomorphological, sedimentological and palynological records. Quat Sci Rev 29(9–10):1138–1155

Xiao X, Haberle SG, Shen J et al (2014) Latest Pleistocene and Holocene vegetation and climate history inferred from an alpine lacustrine record, northwestern Yunnan Province, southwestern China. Quat Sci Rev 86:35–48
Xu H, Song Y, Goldsmith Y et al (2019) Meridional ITCZ shifts modulate tropical/subtropical Asian monsoon rainfall. Sci Bull 64(23):1737–1739
Yan G, Wang FB, Shi GR et al (1999) Palynological and stable isotopic study of palaeoenvironmental changes on the northeastern Tibetan plateau in the last 30,000 years. Palaeogeogr Palaeoclimatol Palaeoecol 153(1–4):147–159
Yang B, Bräuning A, Dong Z et al (2008) Late Holocene monsoonal temperate glacier fluctuations on the Tibetan Plateau. Glob Planet Chang 60(1–2):126–140
Yokoyama Y, Lambeck K, De Deckker P et al (2000) Timing of the Last Glacial Maximum from observed sea-level minima. Nature 406(6797):713–716
Yu G, Tang L, Yang X et al (2001) Modern pollen samples from alpine vegetation on the Tibetan Plateau. Glob Ecol Biogeogr 10(5):503–519
Zhang E, Wang Y, Sun W et al (2016) Holocene Asian monsoon evolution revealed by a pollen record from an alpine lake on the southeastern margin of the Qinghai-Tibetan Plateau. China. Clim past 12(2):415–427
Zhang G, Zhu C, Wang J et al (2010) Environmental archaeology on Longshan culture (4500–4000 BP) at Yuhuicun site in Bengbu. Anhui Province. J Geogr Sci 20(3):455–468
Zhang H, Cheng H, Cai Y et al (2018) Hydroclimatic variations in southeastern China during the 4.2 ka event reflected by stalagmite records. Clim Past 14(11)

Chapter 5
Recession of Gaglu Glacier, Chandra Basin, Western Indian Himalaya

Rupendra Singh, Rajesh Kumar, Syed Umer Latief, Rajesh Kumar, and Mayank Shekhar

Abstract Outside the polar regions, the Himalayan region has a large number of glaciers which are considered the primary source of water for the major perennial rivers of northern India. These glaciers are under threat due to ongoing climate change. The Gaglu glacier located in the Chandra river basin of Western Himalaya was studied using satellite data and ground-based field validation. During 1973–2017, the Gaglu glacier has shown a total retreat of 523.04 m with an average annual rate of 11.89 ma^{-1} and vacated an area of 432×10^3 m^2 in its frontal part. The time-series map prepared using satellite data of the intervening period indicates a variable rate of recession. The continuous recession of the glacier is attributed to the rising mean brightness surface temperature of the order of 0.06°Ca^{-1} during 1989–2015 and mean annual temperature is rising at a rate of 0.021°Ca^{-1} during 1979–2017. A rise in surface temperature causes equilibrium line altitude (ELA) to shift at a higher altitude, which increases the ablation area of a glacier. The terminus of the glacier is also highly sensitive to a rise in temperature. Hence, an increase in mean brightness temperature of the Gaglu glacier and mean annual temperature are the main causal factors for thinning of the glacier, causing enhanced snout retreat and increasing ELA elevation.

Keywords Gaglu glacier · Landsat 8 · Brightness temperature · Chandra basin

R. Singh (✉)
CPS, School of Social Sciences, Jawaharlal Nehru University, New Delhi, India

R. Kumar
Department of Geography, School of Human Sciences, Sikkim University, Tadong, Gangtok, Sikkim, India

S. U. Latief
Post-Graduate Department of Geography, Amar Singh College, Srinagar, Jammu and Kashmir (UT), India

R. Kumar
Department of Environmental Sciences, School of Earth Sciences, Central University of Rajasthan, NH - 8, Bandarsindri, Ajmer, Rajasthan, India

M. Shekhar
Birbal Sahni Institute of Palaeosciences, Lucknow, Uttar Pradesh, India

S. Rani and R. Kumar (eds.), *Climate Change*, Springer Climate,
https://doi.org/10.1007/978-3-030-92782-0_5

Introduction

The Himalayas have one of the largest concentrations of glacier ice and seasonal snowfields. The Himalayan glaciers cover more than 40,000 km^2 of area, representing one of the largest freshwater sources outside the polar regions (Kulkarni and Bahuguna 2001). The Indian Himalayas comprise a sum of 9575 glaciers (Raina and Srivastava 2008). These glaciers provide fresh water to the northern Indian river system, i.e. Ganga and Indus (Bahuguna et al. 2004).

Glaciers and snow are sensitive to climate change (Ahmad et al. 2004; Kulkarni et al. 2007; Rathore et al. 2009). In the North-western Himalayan region, the historical instrumental weather records show an increase in temperature by −1.68 °C over the past century and a decreasing trend in monsoonal precipitation (Bhutiyani et al. 2007, 2010). Nowadays, Himalayan glaciers have been retreating since the 1850s (Majewski and Jaschke 1979). The rate of retreat of Himalayan glaciers ranges from 1 to 60 ma^{-1}. The glaciers of the Baspa basin (Kulkarni et al. 2007; Gaddam et al. 2016) and Uttarakhand Himalaya (Mehta et al. 2011) are retreating over the years. Many studies on the mass balance of most of the Himalayan glaciers have reported a negative mass balance (Kulkarni 2010; Scherler et al. 2011; Kumar et al. 2007; Wagnon et al. 2007; Azam et al. 2012). Contrary to these facts, some large glaciers have been reported to be stable or advancing in the Karakoram region (Scherler et al. 2011; Gardelle et al. 2012; Tangri et al. 2013; Singh et al. 2014). The main reason for their advancement or surging is still not adequately understood; however, considered due to the local climate of the region (Space Application Centre Report 2016).

Some of the Himalayan glaciers are located in remote areas, so the accessibility of such glaciers is challenging and costly. Hence, a significantly fewer number of glaciers are investigated owing to logistical and harsh weather constraints. Under these conditions, remote sensing imageries are helpful to quantify the historical record of the glaciers. Hence, the temporal mapping and monitoring of glaciers are being done using multi-temporal optical remote sensing images (Vohra 1980; Mehta et al. 2011). Most of the glacier's snout monitoring in India is based on the 1960s and 1970s topographic maps of the Survey of India (SoI). However, the snout positions are mapped using toposheets (Raina and Srivastava 2008; Raina 2009). But Corona and Hexagon's images (declassified images) from the 1960s and 1970s are ideal and can be used for snout mapping and comparison with snout position, extracted from the topographic maps (Bhambri et al. 2012). Against the backdrop of these studies, the present study aims to compute the glaciated area, equilibrium line altitude (ELA), brightness temperature and the trend in ERA-Interim mean annual temperature. Moreover, snout retreat and area vacated have been calculated with the help of the SoI topographical map of 1963 (KH-9 of 1973) and Landsat images captured in 1989, 2000 and 2015.

Study Area

The Gaglu glacier lies between 32°21′40″ and 32°28′10″ N latitudes and 77°20′40″ and 77°26′40″ E longitudes situated in the Chandra river valley in Lahul and Spiti district of Himachal Pradesh in the Western Indian Himalaya. The glacier's length is about 5 km in the central part and about 6 km from the left part of the glacier (Kurien and Mush 1959). The total surface area covered by the glacier is 11 km^2. The glacier snout is located at an elevation of 4150 m, while the glacier headwall is situated at 5676 m. The glacier flows in the southeast direction with an uneven gradient. The stream originating from the glacier is known as *Purana Khoksar Nala* and it joins the Chandra river at Chhatru (Fig. 5.1). The lithology of the study area is constituted by shales, phyllites and quartzites (Finkel et al. 2003). The Gaglu glacier is a valley-type glacier and the geomorphology of glaciers indicates the movement of glacier and landform development in the past. In the valley, erosional and depositional landforms are found below the snout of recent times (Fig. 5.2).

The climate in the Chandra valley is characterized by the long and cold winters extending from October to mid-April. The winter season commences in the last week of August or at the beginning of September when the night temperature falls below the freezing point. According to Yang et al. (2008) and Pandey et al. (2011), this region falls in the rain-shadow zone and is primarily fed by the western disturbances. The mean annual temperature at Keylang ranges from –4.8 °C in February to 16.5 °C in July while the mean annual precipitation varies from 9.4 mm in November to 97.5 mm in March (Kurien and Mush 1959).

Material and Methods

In this study, the SoI toposheets, digital elevation models (DEMs) and multi-sensor cloud-free Landsat satellite images have been used for assessing the glacier dynamics. Cloud-free and snow-free images are available in October. A detailed list of data set used is given in Table 5.1. Due to the lack of a reliable observational network, we also included reanalysis data, namely, ERA-Interim for the mean temperature at the surface for 1979–2017 (Dee et al. 2011). The ERA-Interim reanalysis data have been extracted from the European Centre for Medium-Range Weather Forecasts (ECMWF data repository, (http://apps.ecmwf.int/datasets/) at the Gaussian grid resolution of 0.75° X 0.75° (global grid). The nearest grid to the study area (77.22 E-77.26 E and 32.21 N-32.24 N) has been selected.

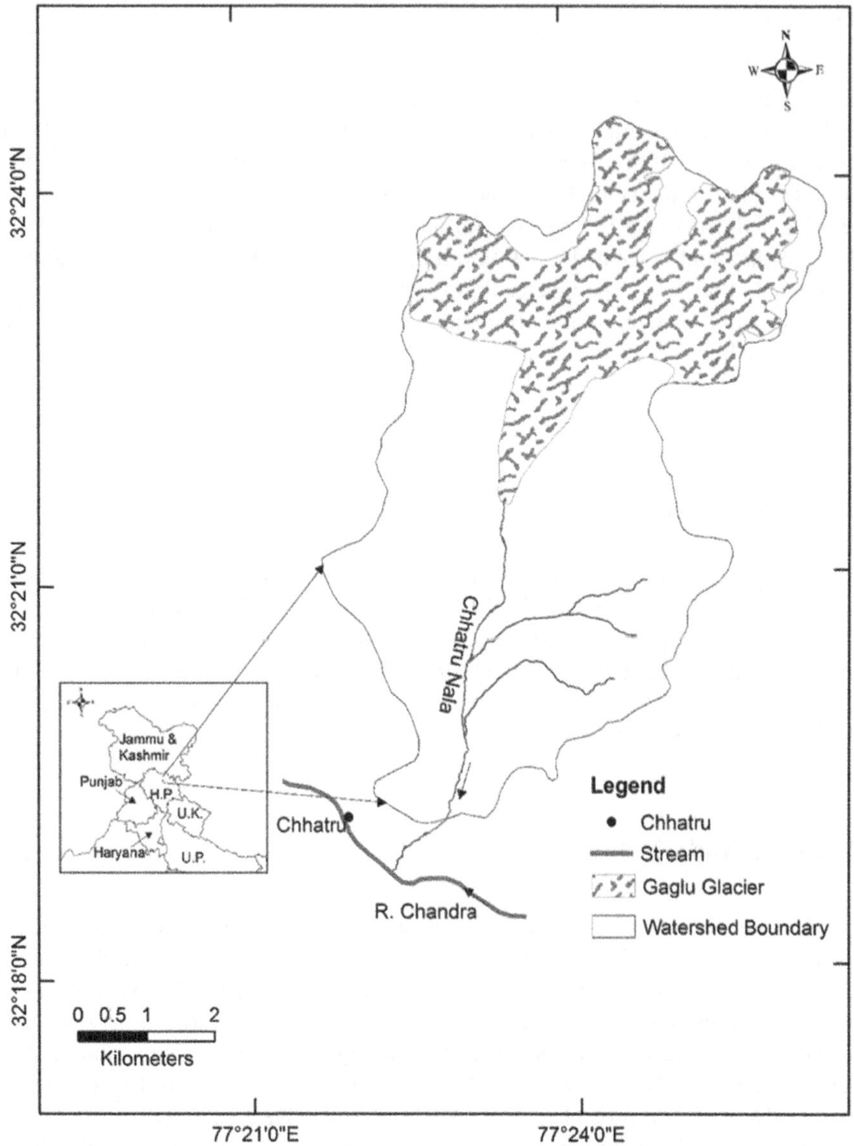

Fig. 5.1 Location map of Gaglu Glacier, Lahul and Spiti district of Himachal Pradesh, Western Indian Himalaya

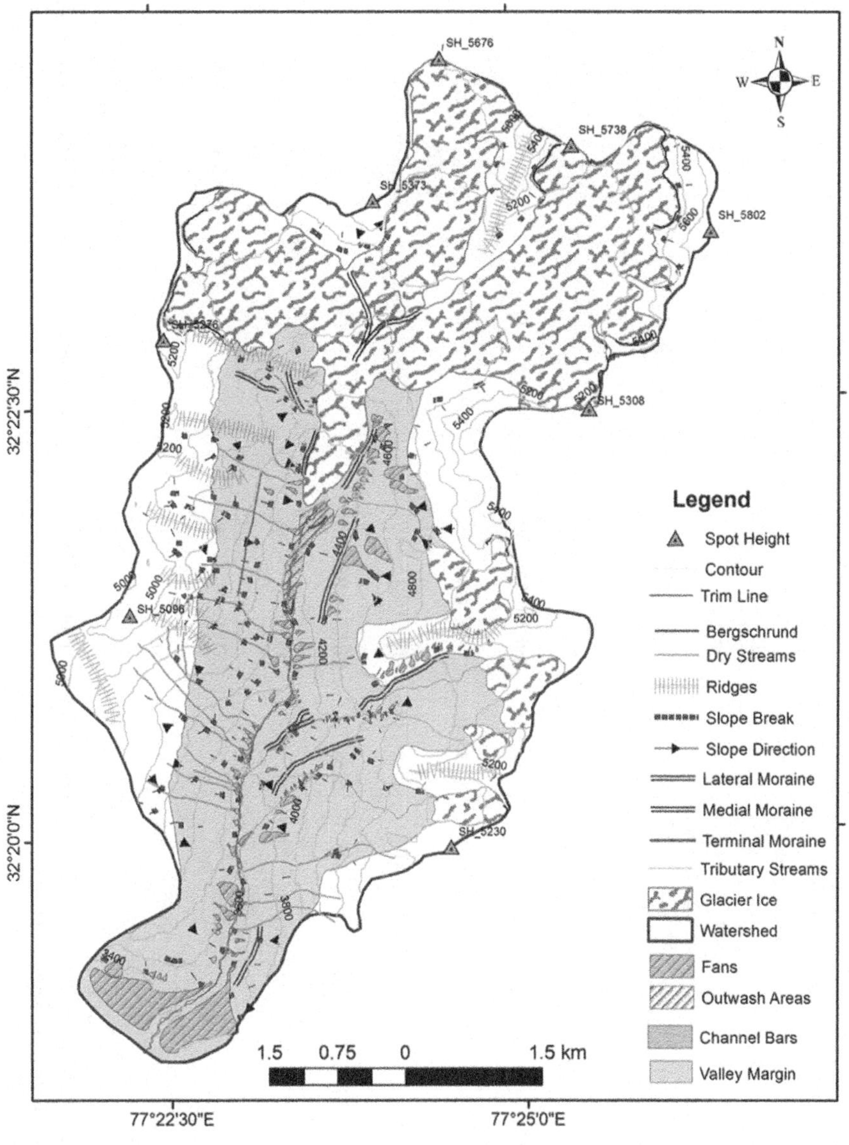

Fig. 5.2 Geomorphological map of Gaglu glacier prepared using Google Earth high-resolution images of 2014

Snout Retreat and Areal Loss of Glacier

Glacier boundaries of different years have been delineated using a visual interpretation of the SoI topographic map and satellite images in the GIS environment. The glacier's length has been derived by drawing lines along the central flow of the

Table 5.1 A detailed list of data set used in the present study

S. No	Scene ID	Agency/satellite name	Sensor	Acquisition date	Resolution (m)	Path and row	Uncertainty (m)	Source of data
1	52H/7 (Sheet No.)	SoI topographical Sheet	–	1963	–	–	–	Survey of India
2	DZB1207-500030L015001	KH-9	Aereal photograph	1973/11/19	7	–	–	Earth explorer United State geological society (USGS)
3	ETP147R38_5T19891009	LANDSAT 5	TM	1989/10/09	30	147–38	31.6	Earth explorer (USGS)
4	EPP147R038_7F20001015	LANDSAT 7	ETM +	2000/10/15	15	147–38	33.54	Earth explorer (USGS)
5	LC81470382015258LGN00	LANDSAT 8	OLI-TIRS	2015/09/15	15	147–38	23.43	Earth explorer (USGS)
6	L1C_T43SGR_A011971_20171007T053044	Sentinel 2	S2A	2017/10/07	10	T43SGR	18.3	Copernicus, sentinels scientific data hub
7	SRTM1N32E077V3	SRTM	–	2000	30	–	–	Earth explorer (USGS)
8	ASTERN32E077V2	ASTER	–	2009	30	–	–	USGS LP DAAC (2016)
9	C1v3_R-1_ 77E32N_i43x 78E31N_h44a	CARTOSAT-1 (version 3-R1)	–	2014	30	–	–	Bhuwan Indian geo-platform of NRSC-ISRO

SoI = Survey of India, TM = Thematic Mapper, ETM + = Enhanced Thematic Mapper Plus, SRTM = Shuttle Radar Topography Mission, ASTER = Advanced Spaceborne Thermal Emission and Reflection Radiometer glacier tongue from head to the terminus. Gaglu glacier had two major tributary glaciers (Left and right tributary). The reflectance of snow, glacier ice and debris shows a distinct difference in visible and near-infrared bands (VNIR) compared to short wave infrared bands (Dozier 1989; Ali et al. 2019). The band combination of 5, 4, 3 is selected for mapping the glacier outline using the Landsat image. Furthermore, the band combination of 4, 3, 2 is applied to map the glacier outline using Landsat TM and ETM + images. The vacated area due to the glacier recession has been computed by overlapping the glacier boundaries of 1973, 1989, 2000, 2015 and 2017.

ELA Estimation and DEM Accuracy Assessment

ELA is an imaginary elevation line that separates the accumulation and ablation zone of a glacier. There are different types of methods to calculate ELA. These methods are namely, area weighted mean (AWM), maximum elevation of the lateral moraine (MELM), the terminus to headwall altitude ratio (THAR) and accumulation area ratio. We also analyzed the ERA-Interim dataset on temperature. The Mann–Kendall (MK) (Mann 1945) test has also been performed using time series of annual mean temperature (1979–2017) and monthly mean temperature for October). In this study, we applied THAR, MELM and AWM methods to estimate ELAs.

Furthermore, we took the average of ELAs computed using these methods to avoid uncertainty in ELA estimation. Due to the lack of appropriate satellite images, we did not calculate ELA using the accumulation area ratio (AAR) method. The ELA establishes a link between climate and glacier health as it is highly influenced by climatic variability (Mehta et al. 2011). The advancement and retreat of glaciers are predominantly controlled by geographical location, climate, slope and aspect.

We have used the SoI topographic map contours for preparing a digital terrain model (DTM) of the Gaglu glacier. The contour interval is 40 m. In addition to this, we checked the Root Mean Square Error (RMSE) in the elevation accuracy of the Shuttle Radar Topography Mission (SRTM), Advanced Spaceborne Thermal Emission and Reflection Radiometer (ASTER version 2) and Cartosat-1 DEMs with the help of spot heights given in the SoI topographic map using Eq. (5.1) (Schumann et al. 2008).

$$\mathrm{RMSE_{DEM}} = \sqrt{\frac{(\mathrm{Topo_{SH}} - \mathrm{DEM_{EL}})^2}{n}} \tag{5.1}$$

where the spot heights of SoI topographic maps ($\mathrm{Topo_{SH}}$) denotes reference elevation data, $\mathrm{DEM_{EL}}$ is elevation values from different SRTM, ASTER and CARTOSAT-1 DEMs and n shows a sum of reference data points.

Root mean square error (RMSE) values for SRTM, ASTER and CARTOSAT-1 DEMs are 69, 276 and 315 m, respectively. The computed absolute maximum elevation error values for SRTM, ASTER and CARTOSAT-1 DEM are 101, 466 and 509 m, respectively. The total maximum elevation error and RMSE values are found to be the least for the SRTM DEM. Hence, SRTM DEM and DTM generated using contours of SoI topographic map have been used in ELAs estimation.

The ELA estimation by THAR method has been computed using Eq. (5.2) (Sharma and Owen 1996).

$$\text{THAR} = [(\text{Altitude at snout} + \text{Headwall altitude})/2] \tag{5.2}$$

The ELA estimation using MELM method is based on the visual interpretation of SoI topographic sheet; Landsat images from TM, ETM + and OLI-TIRS sensors help to find out the starting point of lateral moraines. Further, elevation at the starting point of the lateral moraine was obtained from DTM and SRTM DEM.

The AWM method for ELA estimation is computed using Eq. (5.3) (Sissons 1974; Sharma and Owen 1996; Portar 2001; Mehta et al. 2011). This method assumes that the ablation decreases linearly with an increase in elevation (Sissons 1974).

Area weighted mean (AWM) method (Zamp et al. 2009; Thibert and Vincent 2009; Beedle et al. 2014) is appropriate and straightforward for ELA estimation and has been widely used for the assessment of ELA in the Himalayan glaciers (Sharma and Owen 1996; Benn et al. 2005).

$$\text{AWM} = \left[\sum (A_i \times h_i)/\sum A_i\right] \tag{5.3}$$

where A_i is the area of the glacier in km^2 and H_i denotes midpoint elevation (AMSL) in different contour intervals.

Error Estimation

Data of different spatial resolutions have been used in this study. Hence, area and length computations have some uncertainties. Quantifying the error introduced in the calculation is essential. The co-registration errors always cause an error in area and length computation in the GIS platform. The uncertainties have been computed using Eq. (5.4) (Hall et al. 2003). This formula has been used by Ye et al. (2006) for calculating the resolution error in area and length of the snout change.

$$e = \sqrt{a^2 + b^2 + \text{E}_{\text{reg}}}, \tag{5.4}$$

where 'a' is the resolution of image 1, 'b' is the resolution of image 2 and 'E_{reg}' is the registration error.

Brightness Temperature of Gaglu Glacier

Digital number (DN) of thermal bands of Landsat 5 (Band 6) and Landsat 8 (Band10) images have been converted into radiance using Eq. (5.5) (Sen 1986).

$$R = M_L \times (\mathrm{DN}) + A_L \tag{5.5}$$

where R is the Top of Atmosphere (TOA) radiance in W/(m^2-srad-μm), M_L shows the band-specific multiplicative rescaling factor while the AL is the band-specific additive rescaling factor. M_L and A_L rescaling factors have been obtained from the Metadata files of the scenes.

Equation (5.6) has been applied to convert the radiance values of Landsat 5 TM into temperature, while Eq. (5.7) was used to Landsat 8 thermal image (band 10) (Sen 1986).

$$T = \frac{K2}{\mathrm{In}\left[\frac{K1 \times E}{\mathrm{CV}K1+1}\right]} \tag{5.6}$$

where T = Kelvin (K), CVR1 = Cell value as radiance and ε = emissivity (0.95)

$$T = \frac{K2}{\mathrm{In}\left[\frac{K1 \times E}{\mathrm{CV}K2+1}\right]} \tag{5.7}$$

where T = Kelvin (K) and CVR2 = radiance values.

The brightness temperature in Kelvin has been converted into degree Celsius using Eq. (5.8).

$$\mathrm{Temperature}(^{\circ}\mathrm{C}) = \mathrm{Temperature}(\mathrm{K}) - 273.15 \tag{5.8}$$

Analysis of the ERA-Interim Temperature Dataset

Monotonic upward and downward trends in mean annual and October month temperatures (1979–2017) have been analyzed using a non-parametric Mann–Kendall (MK) test (Mann 1945; Kendall 1975). The sign of Z values indicates positive and negative trends in the data series. The Sen's slope values (Q) give the rate of change (Q) in mean annual and monthly temperatures (Sen 1986).

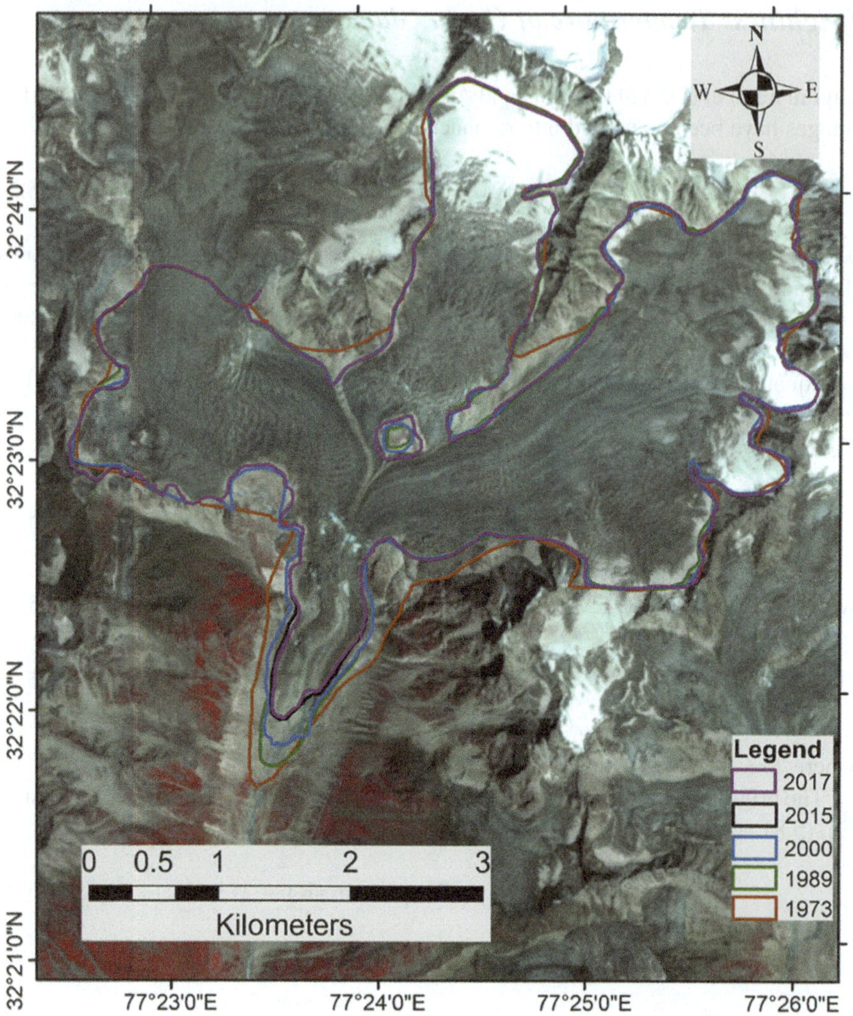

Fig. 5.3 Outer boundaries of the Gaglu glacier in different years are superimposed on the 2017 satellite image to deduce the spatiotemporal changes in the areal extent of the glacier

Results and Discussion

Snout Retreat and Area Vacated

The periodical analysis of the retreat rate and total retreat of the Gaglu glacier snout has been presented in Table 5.2 and Fig. 5.4. This is based on the available satellite data for the given period. The analysis shows that the Gaglu glacier has a total retreat of 523.04 m from 1973 to 2017 (44 years), with an average rate of retreat of

Table 5.2 The periodical retreat of the snout and the rate of retreat of snout during those periods of the Gaglu Glacier

Period	Total retreat (m)	Rate of retreat (m/year)
1963–1973	222.4	22.2
1973–1989	176.5 ± 31.6	11.1 ± 2.0
1989–2000	156.3 ± 33.5	14.2 ± 3.1
2000–2015	166.4 ± 23.4	11.1 ± 1.5
2015–2017	23.84 ± 18.3	11.9 ± 9.1
1973–2017	523.04 ± 1.7	11.89 ± 0.12

Fig. 5.4 Field photographs show changes in the frontal part and Field Photograph for the years 2014 (**a**) and 2017 (**b**)

11.89 ma^{-1}. During 44 years (1973–2017), the Gaglu glacier has lost 11% of its total area (Fig. 5.5 and Table 5.3).

The total area vacated by the glacier is 1576 × 10^3 m^2 from 1973 to 2017, out of which 453 × 10^3 m^2 was due to the frontal recession and the remaining 1123 × 10^3 m^2 from other glaciated regions (Table 5.3, Fig. 5.5). Field photographs (Fig. 5.4) and other studies (Tables 5.5 and 5.6) also show retreat trends in the Chanda river valley. The highest retreat rate has been observed between 1989 and 2000, while the lowest was between 2000 and 2017 (Table 5.3).

Equilibrium Line Altitude (ELA) Changes

The ELA, the imaginary snow line at the end of the ablation season divides the glacier area into two parts (i) the accumulation area and (ii) the ablation area. The ELA has been calculated using several methods for the years 1963 and 2000. There are several methods for the ELA computation like the THAR, MELM and AWM. The ELA estimation by the THAR method has been calculated using Eq. (5.2) and AWM method for ELA has been calculated using Eq. (5.3). In contrast, the MELM method is based on the visual interpretation of the topographic map of SoI, and the Landsat imageries (TM, ETM + and OLI-TIRS sensors). The average ELA of the

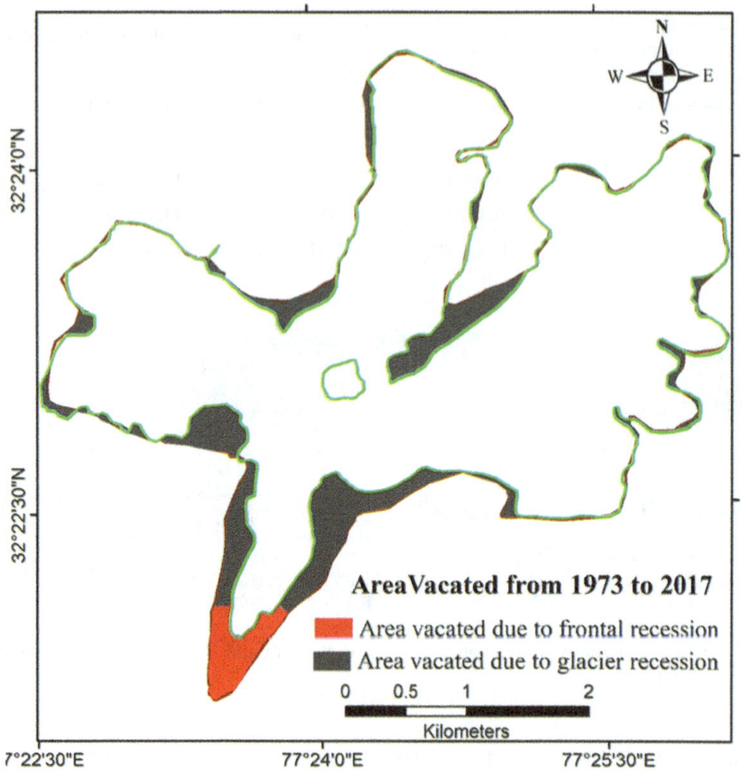

Fig. 5.5 Total area vacated by the Gaglu glacier from 1973 to 2017

Table 5.3 The periodical loss of glacial area of Gaglu Glacier in total as well as an average loss

Period	Total area vacated (10^3 m^2)	Years	Average area vacated per year (10^3 m^2)
1963–1973	178.0	10	17.8
1973–1989	1303.0 ± 15.8	16	81.4 ± 1.0
1989–2000	87.0 ± 17	11	8.0 ± 1.5
2000–2017	186.0 ± 11.7	17	12.6 ± 0.8
1963–2017	1760 ± 1	54	33.9 ± 0.1
1973–2017	1576 ± 1	44	37.65 ± 0.1

Table 5.4 Presentation of equilibrium line altitude (ELA) for 1963 and 2000 calculated by various methods

S. No	Method	ELA (m) 1963	ELA (m) 2000
1	THAR	4825	4845
2	AWM	4786	4943
3	MELM	4640	4720
	Average	4750	4836

Table 5.5 Retreat trends of glacier snouts in the Himalayan region during different

Glacier	Observation period	No. of years	Average rate of retreat (ma^{-1})	length (km.)	Aspects	Reference
Chhota Shigri (HP)	1962–2010	48	7.0	9	North	Azam et al. (2012)
Bara Shigri (HP)	1906–1995	99	29.8	27	North West	GSI Report (1999)
Gepang Gath (HP)	1965–2012	47	19.7	5.51	North West	Mukhtar and Prakash (2013)
Hamta	1963–2010	21	5.5	6	North West	Pandey et al. (2011)
Batal	1980–2011	31	7.5	4.9	East	SAC-ISRO report (2016)
Miyar	1980–2013	50	19.34	27	South	SAC-ISRO report (2016)
Menthosa	1965–2013	48	4.5	3.86	South East	SAC-ISRO report (2016)
Gaglu (HP)	1973–2017	44	11.89	9	South	Present Study

Gaglu glacier based on these methods is presented in Table 5.4 and Fig. 5.6. The ELA rages from 4640 to 4943 m for the years 1963 and 2000. The lowest value of ELA was 4640 m in 1963 by MELM method and the highest is 4825 m by THAR methods. However, the ELA in 2020 was represented lowest (4720 m) by MELM method and highest (4943 m) by AWM method. The average ELA of the Gaglu glacier was 4750 m in 1963 and 4836 m in 2000, which indicates upward shifting of the ELA while advancing years. This may be related to rising temperature and varying precipitation under the changing climatic behaviour.

An average ELA depression was −86 m from 1963 (4750 m) to 2000 (4836 m) has been observed. In a similar pattern, the response of the terminus variability of the Gaglu glacier was observed. The terminus altitude was at 4150 m in the year 1963, which moved upward to the altitude of 4240 m in the year 2000.

Brightness Temperature of Gaglu Glacier

Since the last century, many glaciers are showing a retreating trend in most parts of the Himalayan region due to the increased temperature (Table 5.5). There has been

Table 5.6 Observational characteristics of studied Glaciers in Chandra Valley

Glacier	Observation period	No. of years	Average rate of retreat (ma^{-1})	Reference
Kolahoi (J&K)	1962–2010	48	11.97	Latief et al. (2016)
Gangotri (UK)	1842–2010	168	11.0	Auden (1937),; Bhambri et al. (2012)
Dokriani (UK)	1962–2007	45	16.7	Dobhal et al. (2008)
Chaurabari Glacier (UK)	1962–201	48	6.8	Swaroop et al. (2001), Dobhal et al. (2013)
Satopanth (UK)	1968–2013	45	9.86	Tangri et al. (2014)
Tipra Bamak (UK)	1962–2008	46	14.4	Mehta et al. (2011)
Pindari (UK)	1845–2010	165	18.7	Tewari (1973); Bali et al. (2013)
Milam (UK)	1849–2006	157	17.0	Cotter and Brown (1907); Govindharaj (2011)
Rathong (Sikkim)	1976–2005	29	18.2	Raina (2009)
S. Lhonak (Sikkim)	1962–2008	46	42.2	Govindharaj et al. (2013)
Zemu (Sikkim)	1909–2005	96	9.0	Raina (2009)
Khumbu (Nepal)	1960–2006	46	18.3	Bajracharya and Mool (2009)

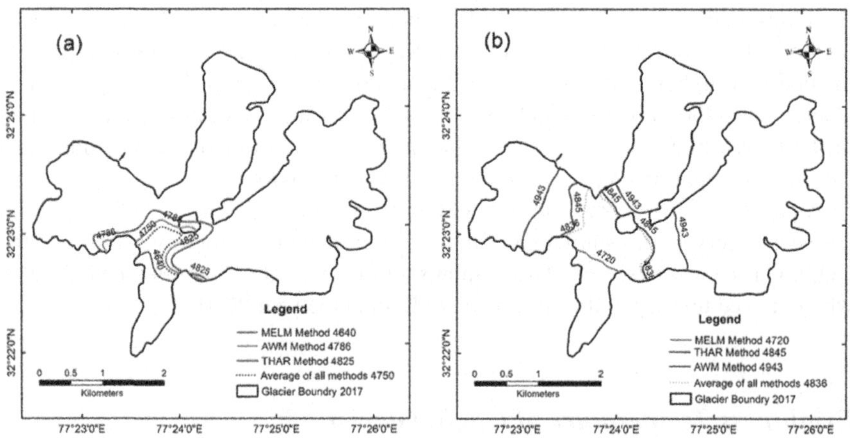

Fig. 5.6 **a** Altitudinal position of ELA in the year 1963. **b** Altitudinal position of ELA in the year 2000

a linkage between increased temperature and glacier retreats (Haeberli et al. 2005; Kaser et al. 2006; Mehta et al. 2011). A study carried out by Dimri and Mohanty (2007) pointed out that the temperature of winter months (December, January and February) increased by 2 °C during 1974–2006 in the North-western Himalayan region, whereas another study showed that the annual mean temperature increased by 1.49 ºC in the same region during 1985–2009 (Kumar et al. 2014). Because of an increase in temperature, precipitation converts into liquid from solid, which is the main cause for rainfall to the higher elevation than earlier reported in the Himalaya (Dimri and Mohanty 2007; Thayyen et al. 2005). Moreover, many authors have reported temperature rise in different parts of Himalaya (Seko and Takahashi 1991; Borgaonkar et al. 1996; Shrestha and Wake 1999; Bhutiyani et al. 2007). From 1962 to 2004, North-western Himalaya lost 21% of the glacier area due to temperature increases. It varies in different watersheds and the retreat of small glaciers is faster than the large glaciers (Kulkarni et al. 2007; Kumar et al. 2009). The 1990s was the warmest decade of the century (IPCC 2001a, b). Furthermore, the warmest period was 1997 and 1998 due to El Niño-Southern Oscillation (ENSO) (Trenberth et al. 2002, Latief et al. 2016). The retreat of glaciers also affects the location's characteristics and landscape, hence changing the albedo pattern and energy balance of the region. The size and orientation of glaciers are directly related to the rate of glacier retreat. Big size glaciers and south, south-west facing glaciers receive more energy compared to glaciers facing other sides. Therefore, the rate of retreat is high in these glaciers (IPCC 2001a, b) (Table 5.6).

The minimum brightness temperature of the Gaglu glacier has increased by −3 °C at the frontal part of the glacier between 1989 and 2015 (Fig. 5.7). Saurabh and Matthias (2016) researched the Lahaul-Spiti valley and defined three thinning patterns. Gaglu glacier falls under the first pattern (i.e. <10% debris cover type-1 glacier), which denotes the maximum thinning rate at the glacier terminus.

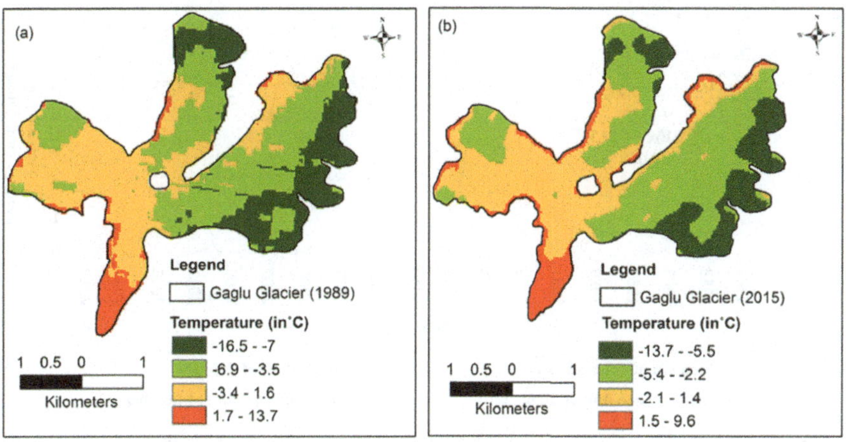

Fig. 5.7 Distribution of land surface temperature (°C) of Gaglu glacier during **a** 1989 and **b** 2015

Table 5.7 Distribution of minimum, maximum and mean brightness temperature (°C) along with standard deviation (SD) for the Gaglu glacier

Year	Min	Max	Mean	SD
09 October 1989	−16.5	13.7	−4.2	3.9
19 October 2015	−13.7	9.6	−2.6	3.2

By standard deviation interpretation, it is observed that the variation in brightness temperature was high in 1989 as compared to 2015. The range of temperature was about 30.2 °C in 1989, while it decreased to 23.3 °C in 2015. The mean brightness temperature of the Gaglu glacier has increased by 1.6 °C between 1989 and 2015. Thus, the mean brightness temperature has been rising at 0.06 °C/year since 1989 (Table 5.7 and Fig. 5.7). Such temperature rise is in agreement with the study done by Bhutiyani et al. (2007, 2010) and Kumar et al. (2014) in the north-western Himalaya region.

The ERA-Interim Temperature

The ERA mean annual temperature (MAT) has been found significant at the 0.01 level. The positive Z value (+2.85) indicates a significant increasing trend in MAT. The rate of increase in MAT is only 0.021°Ca^{-1}. On the other hand, the mean temperature of October month shows an insignificant positive Z value (1.19) along with a rate of change of 0.016°Ca^{-1} (Fig. 5.8). Hence, an increase in mean brightness temperature of the Gaglu glacier and mean annual temperature are the main contributing

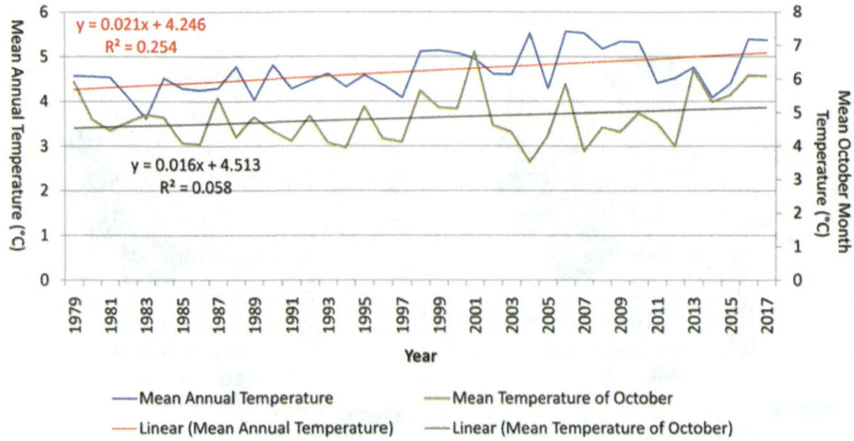

Fig. 5.8 Mean annual and mean October month temperature of the Gaglu Glacier between 1979 and 2017

factors for thinning of the glacier, causing enhanced snout retreat and upward shifting of the ELA.

Conclusions

The Gaglu glacier is highly sensitive to subtle changes in climate. Increasing temperature has seriously impacted the health of the Gaglu glacier. The mean brightness temperature of the Gaglu glacier has increased by 1.6 °C between 1989 and 2015. The time-series satellite image analysis has shown that the Gaglu glacier has retreated by 523.04 m in just 44 years at an average annual rate of 11.89 ma^{-1} and the total area vacated is 1576×10^3 m^2. The temperature rise has also caused thinning of the Gaglu glacier and ELA to shift towards higher altitudes and therefore, an increase in the ablation area has been observed. Therefore, it can be inferred that if the present scenario continues, the Gaglu glacier will not survive for long. However, knowing the complexity of Himalayan climate, terrain and glacier behaviour, more research needs to be carried out in the study area to understand the paleoclimate and accordingly predict the future behaviour of the Gaglu glacier.

Acknowledgements We are grateful to the United States Geological Survey (USGS) for making available Landsat and Orb View data and the European Space Agency ESA for Sentinel two satellite images. Rupendra Singh is thankful to Dr. Satya Prakash Shukla, Director- Polar Studies division, Geological Survey of India, Faridabad, Haryana, India and Dr. S. Nawaz Ali, Scientist of Birbal Sahni Institute of Palaeosciences (BSIP), Lucknow, Uttar Pradesh, India for thoughtful discussion on glacier dynamics. We are also thankful to Dr. Vijendra Kumar Pandey, Kirori Mal College, University of Delhi, New Delhi, India for providing the field photographs.

References

Ahmad S, Hasnain SI, Selvan MT (2004) Morpho-metric characteristics of Glaciers in the Indian Himalayas. Asian J Water Environ Pollut 1:109–118

Ali SN, Singh R, Pandey P (2019) Estimation of the frontal retreat rate of the Pindari Glacier, Central Himalaya using remote sensing technique. Earth Sci India 12(3):146–157

Auden JB(1937) The snout of the Gangotri glacier, Tehri Garhwal. Records Geol Surv India 72(II):135–140

Azam MF, Wagnon P, Ramanathan AL et al (2012) From balance to imbalance: A shift in the dynamic behaviour of Chhota Shigri glacier western Himalaya India. J Glaciol 58:315–324. https://doi.org/10.3189/2012JoG11J123

Bahuguna IM, Kulkarni AV, Nayak S (2004) DEM from IRS-1C PAN stereo coverages over Himalayan glaciated region—accuracy and its utility. Int J Remote Sens 25:4029–4041. https://doi.org/10.1080/01431160310001652376

Bajracharya SR, Mool P (2009) Glaciers, glacial lakes and glacial lake outburst floods in the Mount Everest region, Nepal. Ann Glaciol 50(53):81–86. https://doi.org/doi.org/10.3189/172756410790595895

Bali R, Ali SN, Agarwal KK, Rastogi SK et al (2013) Chronology of late quaternary glaciation in the Pindar valley, Alaknanda Basin, Central Himalaya (India). J Asian Earth Sci 66:224–233. https://doi.org/doi.org/10.1016/j.jseaes.2013.01.011

Beedle MJ, Menounos B, Wheate R (2014) An evaluation of mass-balance methods applied to Castle Creek Glacier, British Columbia, Canada. J Glaciol 60:262–276. https://doi.org/10.3189/2014JoG13J091

Benn DI, Owen LA, Osmaston HA et al (2005) Reconstruction of equilibrium-line altitudes for tropical and sub-tropical glaciers. Quat Int 138(139):8–21. https://doi.org/10.1016/j.quaint.2005.02.003

Bhambri R, Bolch T, Chaujar RK (2012) Frontal recession of Gangotri Glacier, Garhwal Himalayas, from 1965 to 2006, measured through high-resolution remote sensing data. Curr Sci 102:489–494. https://doi.org/10.5167/uzh-59630

Bhutiyani MR, Kale VS, Pawar NJ (2007) Long-term trends in maximum, minimum and mean annual air temperatures across the Northwestern Himalaya during the twentieth century. Clim Change 85:159–177. https://doi.org/10.1007/s10584-006-9196-1

Bhutiyani MR, Kale VS, Pawar NJ (2010) Climate change and the precipitation variations in the north-western Himalaya: 1866–2006. Int J Climatol 30:535–548. https://doi.org/10.1002/joc.1920

Borgaonkar HP, Pant GB, Rupa KK (1996) Ring width variation in Cedrus deodara and its climatic response over the Western Himalaya. Int J Climatol 16:1409–1422. https://doi.org/10.1002/(SICI)1097-0088(199612)16:12%3c1409::AID-JOC93%3e3.0.CO;2-H

Cotter GD, Brown ZC (1907) Notes on certain glaciers in Kumaon. Rec Geol Surv India 35(4):148–157

Dee Dick P, Uppala SM, Simmons AJ, Berrisford P, Poli P, Kobayashi S, Andrae U et al (2011) The ERA-Interim reanalysis: configuration and performance of the data assimilation system. Q J R Meteorol Soc 137(656):553–597

Dimri AP, Mohanty UC (2007) Location-specific prediction of maximum and minimum 93:79–93. https://doi.org/10.1002/met

Dobhal DP, Gergan JG, Thayyen RJ (2008) Mass balance studies of the Dokriani Glacier from 1992 to 2000, Garhwal Himalaya, India. Bull Glaciol Res 25:9–17

Dobhal DP, Mehta M, Srivastava D (2013) Influence of debris cover on terminus retreat and mass changes of Chorabari Glacier Garhwal Region, Central Himalaya, India. J Glaciol 59(217):961–971. https://doi.org/doi.org/10.3189/2013JoG12J180

Dozier J (1989) Spectral signature of alpine snow cover from the Landsat thematic mapper. Remote Sens Environ 28:148–157

Finkel RC, Owen LA, Barnard PL, Caffee MW (2003) Beryllium-10 dating of Mount Everest moraines indicates a strong monsoon influence and glacial synchroneity throughout the Himalaya. Geology 31:561–564. https://doi.org/10.1130/0091-7613(2003)031%3c0561:BDOMEM%3e2.0.CO;2

Gaddam VK, Kulkarni AV, Kumar GA (2016) Estimation of glacial retreat and mass loss in Baspa basin, Western Himalaya. Spat Inf Res 24:257–266. https://doi.org/10.1007/s41324-016-0026-x

Gardelle J, Berthier E, Arnaud Y (2012) Slight mass gain of Karakoram glaciers in the early twenty-first century. Nat Geosci 5:322–325. https://doi.org/10.1038/ngeo1450

Geological Survey of India (1999) Annual general report 8(133)

Govindharaj KB (2011) Recession and reconstruction of Milam glacier, Kumaon Himalaya, observed with satellite imagery. Curr Sci 100(9):1420–1425

Govindharaj KB, Remya SN, Kumar KV (2013) Remote sensing-based hazard assessment of glacial lakes in Sikkim Himalaya. Curr Sci 104:3

Haeberli W, Zemp M, Hoelzle M et al (2005) Fluctuations of Glaciers 1995–2000, vol 8. Paris, UNESCO. Available at: http://www.geo.uzh.ch/microsite/wgms/fog/fog8.pdf

Hall DK, Bayr KJ, Schöner W et al (2003) Consideration of the errors inherent in mapping historical glacier positions in Austria from the ground and space (1893–2001). Remote Sens Environ 86:566–577. https://doi.org/10.1016/S0034-4257(03)00134-2

Intergovernmental Panel on Climate Change (IPCC) (2001a) Climate change: the scientific Basis. In: Contribution of Working Group I to the Third Assessment Report of the Intergovernmental Panel on Climate Change. Cambridge University Press, Cambridge
Intergovernmental Panel on Climate Change (IPCC) (2001b) Climate change impacts, adaptation, and vulnerability. In: Contribution of Working Group II to the Third Assessment Report of the Intergovernmental Panel on Climate Change. Cambridge University Press, Cambridge
Kaser G, Cogley JG, Dyurgerov MB et al (2006) Mass balance of glaciers and ice caps : consensus estimates for 1961–2004. 33:1–5. https://doi.org/10.1029/2006GL027511
Kendall MG (1975) Rank correlation methods, 4th edn. Charles Griffin, London
Kulkarni AV (2010) Monitoring Himalayan cryosphere using remote sensing techniques. J Indian Inst Sci 90:457–469
Kulkarni AV, Bahuguna IM, Rathore BP et al (2007) Glacial retreat in Himalaya using Indian remote sensing satellite data. Curr Sci 92:69–74. https://doi.org/10.1117/12.694004
Kulkarni AV, Bahuguna IM (2001) Role of satellite images in snow and glacial investigations. In: Geological Survey of India Special Publication 53, Publications Division, Geological Survey of India, Lucknow, India, pp 233–240
Kumar R, Areendran G, Prakash R (2009) Witnessing change: Glaciers in the Indian Himalayas, WWF, 2009. Impress, India
Kumar R, Singh S, Randhawa SS, Kumar K, Rana JC (2014) Comptes rendus geoscience temperature trend analysis in the glacier region of Naradu. Comptes Rendus Geosci 346:213–222. https://doi.org/10.1016/j.crte.2014.09.001
Kumar R, Hasnain S, Wagnon P et al (2007) Climate change signal detected through mass balance measurement on benchmark glacier, Himachal Pradesh, India. Climate and anthropogenic impacts on the variability of water resources. Technical Documents in Hydrology 80. UNESCO/UMR 5569, Hydro Sciences Montpellier, Paris, pp 65–74
Kurien TK, Mush MM (1959) Report on the survey of the gaglu glacier of Lahul in the Kangra district of Punjub, Geological Survey of India, GSI-CHQ-1164
Latief SU, Rashid SM, Singh R (2016) Impact analysis of climate change on Kolahoi Glacier in Liddar Valley, north-western Himalayas. Arab J Geosci 9:705. https://doi.org/10.1007/s12517-016-2724-z
Mann HB (1945) Nonparametric tests against trend. Econometrica 13:245–259
Mayewski PA, Jeschke PA (1979) Himalayan and Trans-Himalayan Glacier Fluctuations Since AD 1812. Arct Alp Res 11:267. https://doi.org/10.2307/1550417
Mehta M, Dobhal DP, Bisht MPS (2011) Change of Tipra Glacier in the Garhwal Himalaya, India, between 1962 and 2008. Prog Phys Geogr 35:721–738. https://doi.org/10.1177/0309133311411760
Mukhtar MA, Prakash O (2013) Glacio-geomorphological studies in the pro-glacial region of Gepang Gath Glacier, Lahaul and Spiti district, Himachal Pradesh on expedition basis 2012–13, Geological Survey of India
Pandey P, Venkataraman G, Shukla SP (2011) Study of retreat of Hamtah glacier, Indian Himalaya, using remote sensing technique. In: Geoscience and remote sensing symposium (IGARSS), 2011 IEEE international, 2011, pp. 3194–3197
Porter SC (2001) Snowline depression in the tropics during the Last Glaciation 20
Raina VK, Srivastava D (2008) Glacier Atlas of India. Geological Society of India, Bangalore
Raina V K (2009) Himalayan glaciers, A state-of-art review of glacial studies, glacial retreat and climate change, Ministry of Environment and Forests, India, http://go.nature.com/pLgJ6D
Rathore BP, Kulkarni AV, Sherasia NK (2009) Understanding future changes in snow and glacier melt runoff due to global warming in Wangar Gad Basin, India. Curr Sci 97:1077–1081
Saurabh V, Matthias B (2016) Elevation change rates of Glaciers in the Lahaul-Spiti (Western Himalaya, India) during 2000–2012 and 2012–2013. Remote Sens 1–16. https://doi.org/10.3390/rs8121038

Scherler D, Bookhagen B, Strecker MR (2011) Spatially variable response of Himalayan glaciers to climate change affected by debris cover. Nat Geosci 4:156–159. https://doi.org/10.1038/ngeo1068

Schumann G, Matgen P, Cutler MEJ et al (2008) Comparison of remotely sensed water stages from LiDAR, topographic contours and SRTM. ISPRS J Photogramm Remote Sens 63:283–296. https://doi.org/10.1016/j.isprsjprs.2007.09.004

Seko K, Takahashi S (1991) Characteristics of winter precipitation and its effect on glacier in the Nepal Himalaya. Bul Glacier Res 9:9–12

Sen PK (1986) Estimates of the regression coefficient based on Kendall's tau. J Am Stat Assoc 63(324):1379–1389

Sharma MC, Owen LA (1996) Quaternary glacial history of NW Garhwal, Central Himalayas. Quat Sci Rev 15:335–365. https://doi.org/10.1016/0277-3791(95)00061-5

Shrestha AB, Wake CP (1999) Maximum temperature trends in the himalaya and its vicinity : an analysis based on temperature records from Nepal for the period 1971–94 maximum temperature trends in the himalaya and its vicinity : an analysis based on temperature records from Nepal for the period 12:2775–2786

Singh R, Ramchandra, Tangri AK (2014) Surging Glaciers of upper Shyok Valley, Eastern Karakoram Himalayan attempt at their temporal monitoring using Remote Sensing & GIS techniques. International Geographical Union (IGU), 4-5 June 2014 at Department of Geography and Regional Development, University of Kashmir Srinagar-190006, J&K, India.

Sissons JB (1974) A late-glacial ice cap in the central Grampians, Scotland. Trans Inst Br Geogr 62:95–113

Space Application Centre (SAC) Indian Space Research Organization (ISRO) (2016) Monitoring snow and glaciers of Himalayan Region, SAC-ISRO, Ahmedabad, 146–148. Available at: https://vedas.sac.gov.in/vedas/downloads/SAC_Snow_Glacier_Book.pdf

Swaroop S, Oberoi LK, Srivastava D, Gautam CK (2001) Recent fluctuation in snout front of Dunagiri and Chaurabari glaciers, Dhauliganga and Mandakini—Alaknanda basins, Chamoli district, Uttar Pradesh. In: Symposium on Snow, Ice and Glaciers—A Himalayan Perspective, vol 53. Lucknow, 9–11 March 1999, pp 77–81

Tangri AK, Ramchandra, Singh R (2014) Long term monitoring of the deteriorating health of Satopanth and Bhagirath Kharak Glaciers in Upper Alaknanda valley, Chamoli District, Uttarakhand Himalaya, through integration of conventional and remote sensing techniques. In: National Conference on Himalayan Glaciology (NCHG-2014) 30–31 October, 2014, pp 73–74. http://www.hpccc.gov.in/PDF/Abstract%20Final.pdf

Tangri AK, Chandra R, Yadav SKS (2013) Signatures and evidence of surging glaciers in the Shyok valley, Karakoram Himalaya, Ladakh region, Jammu & Kashmir State, India. In: Sinha R, Ravindra R (eds) Earth system processes and disaster management, society of earth scientists series, vol 1. Springer Berlin Heidelberg, pp 37–50.https://doi.org/10.1007/978-3-642-28845-6_4

Tewari AP (1973) Recent changes in the snout of Pindari glacier (Kumaun Himalayas), the role of snow and ice in hydrology. In: Proceedings of Banff Symposia, September 1972, UNESCO-WHOIAHS

Thayyen RJ, Gergan JT, Dobhal DP (2005) Slope lapse rates of temperature in Din Gad (Dokriani Glacier) catchment. Bull Glaciolo Res 31–37

Trenberth KE (2002) Evolution of el niño–southern oscillation and global atmospheric surface temperatures. J Geophys Res 107(D8). https://doi.org/10.1029/2000JD000298

Vohra CP (1980) Some problems of glacier inventory in the Himalayas. IAHS Publication. (Riederalp Workshop 1978–World Glacier Inventory). 126:67–74

Wagnon P, Linda A, Arnaud Y et al (2007) Four years of mass balance on Chhota Shigri Glacier, Himachal Pradesh, India, a new benchmark glacier in the western Himalaya. J Glaciol 53:603–611. https://doi.org/10.3189/002214307784409306

Yang X, Zhu B, Wang X et al (2008) Late quaternary environmental changes and organic carbon density in the Hunshandake Sandy Land, eastern Inner Mongolia. China 61:70–78. https://doi.org/10.1016/j.gloplacha.2007.08.007

Ye Q, Kang S, Chen F, Wang J (2006) Monitoring glacier variations on Geladandong mountain, central Tibetan Plateau , from 1969 to 2002 using remote-sensing and GIS technologies 52:537–545

Zamp M, Hoelzel M, Haeberli W (2009) Six decades of glacier mass-balance observation: A review of the worldwide monitoring network. Ann Glaciol 50:101–111

Chapter 6
Holocene Climate and Glacial Extents in the Gangotri Valley, Garhwal Himalaya, India: A Review

Parminder Singh Ranhotra, Mayank Shekhar, Ipsita Roy, and Amalava Bhattacharyya

Abstract The Himalayan glaciers being influenced by different precipitation regimes, needs an understanding of their response to climatic changes. Instrumental monitoring of some Himalayan glaciers shows a continuous negative mass balance and retreat due to recent warming. Here, the study discusses the previously established glacial stages within the Gangotri valley (Western Indian Himalaya) concerning climatic changes during Holocene. Palynology-based climate reconstructions from the Gangotri valley showed intermittent cool phases ca. 8.3 ka, 2 ka, 1 ka and 0.8–0.2 ka. During the early Holocene, the Gangotri glacier snout was probably much downstream from its present position (~3980 m above mean sea level (AMSL)), but the lateral height of the glacier remained lower than the sampling point altitudes (~4300 and ~4000 m AMSL). Documented Kedar Glacial Stage ca. 7 ka might be the end stage of a transient cool event ca. 8.3–7.1 ka. Subsequent paraglacial activity coincided with the warm phase. Shivling advance ca. 5 ka needs to be resolved climatically due to poor time resolution and age constraints. However, studies from other regions analogue the glacier extent at ~3300 to 3500 m AMSL between 7 and 5 ka. Two short-lived glacial stages, Gangotri and Bhojbasa, show respective concomitance with warm and cool conditions ca. 1 ka and 0.8–0.2 ka, later coincide with Little Ice Age (LIA). Recorded glacial stages in the valley fairly correspond with cool phases when winter westerlies were strong.

Keywords Palynology · Palaeoclimate · Western Himalayas · Glacial stages · Gangotri Glacier

Introduction

Outside Polar circles, the Himalaya and Tibetan plateau are the most glaciated regions in the world and play an important role in the dynamics of the Asian summer monsoon system and winter westerlies (Flohn 1957; Hahn and Manabe 1975; Kutzbach et al. 1993) on which the glacial system of the entire highland depends. Glacier dynamism

P. S. Ranhotra (✉) · M. Shekhar · I. Roy · A. Bhattacharyya
Birbal Sahni Institute of Palaeosciences, 53 University Road, Lucknow, Uttar Pradesh, India

S. Rani and R. Kumar (eds.), *Climate Change*, Springer Climate,
https://doi.org/10.1007/978-3-030-92782-0_6

is readily linked to the changes in hydroclimatic conditions. To better understand the relationship between the glacial and hydroclimatic systems, the knowledge of the duo during much past, i.e. beyond the instrumental records is important. This could be available only through proxy data analyses from various geomorphological features within the glaciated valleys. Many late Quaternary climatic reconstruction studies are available from the Himalayan and Trans-Himalayan regions (Bhattacharyya et al. 2006, 2011; Ranhotra et al. 2007; Demske et al. 2009) albeit needs more data flux from the glaciated regions. In last two decades many reconstructions on the glacier dynamics since the Last Glacial Maxima (LGM) till late Holocene are available from the Himalayan regions (Sharma and Owen 1996; Owen et al. 2002a, b; Bhambri and Bolch 2009; Mehta et al. 2012, 2014; Sharma and Shukla 2018; Ali et al. 2019) and Tibetan (Shi 2002; Owen et al. 2005; Jin et al. 2008; Yang et al. 2008; Owen 2009; Zhu et al. 2013; Loibl et al. 2014) based on the field mapping and absolute dating of the glacio-geomorphic features. It is therefore of interest to compare the reconstructed climatic phases with the glacier dynamics in the valleys.

There exists spatial variation in annual precipitation over the Indian subcontinent due to the seasonal shifting of the Inter-Tropical Convergence Zone (ITCZ). The northward shifting of ITCZ controls the Indian summer monsoon (ISM) by drawing the moisture from the Bay of Bengal (BOB) and Arabian Sea (AS). The climate of Eastern Himalaya has a strong summer monsoon influence through the BOB branch. The central and western Himalaya are also influenced by the extra-tropical Western Disturbance (WD) during the winter months (Dimri and Mohanty 2009; Dimri 2013; Polanski et al. 2014). Previous studies are of debate on whether the ISM precipitation or strong winter westerlies (WDs) as a major driver to glacial advancements in the Himalaya during Late Pleistocene—Holocene. Few discussed ISM as a major source of moisture resulting in the glacial accumulation (Yang et al. 2013; Loibl et al. 2014) and advancement during the marine isotope stages (MIS) 3 and 4 (Owen et al. 2002a, b; Finkel et al. 2003; Yang et al. 2008; Owen 2009; Ali and Juyal 2013; Murari et al. 2014). However, recent studies have attributed strengthened mid-latitude westerlies and low temperatures contributing to the glacier response during the MIS-2 (LGM) phase (Ganopolski et al. 1998; Nagar et al. 2013; Eugster et al. 2016; Sharma et al. 2016; Ganju et al. 2018; Ali et al. 2019).

Within this approach, the Gangotri valley holding the largest glacier of the Garhwal Himalaya has also been widely studied for the glacier dynamics including reconstruction of glacier extent beyond LGM (Sharma and Owen 1996; Barnard et al. 2004; Owen 2009) as well in terms of past climate changes (Ranhotra et al. 2001; Kar et al. 2002; Ranhotra and Kar 2011). But a comprehensive scenario of the glacial dynamics concerning climate change is still lacking for the Gangotri valley. Here we have attempted to relate the identified glacial stages with the reconstructed climatic events for the Holocene period and to generalize the temporal extent of the Gangotri Glacier in the valley. To meet out the objectives we used previously available records of the reconstructed glacial stages based on the geomorphological evidence and the palynology-based Holocene climate reconstructions from the Gangotri valley.

Study Area

The present study has been conducted in the Gangotri valley of the Western Indian Himalaya (Garhwal Himalaya) (Fig. 6.1). The upper reaches of Bhagirathi valley

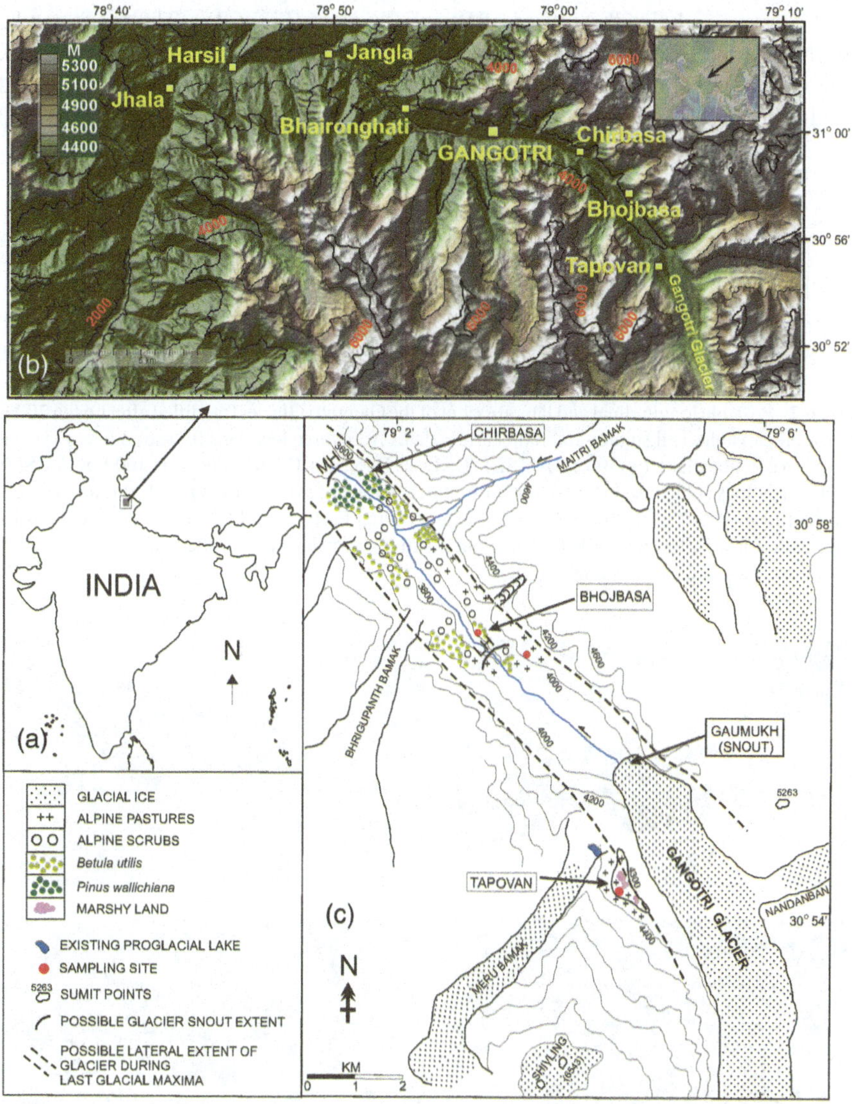

Fig. 6.1 Gangotri Glacier valley showing, **a** location of the study site. **b** study sites and sampling points. **c** vegetation distribution, possible longitudinal extent of glacial snout during mid Holocene (MH) and late Holocene (LH) and lateral extent of glacier body till LGM (Modified based on Owen 2009)

in Uttarkashi district of Uttarakhand (Fig. 6.1a–c) hold settlements by the name Gangotri township is situated at an altitude around 3000 m AMSL. The valley is characterized by ~30 km long Gangotri glacier that starts from the base of Chaukhamba peaks at the elevation of ~7000 m AMSL. The Gangotri glacier forms the main trunk glacier of the valley with its snout positioned at ~ 3970 m AMSL (Fig. 6.2 and Fig. 6.3b, June 2012 survey) and ~ 3980 m AMSL (Figs. 6.2 and 6.3c, d, October 2015 survey) and also the source of Bhagirathi river (Fig. 6.1c). The snout can be reached following the ~18 km of mule track upstream from the Gangotri township. The valley is well marked by the geomorphic features of glacial and glacio-fluvial origin. From the snout position till 4 km downstream, the valley slopes gently and

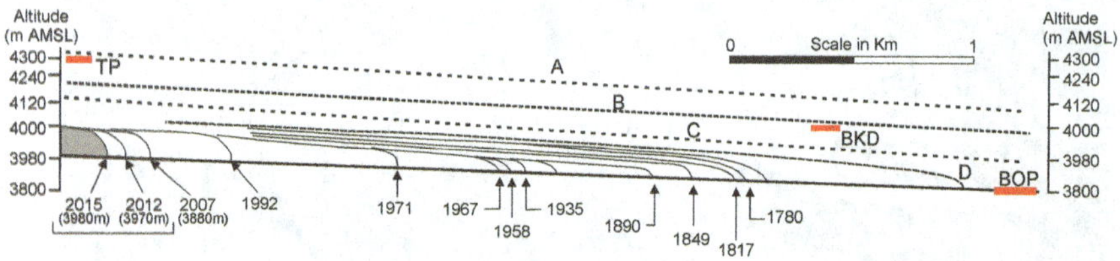

Fig. 6.2 Possible longitudinal and lateral extent of the Gangotri Glacier (modified after Owen 2009) based on palynological studies and radiocarbon dates from three locations (red solid lines) viz. TP = Tapovan Palaeolake (~4300 m AMSL), BKD = Bhojbasa Kame Deposit (~4000 m AMSL), BOP = Bhojbasa Outwash Plain (~3800 m AMSL). **A** Lateral extent during LGM. **B** Lateral extent during Local BGS (Owen 2009). **C** Lateral extent during Early to Mid Holocene. **D** Extent during later part of late Holocene (~400 C.E.). Snout positions and elevation for the years 2012 and 2015 C.E. were surveyed by one author (PSR)

Fig. 6.3 Photographs of Gangotri Glacier snout positions surveyed in different years. **a** May 2002. **b** June 2012. **c** and **d** October 2015

appears as perfect U-shaped with distinctly traceable older and recently exposed fresh lateral as well as ground moraines. The fresh moraines in this region are more confined within a 2 km distance downstream of glacier snout. Further downstream, the valley becomes comparatively narrow and more influenced by the fluvial activities besides the reworking of older glacial deposits (Nainwal et al. 2004; Srivastava 2012).

Climatically the valley is dry, receives around 500 mm of precipitation during May–September through summer monsoon (Singh et al. 2005) and 200–300 cm of snowfall during winter (October to April) through western disturbances (Raina and Srivastava 2008). The Gangotri valley lying at a sharp right-angle turn to summer monsoon front (Sharma and Owen 1996) remains devoid of summer rain penetrating the valley. Monthly mean minimum and maximum temperature range between −10 and 12 °C. The gridded climate data record for the Gangotri valley (Fig. 6.4) obtained from the Climatic Research Unit; University of East Anglia (CRU-TS v. 3.23; (0.5 × 0.5°) grid; Harris and Jones 2015) explains the mean annual temperature (MAT) and mean annual precipitation (MAP). The summer months (June–September) contribute ~71% to the annual rainfall with July (~15 mm) being the wettest month. Winter months (December-February) contribute only around ~12% of the total rainfall with November being the driest month with ~10 mm rainfall. The MAT for the area is ~7.7 °C with the highest in July (14.4 °C) and the lowest in January (−0.7 °C).

The distribution of vegetation in the valley is also quite distinct with the altitude and characteristics of environments compared to the lower reaches of the Bhagirathi valley. Temperate conifers (*Abies, Cedrus, Picea, Pinus*) and broad-leaved taxa (*Quercus, Alnus, Corylus, Carpinus, Ulmus, Juglans,* etc.) grow at the lower elevations around Gangotri town (~3000–3200 m AMSL). *Pinus wallichiana* (blue pine)

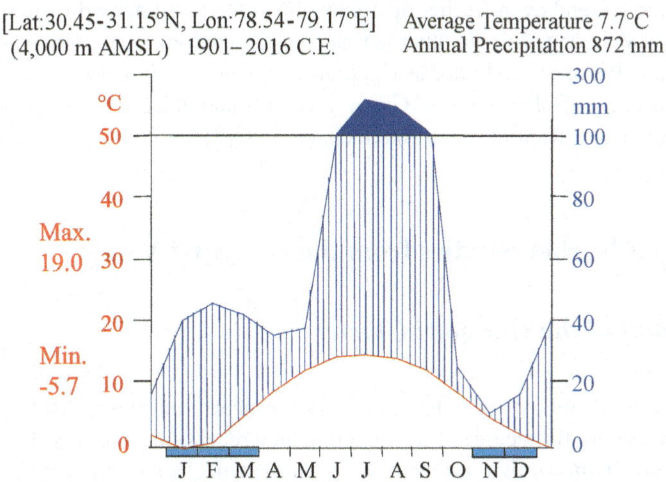

Fig. 6.4 The Walter-Leith climate diagram showing temperature and precipitation values (CRU-TS v. 3.22; (0.5 × 0.5 degree) data for the years 1901–2016 C.E.) for Gangotri valley

Fig. 6.5 Photographs of the palynological sampling sites in the Gangotri valley. **a** Tapovan Palaeolake (TP) at ~4300 m AMSL. **b** Bhojbasa Kame Deposit (BKD) at ~4000 m AMSL. **c** Bhojbasa Outwash Plain (BOP) at ~3800 m AMSL. **d** Growth of *Betula utilis* on moraine deposits at Bhojbasa

forms the upper limit of conifer forest at the altitude of ~ 3600 m AMSL with dense growth along with a broad-leaved sub-alpine taxa *Betula utilis* (birch) growing on the older stabilized morainic and fan deposits (Figs. 6.1c and 6.5d). Upstream growth of the birch on the lateral and ground moraines forms the tree line at the altitude of ~3800 m AMSL along with the other shrubby and scrubby taxa viz. Rosaceae, *Juniperus, Lonicera, Ephedra,* etc. Few scrubby taxa such as *Juniperus, Lonicera, Ephedra,* etc. extend even further upstream above the tree line and form patches on the valley walls along with the other herbaceous taxa belonging to widely distributed families like Poaceae, Asteraceae (including *Artemisia*), Amaranthaceae, Brassicaceae, Rosaceae, Polygonaceae, Cyperaceae, Ranunculaceae, Saxifragaceae, etc. and Pteridophytes (Ranhotra and Bhattacharyya 2013).

Preview of Earlier Studies from the Gangotri Valley

Past Extent of the Gangotri Glacier

In the Himalayan region, very few glaciers have nearly more than 100 years (mid-nineteenth century till recent) of well-documented recessional records (Bhambri and Bolch 2009). It shows that amongst the documented glaciers, the oldest recession records are available for the Gangotri glacier that may be due to its mythological importance for centuries and its accessibility. This glacier has been continuously explored by various scientific groups for various aspects viz. hydrology (Kumar et al. 2002; Singh et al. 2008), glacier mass balance (Gantayat et al. 2014; Kulkarni and Karyakarte 2014), geomorphology (Singh and Mishra 2002; Bali et al. 2003; Srivastava 2012) and past climate (Ranhotra et al. 2001; Kar et al. 2002; Ranhotra and Kar 2011). In the year 1935 Auden (1937) initiated the systematic snout mapping of the Gangotri glacier, which was followed by several recessional and snout monitoring studies by the Geological Survey of India (Jangpangi 1958; Tewari 1972). According to Sharma and Owen (1996) and Naithani et al. (2001) the Gangotri glacier has been retreating since 1780 C.E. at various rates. However, Mukherjee and Sangewar (2001) calculated the snout advancement of about 500 m until 1935 C.E. based on the sketch of the Gangotri glacier snout position prepared by Griesbach in 1889 C.E. (Bhambri and Bolch 2009). The available documented records show that from the last phase of the Little Ice Age (LIA) time (1780 C.E.) till recent (2006 C.E.) the Gangotri glacier front has shown a cumulative recession of around 2300 m (Bhambri and Bolch 2009; Bolch et al. 2012) at various rates (Fig. 6.2, modified after Owen 2009).

The geomorphological features within the glacier valley are the best archives to reconstruct the glacier extent, especially beyond the available documented records in the last several thousand years. Sharma and Owen (1996) recognized three glacial stages based on optically stimulated luminescence (OSL) dating of the sediment landforms within the Bhagirathi valley. During the late Pleistocene, the maximum extent of ice occurred ca. 63 ka BP (~MIS 3), which is considered as Bhagirathi Glacial Stage (BGS) when the snout extent of the Gangotri glacier was documented at ~40 km downstream to Jhala (~2300 m AMSL) concerning present snout position (Fig. 6.1a). The snout extent of most of the Himalayan glaciers during LGM ca. 24–18 ka BP was considered much limited (~10–15 km down than their present position). The other two glacial advances were recorded during mid to late Holocene when glaciers were around 3–1.5 km beyond the present ice fronts, named as Shivling Glacial Advance (SGA) ca. 5 ka, followed by Bhojabasa Glacial Advance (BGA) of 1 to 2 km than present ice fronts during latter phase of LIA (200–300 years BP). With the addition of more dates through Cosmogenic Radionuclide (CRN) dating of glacial and paraglacial landforms, Barnard et al. (2004) reconstructed five glacial stages within the Bhagirathi valley (Table 6.1).

Palaeoclimatic Scenario of the Gangotri Area

During the last few years, palynological studies of subsurface sedimentary profiles collected from various geomorphic features at different altitudes (Figs. 6.1c and 6.4) have been carried out and the reconstruction of Holocene vegetation and climate of the Gangotri area was provided (Ranhotra et al. 2001; Kar et al. 2002; Bhattacharyya and Ranhotra 2003; Ranhotra and Bhattacharyya 2004; Ranhotra and Kar 2011). The identified climatic events supported by the radiocarbon chronology (^{14}C ages) are provided in Table 6.1. The calibrated ages are mentioned in 'ka BP' in the table and text.

Table 6.1 Correlation of climatic phases with the different glacial stages within the Bhagirathi Valley

Sampling site Geomorphology	Calibrated age range (ka BP)[a]	Palynology-based Climatic phases	Glacial stages with approximate timing (Barnard et al. 2004)	
Bhojbasa Outwash Plain (BOP)	Recent	Warm—Moist		
	~0.8 to ~0.2	Cool—Dry	Bhojbasa stage	~ 0.3–0.2 ka
	~1.7 to ~0.8	Warm—Moist	Gangotri stage	~1 ka
	~2 to ~1.7	Cool—Moist		
Tapovan Palaeolake (TP)	~5.4 to ~2	Less Arid		
Bhojbasa Kame Deposit (BKD)	~6.8 to ~5.4	Varied climate mostly arid	Shivling advance	~5 ka
	~7.8 to ~ 6.8	Warm—Moist	Kedar stage	~7 ka
	~9 to ~7.8	Cool—Moist		
	~10 to ~9	Warm—Moist		
			Bhagirathi stage	~63–11 ka

[a]All dates in "ka BP" are based on 1950 C.E

Study Sites in the Gangotri Valley

The reconstructed climatic events are available by the previously analysed subsurface sedimentary sequences from the three sites within the Gangotri glacier valley, in the upper reaches of Bhagirathi valley, Uttarakhand, western Himalaya (Fig. 6.1b, c). The three sites are located within the altitudinal range of around 500 m and cover the transect of ~4 km from Bhojbasa to Tapovan Palaeolake upstream (Fig. 6.2). They represent different glacio-geomorphological features and ecological regimes. The site, Tapovan Paleolake (TP) is a sediment-filled ablation valley at the altitude of ~ 4300 m AMSL and above the present glacier snout position (Figs. 6.1c, 6.2 and 6.5). This palaeolake surrounded by the base of Shivling peak, right lateral moraines of Meru tributary glacier and left lateral moraines of the main trunk glacier. The meltwater streams from the Shivling ice melts and Meru Bamak glacier (a left tributary glacier) join the Bhagirathi stream near Gangotri glacier snout. The centre part of the palaeolake holds marshy conditions supporting the good growth of Cyperaceae, whereas the steppe elements are growing along the dry margins of the palaeolake and mountain slopes. Bhattacharyya and Ranhotra (2003) and Ranhotra and Bhattacharyya (2004) carried out a palynological analysis on the sedimentary sequence of this palaeolake and provided the climatic events covering the Holocene time. The second site is Bhojbasa outwash plain (BOP), which is around 4 km downstream of the glacier snout (Figs. 6.1c, 6.2 and 6.5). Bhojbasa, also a base camp is an outwash plain at an altitude of ~3800 m AMSL supporting the sparse birch forest associated with the scrubby and steppe taxa. The third site is located, ~500 m upstream of Bhojbasa, is the top of the kame deposit (BKD) at the right lateral moraine (~4000 m AMSL) with alpine steppe elements and grass cover (Figs. 6.1c, 6.2 and 6.5). Subsurface sediment profiles, one each from the sites BOP and BKD also provided palynology-based Holocene climate reconstruction (Ranhotra et al. 2001 (BKD); Kar et al. 2002 (BOP)).

Chronology of Sedimentary Sequences

From TP, the chronology and climate records are available from two sedimentary sequences. A 2.4 m deep sedimentary sequence from the TP deposits consisted of clay to coarse sand and grit size sediments, indicating a fluvio-lacustrine depositional environment. The base of the profile was dated as 9 ka BP by using the radiocarbon chronology (Ranhotra and Bhattacharyya 2004). A short 70 cm deep sequence near the base of Shivling peak contains organic-rich silt and clay sequence and covers the late Holocene time frame since ~3 ka (Ranhotra and Bhattacharyya 2004). The 1.4 m deep profile from the top of Kame terrace near Bhojbasa (BKD site) consists of silt and fine to coarse sand and grit with the bands of organic-rich clay at certain depths. Three radiocarbon dates are available for the sediment samples at various depths, with the date of ~8.73 ka BP near the base of the profile (Ranhotra et al. 2001) made the sequence represents Holocene time.

The sediment profile from the BOP is 1.25 m deep and consists of humic clay with rootlets at the top 40 cm depth from the surface. The underlying sediments mostly contain fine to coarse-grained sand with gravels with occasional clay lenses and intercalations. The samples were sequentially dated as 0.6 ± 0.09 ka BP (BS-1750) and 1.59 ± 0.25 ka BP (BS-1788) and extrapolated to 2 ka BP till the depth of profile (Kar et al. 2002). Thus, the sedimentary profiles from the TP and BKD sites covered almost the entire Holocene whereas the profile from Bhojbasa outwash plain covered the late Holocene part. The dates of all the sedimentary sequences from three sites are provided in Table 6.2.

Vegetation vis-à-vis Climate Reconstruction

The palynological record of Holocene using sedimentary profiles from this region revealed abundant pollen-spores of herbaceous steppe and marshy taxa, which are most representative of alpine vegetation of this region (Ranhotra et al. 2001; Kar et al. 2002; Bhattacharyya and Ranhotra 2003; Ranhotra and Bhattacharyya 2004). The pollen of temperate tree taxa, presently not growing within the vicinity of sampling sites, are also present in a good amount. These pollen grains are transported from the lower altitudes by upthermic winds in which the pollen of *Pinus* (conifer) dominates despite this taxon presently forming its upper growth limit at the altitude (~3600 m AMSL) which is lower than the altitudes of study sites (Fig. 6.1c). The winged

Table 6.2 Radiocarbon chronologies of sedimentary sequences from different geomorphic locations in Gangotri valley

Sample number	Depth range from surface (cm)	^{14}C dates (Yr B.P.)	Calibrated ages (cal Yr B.P.)	BS (BSIP) number
A				
TP-1	20–30	1910 ± 270	1868	1876
TP-2	120–130	1710 ± 90	1609	1863
TP-3	230–240	9000 ± 450	10,189	1859
B				
BKD-1	20–30	1030 ± 90	953	1854
BKD-2	50–54	5990 ± 120	6835	1730
BKD-3	120–124	8730 ± 170	9730	1729
C				
BOP-1	30–40	600 ± 90	614	1750
BOP-2	80–90	1590 ± 250	1513	1788

A = Tapovan Palaeolake (TP); B = Bhojbasa kame deposit (BKD); C = Bhojbasa outwash plain (BOP)

pollen of *Pinus* is produced profusely and transported long-distance by the winds (Ranhotra and Bhattacharyya 2013; Roy et al. 2018). The pollen of other conifers (*Abies, Picea and Cedrus*) and the temperate broad-leaved taxa, which are growing further downstream, were also sparsely present. Pollen of *Betula*, the only broad-leaved sub-alpine tree line taxa, has been found in a fair amount in the samples from all three sites.

The climate reconstruction from the three sites shows warm and moist conditions during the early Holocene, i.e. ca. 9000 to 8300 yrs BP (10.4–9.25 ka BP) indicated by the presence of a fair amount of pollen of conifers, temperate broad-leaved taxa and local herbaceous taxa. This was followed by a cool phase till ~7100 yrs BP (~7.8 ka BP) with a decline in conifers and temperate broad-leaved taxa. The other intermittent dry phases around 6000-5000 yrs BP (6.8–5.4 ka BP), ~3000 yrs BP (3.1 ka BP) and ~1 ka BP were noticed. The late Holocene part, i.e. since ca. 2 ka BP till recent was better shown by the study from the BOP site (Kar et al. 2002) depicting warm climate between ca. 1.7–0.8 ka BP that can be correlated with the Medieval Warm Period (MWP) followed by cool conditions from ca. 0.8 ka BP till ~0.2 ka BP corresponding to LIA (Table 6.1).

Climate Versus Glacier Extent

Pre-Holocene

Glaciers are sensitive to climatic changes and show major or minor oscillations in response to the intensity of climate change. These oscillations accelerate various paraglacial activities that frequently modify the glacial landscapes. Hence, it is a challenge to quantify the glacial landforms chronologically, to understand the glacier dynamics much beyond the instrumental records. Within the Bhagirathi valley, the late Pleistocene extensive glaciation (BGS) ca. 63 ka, (Sharma and Owen 1996) corresponds to the marine isotope stage 3 (MIS 3), when the high South Asian Summer Monsoon, due to increased insolation, might have resulted to more snowfall at the high altitudes of the Himalayan region to further north producing positive mass balance. Subsequently during the LGM ca. 24–18 ka, the weaker monsoon due to reduced insolation might have kept Himalayan glaciers restricted to higher altitudes, generally extending around 10–15 km from present ice margins (Owen et al. 2002a). Besides the longitudinal extent of the Gangotri glacier, the lateral extent or the thickness of the glacier mass during BGS and LGM are not well documented at the lower reaches due to the reworking of the glacial deposits. However, at the higher reaches, i.e. upstream of Gangotri town the high extent of the lateral moraines is still preserved and is more prominent near the vicinity of the present glacier snout. Owen et al. (2002a) documented the height of these preserved moraines as the maximum thickness during BGS (MIS 3) (Fig. 6.2) when the glacier snout was ~40 km downstream from its contemporary position.

The palynology-based climatic studies carried out on the sediments trenched from the Tapovan Palaeolake (TP) between the left lateral moraine of the main valley (Ranhotra and Bhattacharyya 2004) and the top of kame deposit near Bhojbasa (BKD) at the right lateral moraine of the main valley (Ranhotra et al. 2001) have provided respective radiocarbon dates of ~10 and ~9.7 ka BP. Presently the site TP is at an altitude of ~4300 m AMSL, which is ~300 m vertically higher than the present valley bed. Also, the top of kame deposit (BKD) at ~4000 m AMSL on right lateral moraine ~3.5 km downstream from glacier snout is ~ 200 m vertically higher than the present river bed (sites TP and BKD in Figs. 6.1c and 6.2). High pollen frequency of alpine herbaceous taxa from near bottom samples of both the profiles suggests that at least since early Holocene both the sites might not be covered by the glacial ice. The lithology of the sediments indicates the shallow marshy to lacustrine or fluvial conditions at sites TP and BKD. The availability of open grounds along the lake margins and mountainous slopes might have supported the meadow-type vegetation comprising the herbaceous and steppe taxa under the early Holocene warm-moist conditions. Thus during the early Holocene, the thickness of the glacial ice, under retreating conditions, was lower than the heights of sampling points TP and BKD (Fig. 6.2). The CRN dates of 15 ka, 11 ka and 8 ka BP respectively at the higher and lower levels of the right lateral moraine deposits (Barnard et al. 2004) between Bhojbasa and present snout position also suggested the lateral thinning of glacier ice during the commencement of early Holocene warm conditions after the Younger Dryas stadial phase of ~12.7–11.5 ka.

Holocene

The Holocene interglacial phase is generally represented by the warm climatic conditions when the conditions during early Holocene, mid-Holocene optima and MWP of late Holocene were warmer and more moist. The resultant mass balance of most of the glaciers might be negative during these phases. Barnard et al. (2004) documented Kedar Glacial Stage ca. 7 ka BP, which brackets in the warm-moist phase (Table 6.1) might be the end stage of a transient 8.2 ka cool event recorded globally (Alley and Ágústsdóttir 2005; Thomas et al. 2007; Matero et al. 2017) and also experienced in the regions of Himalaya (Demske et al. 2009; Rawat et al. 2015; Srivastava et al. 2017; Ranhotra et al. 2018) when the WD was strong and monsoon precipitation was weak. This was followed by the paraglacial activity ca. 6.6 ka BP (Barnard et al. 2004) coinciding with the subsequent warm phase allowing the reduction of glacial mass balance and glacier retreat. The period during the middle Holocene cannot be well resolved climatically due to the low time resolution of the palynological samples. But the mid-Holocene time ca. 5–4 ka was reported dry (Ranhotra et al. 2001; Chakraborty et al. 2006; Ranhotra and Bhattacharyya 2010) with the weaker summer monsoon precipitation over the Indian subcontinent (Berkelhammer et al. 2012; Dixit et al. 2014; Giesche et al. 2019) and also covers the Shivling advance documented ca. 5 ka. The recent efforts on the reconstruction of glacier extent from the other Himalayan valleys viz. Kedarnath in Uttarakhand (Mehta et al. 2012), Tons valley (Scherler et al. 2010), Dokriani valley (Shukla et al. 2018), Bangni Glacier, Central Himalaya (Sati et al. 2014) and Rukti in Sangla, Himachal Pradesh (Ranhotra and Bhattacharyya 2010) suggested the extent of glacier snouts at the altitudes of 3200–3500 m AMSL during early to mid-Holocene period respectively. Considering the Mid-Holocene longitudinal extent of glacial bodies within the Rukti and Chorabari Valleys as an analogue, the snout position of the glacier within Gangotri valley during mid Holocene also might have been at the altitudes ~3500–3600 m AMSL, i.e. ~9 km downstream from the present position (Fig. 6.1c). During Late Holocene the two short-lived glacial stages viz. Gangotri and Bhojbasa stages identified ca. 1 ka and 0.3–0.2 ka BP coincide with the MWP and LIA climatic phases. Palynological investigation of a sedimentary sequence from BOP revealed the strong ISM phase ca. 1.7–0.8 ka and strong WD phase ca. 0.8–0.2 ka (Sharma and Owen 1996; Barnard et al. 2004).

Correlation with Glacial Records from Other Valleys and Climate-Glacial Dynamics

Other regions of the Himalayas also provided evidence of glacial episodes comparable to the glacial stages from the Gangotri valley. From Nubra valley, Karakoram Himalaya, Ganju et al. (2018) reported major and minor glacial advances during pre-Holocene and Holocene time based on the optical chronologies of lateral moraines. The oldest glacial advance was dated ca. 60.4 ka corresponding to MIS-4 followed by the decrease in ice volume and vertical shrinking of ice mass till ca. 42 ka. The subsequent major glacial advances were dated ca. 30 ka (beginning of MIS-2) and ca. 18 ka (LGM). The minor glacial advancement episodes were recorded during the mid Holocene (ca. 6.8 ka) and late Holocene (ca. 1 ka and between 0.5 and 0.2 ka). Based on ^{10}Be dating of moraines in the upper Tons valley, Uttarakhand western Himalaya, Scherler et al. (2010) identified five glacial episodes after the LGM. An extensive glaciation was dated at ca. 16 ka when the glacier descended to ~2500 m AMSL. Thereafter, the reconstructed glaciations of decreasing magnitude were around 11–12 ka, 8–9 ka, ~5 ka and <1 ka. Correspondingly, from the Chorabari glacier valley, Kedarnath, (Uttarakhand, western Himalaya) Mehta et al. (2012) reported four major glacial stages of decreasing magnitude namely Rambara ca. 13 ka, Ghindurpani ca. 9 ka, Garuriya ca. 7 ka and Kedarnath ca. 5 ka. Both temperature and precipitation have significant control on the dynamics of temperate glaciers (Anderson and Mackintosh 2006). However, the role of ISM and WD in the advancement of glaciers under different precipitation regimes remains debated. Previously the researchers suggested the role of active monsoon system towards positive mass balances to the glaciers present in the monsoon dominant regions (Phillips et al. 2000; Owen et al. 2008, 2005, 2006; Finkel et al. 2003; Rupper et al. 2009). This argument was based on the major periods of glaciations corresponding well with the period of intensified monsoon (Ali and Juyal 2013). Contrarily, the subsequent investigations on the glacial extent in the valleys of different regions bring to the view that the glaciers responded positively during the strong WDs and weak ISM episodes (Owen and Dorch 2014; Bisht et al. 2015; Hu et al. 2015; Eugster et al. 2016; Shukla et al. 2018; Ali et al. 2019). The identified glacial stages within the Gangotri and other valleys almost represent the end phase of the cool episodes when the WDs were strong. The low temperatures and enhanced winter precipitation might have resulted in the positive mass balance and the commencement of subsequent warming resulted in the melting of glacial ice with retreat. Moreover, the response time of glaciers towards the climate anomalies also needs to be verified by generating the climate datasets of finer temporal resolution.

Inferences

Within the Gangotri glacier valley the tree line during the early Holocene might have been at lower altitudes than its present-day distribution of ~3600 m AMSL (by *Pinus wallichiana*) and ~3800 m AMSL (by *Betula utilis*). Accordingly, the snout of the glacier might also be at lower elevations (probably below 3300 m AMSL) than its present-day position of ~3980 m AMSL, though under retreating conditions due to early Holocene warm climate. The Kedar Glacial Stage at ~7 ka BP might be the end stage of a cool event coincided with the 8.2 ka North Atlantic cool event with strong winter westerlies (WD). During the middle Holocene (~6–5 ka) the Gangotri glacier receded to higher altitudes with snout position probably between ~3300 and 3500 m AMSL, as analogous to the studies from Sangla (Ranhotra and Bhattacharyya 2010) and Chorabari (Mehta et al. 2012, 2014) valleys. The continuous retreat of the glacier snout during late Holocene (i.e. since ~2 ka) is evident by the increase in the pollen content of temperate conifers and broad-leaved taxa indicating a shift of tree line to higher altitudes. During the late Holocene, the Gangotri Glacial Stage ca. 1 ka though coincides with the warm phase also globally known as MWP ca. 900–1200 C.E. (Lamb 1965) that might be the result of added moisture to the glacier during cool-moist conditions preceding MWP. The recent last Bhojbasa Glacial Stage 0.3–0.2 ka BP can be bracketed within the cool episode from ~0.8 to 0.2 ka BP that integrates the well-known LIA phase. The low temperatures due to enhanced westerlies and the weak ISM could have been crucial in the positive mass balance of the Himalayan glaciers, however, needs further investigations. The present palynological studies are not substantial and cover the climatic scenario at a low resolution of centennial to millennial-scale. For the detailed analyses, efforts are required towards a quantitative approach on decadal to centennial-scale development of the climatic and glacier scenario. This requires the development of the modern Pollen-Tree Line and Tree Line-Glacier relationships based on the pollen of tree line forming taxa and their distribution spatiotemporally. Also, more absolute dates of the glacio-geomorphic features from the upper reaches of glacial valleys will further augment the available chronologies for the robust reconstruction of glacial episodes in relation to climate variabilities.

Acknowledgements The authors are thankful to the Director, Birbal Sahni Institute of Palaeosciences (BSIP), Lucknow, Uttar Pradesh, India for the kind encouragement towards the completion of this manuscript and permission to publish it (BSIP/RDCC No.14/2021-22). High gratitude is towards the Science and Engineering Research Board (SERB (Department of Science and Technology DST)), New Delhi, India for the financial support vide SERB sponsored project number SR/DGH/56/2013 and the in-house (BSIP) project funding. The author IR is thankful to the DST Innovation in Science Pursuit for Inspired Research (INSPIRE) program (IF150998) of the Ministry of Science and Technology for funding her fellowship during the preparation of this manuscript.

References

Ali SN, Dubey J, Morthekai P et al (2019) Climate forcing and the initiation of glacier advance during MIS-2 in the North Sikkim Himalaya, India. J Asian Earth Sci 174:381–388

Ali SN, Juyal N (2013) Chronology of late quaternary glaciations in Indian Himalaya: a critical review. J Geol Soc India 82(6):628–638

Alley RB, Ágústsdóttir AM (2005) The 8k event: cause and consequences of a major Holocene abrupt climate change. Quat Sci Rev 24(10–11):1123–1149

Anderson B, Mackintosh A (2006) Temperature change is the major driver of late-glacial and Holocene glacier fluctuations in New Zealand. Geology 34(2):121–124

Auden JB (1937) The snout of the Gangotri glacier, Tehri Garhwal. Records of the Geol Surv India 72(II):135–140

Bali R, Awasthi DD, Tiwari NK (2003) Neotectonic control on the geomorphic evolution of the Gangotri Glacier Valley. Garhwal Himalaya Gondwana Res 6(4):829–838

Barnard PL, Owen LA, Finkel RC (2004) Style and timing of glacial and paraglacial sedimentation in a monsoon-influenced high Himalayan environment, the upper Bhagirathi Valley, Garhwal Himalaya. Sediment Geol 165:199–221

Berkelhammer M, Sinha A, Stott L et al (2012) An abrupt shift in the Indian monsoon 4000 years ago. Geophys Monogr Ser 198:75–87

Bhambri R, Bolch T (2009) Glacier mapping: a review with special reference to the Indian Himalayas. Progress Phys Geogr 33(5):672–704

Bhattacharyya A, Ranhotra PS (2003) Late Holocene vegetation and climatic changes around Tapoban, Gangotri glacier, Garhwal Himalaya. In: PAGES workshop and training program French Institute Pondicherry, vol 42, pp 243–246

Bhattacharyya A, Ranhotra PS, Shah SK (2006) Temporal and spatial aspects of western Himalayan climate and their implication to monsoon dynamics. J Geol Soc India 68:507–515

Bhattacharyya A, Ranhotra PS, Shah SK (2011) Spatio-temporal variation of alpine vegetation vis-à-vis climate during Holocene in the Himalaya. Mem Geol Soc India 77:309–319

Bisht P, Ali SN, Shukla AD et al (2015) Chronology of late quaternary glaciation and landform evolution in the upper Dhauliganga valley,(Trans Himalaya), Uttarakhand, India. Quat Sci Rev 129:147–162

Bolch T, Kulkarni A, Kääb A et al (2012) The State and Fate of Himalayan Glaciers. Science 336:310–314

Chakraborty S, Bhattacharyya SK, Ranhotra PS et al (2006) Palaeoclimatic scenario during Holocene around Sangla valley, Kinnaur, North-west Himalaya, based on multiproxy records. Curr Sci 91:777–782

Demske D, Tarasov PE, Wünnemann B et al (2009) Late glacial and Holocene vegetation, Indian monsoon and westerly circulation in the Trans-Himalaya recorded in the lacustrine pollen sequence from TsoKar, Ladakh, NW India. Palaeogeogr Palaeoclimatol Palaeoecol 279:172–185

Dimri AP (2013) Interannual variability of Indian winter monsoon over the Western Himalayas. Glob Planet Change 106:39–50

Dimri AP, Mohanty UC (2009) Simulation of mesoscale features associated with intense western disturbances over western Himalayas. Meteorol Appl J Forecast Pract Appl Training Tech Model 16(3):289–308

Dixit Y, Hodell DA, Petrie CA (2014) Abrupt weakening of the summer monsoon in northwest India ~4100 yr ago. Geology 42(4):339–342

Eugster P, Scherler D, Thiede RC et al (2016) Rapid Last Glacial Maximum deglaciation in the Indian Himalaya coeval with midlatitude glaciers: New insights from 10Be-dating of ice-polished bedrock surfaces in the Chandra Valley NW Himalaya. Geophys Res Lett 43(4):1589–1597

Finkel RC, Owen LA, Barnard PL (2003) Beryllium-10 dating of Mount Everest moraines indicates a strong monsoon influence and glacial synchroneity throughout the Himalaya. Geology 31(6):561–564

Flohn H (1957) Large-scale aspects of the "summer monsoon" in South and East Asia. J Meteorol Soc 35:180–186
Ganju A, Nagar YC, Sharma LN (2018) Luminescence chronology and climatic implication of the late quaternary glaciation in the Nubra valley, Karakoram Himalaya, India. Palaeogeogr Palaeoclimatol Palaeoecol 502:52–62
Ganopolski A, Rahmstorf S, Petoukhov V et al (1998) Simulation of modern and glacial climates with a coupled global model of intermediate complexity. Nature 391(6665):351–356
Gantayat P, Kulkarni AV, Srinivasan J (2014) Estimation of ice thickness using surface velocities and slope: case study at Gangotri Glacier, India. J Glaciol 60 (220)
Giesche A, Staubwasser M, Petrie CA et al (2019) Indian winter and summer monsoon strength over the 4.2 ka BP event in foraminifer isotope records from the Indus River delta in the Arabian Sea. Clim Past 15(1):73–90
Hahn DG, Manabe S (1975) The role of mountains in the south Asian monsoon circulation. J Atmos Sci 32(8):1515–1541
Harris IC, Jones PD (2015) CRU TS3.23: Climatic Research Unit (CRU) Time-Series (TS) Version 3.23 of High Resolution Gridded Data of Month-by-month Variation in Climate (Jan 1901-Dec 2014). Centre for Environmental Data Analysis, University of East Anglia Climatic Research Unit. https://doi.org/10.5285/4c7fdfa6-f176-4c58-acee-683d5e9d2ed5
Hu G, Yi CL, Zhang JF et al (2015) Luminescence dating of glacial deposits near the eastern Himalayan syntaxis using different grain-size fractions. Quat Sci Rev 124:124–144
Jangpangi BS (1958) Report on the survey and glaciological study of the Gangotri glacier, Tehri Garhwal District: Glacier No. 3, Arwa Valley: Satopanth and Bhagirath Kharak Glaciers, Garhwal District, Uttar Pradesh. Mem Geol Surv India pp 18
Jin L, Peng Y, Chen F et al (2008) Modeling sensitivity study of the possible impact of snow and glaciers developing over Tibetan Plateau on Holocene African-Asian summer monsoon climate. Clim past Discuss 4:1265–1287
Kar R, Ranhotra PS, Bhattacharyya A et al (2002) Vegetation vis-à-vis climate and glacial fluctuation of the Gangotri glacier since 2000 years. Curr Sci 82:347–351
Kulkarni AV, Karyakarte Y (2014) Observed changes in Himalayan glaciers. Curr Sci 237–244
Kumar K, Miral MS, Joshi V et al (2002) Discharge and suspended sediment in the meltwater of Gangotri Glacier, Garhwal Himalaya India. Hydrol Sci J 47(4):611–619
Kutzbach JE, Prell WL, Ruddiman WF (1993) Sensitivity of Eurasian climate to surface uplift of the Tibetan Plateau. J Geol 101(2):177–190
Lamb HH (1965) The early medieval warm epoch and its sequel. Palaeogeogr Palaeoclimatol Palaeoecol 1:13–37
Loibl D, Lehmkuhl F, Grießinger J (2014) Reconstructing glacier retreat since the Little Ice Age in SE Tibet by glacier mapping and equilibrium line altitude calculation. Geomorphology 214:22–39
Matero ISO, Gregoire LJ, Ivanovic RF et al (2017) The 8.2 ka cooling event caused by Laurentide ice saddle collapse. Earth Planet Sci Lett 473:205–214
Mehta M, Dobhal DP, Kesarwani K et al (2014) Monitoring of glacier changes and response time in Chorabari Glacier, Central Himalaya, Garhwal, India. Current Sci 281–289
Mehta M, Majeed Z, Dobhal DP et al (2012) Geomorphological evidences of post-LGM glacial advancements in the Himalaya: a study from Chorabari Glacier, Garhwal Himalaya India. J Earth Syst Sci 121(1):149–163
Mukherjee BP, Sangewar CV (2001) Recession of Gangotri glacier through 20th century. Geol Surv India Spl Pub 65:1–3
Murari MK, Owen LA, Dortch JM et al (2014) Timing and climatic drivers for glaciation across monsoon-influenced regions of the Himalayan-Tibetan orogen. Quat Sci Rev 88:159–182
Nagar YC, Ganju A, Satyawali PK et al (2013) Preliminary optical chronology suggests significant advance in Nubra valley glaciers during the Last Glacial Maximum. Curr Sci 96–101
Nainwal HC, Naithani AK, Kumar D et al (2004) Glacial morphology of the Gangotri group of glaciers in Garhwal Himalaya, Uttaranchal India. Geol Surv India Spl Pub. 80:87–103

Naithani AK, Nainwal HC, Sati KK et al (2001) Geological evidences of retreat of Gangotri glacier and its characteristics. Curr Sci 80:87–94

Owen LA (2009) Latest Pleistocene and Holocene glacier fluctuations in the Himalaya and Tibet. Quat Sci Rev 28:2150–2164

Owen LA, Caffee MW, Bovard KR et al (2006) Terrestrial cosmogenic nuclide surface exposure dating of the oldest glacial successions in the Himalayan orogen: Ladakh Range, northern India. Geol Soc Am Bull 118(3–4):383–392

Owen LA, Caffee MW, Finkel RC et al (2008) Quaternary glaciation of the Himalayan-Tibetan orogen. J Quat Sci: Published for the Quat Res Assoc 23(6–7):513–531

Owen LA, Dortch JM (2014) Nature and timing of Quaternary glaciation in the Himalayan-Tibetan orogen. Quat Sci Rev 88:14–54

Owen LA, Finkel RC, Barnard PL et al (2005) Climatic and topographic controls on the style and timing of Late Quaternary glaciation throughout Tibet and the Himalaya defined by ^{10}Be cosmogenic radionuclide surface exposure dating. Quat Sci Rev 24:1391–1411

Owen LA, Finkel RC, Caffee MW (2002) A note on the extent of glaciation throughout the Himalaya during the global Last Glacial Maximum. Quat Sci Rev 21(1–3):147–157

Owen LA, Finkel RC, Caffee MW et al (2002) Timing of multiple late Quaternary glaciations in the Hunza Valley, Karakoram Mountains, northern Pakistan: defined by cosmogenic radionuclide dating of moraines. Geol Soc Am Bull 114(5):593–604

Phillips WM, Sloan VF, Shroder JF Jr et al (2000) Asynchronous glaciation at Nanga Parbat, northwestern Himalaya Mountains. Pakistan. Geology 28(5):431–434

Polanski S, Fallah B, Befort DJ et al (2014) Regional moisture change over India during the past Millennium: A comparison of multi-proxy reconstructions and climate model simulations. Glob Planet Change 122:176–185

Raina VK, Srivastava D (2014) Glacier atlas of India. GSI Publications vol 7, no 1

Ranhotra PS, Bhattacharyya A (2004) Vegetation, Climate and Lake History of Tapoban, Gangotri Glacier, Garhwal Himalaya. Geol Surv India Spl Pub 80:215–221

Ranhotra PS, Bhattacharyya A (2010) Holocene palaeoclimate and glacier fluctuations within Baspa Valley, Kinnaur, Himachal Pradesh. J Geol Soc India 75:527–532

Ranhotra PS, Bhattacharyya A (2013) Modern vegetational distribution and pollen dispersal study within Gangotri Glacier Valley, Garhwal Himalaya. J Geol Soc India 82:133–142

Ranhotra PS, Bhattacharyya A, Kar R et al (2001) Vegetational and climatic changes around Gangotri glacier during Holocene. Geol Surv Indian Spl Pub 65(3):67–71

Ranhotra PS, Bhattacharyya A, Kotlia BS (2007) Vegetation and climatic changes around Lamayuru, Trans-Himalaya during the last 35 ky B.P. Palaeobotanist 56:117–126

Ranhotra PS, Kar R (2011) Palynological study of glacio-geomorphic features and its relevance to Quaternary palaeoclimate and glacial history. Curr Sci 100:641–647

Ranhotra PS, Sharma J, Bhattacharyya A et al (2018) Late Pleistocene-Holocene vegetation and climate from the palaeolake sediments, Rukti valley, Kinnaur, Himachal Himalaya. Quat Int 479:79–89

Rawat S, Gupta AK, Sangode SJ et al (2015) Late Pleistocene-Holocene vegetation and Indian summer monsoon record from the Lahaul, northwest Himalaya, India. Quat Sci Rev 114:167–181

Roy I, Ranhotra PS, Shekhar M et al (2018) Over-representation of some taxa in surface pollen analysis misleads the interpretation of fossil pollen spectra in terms of extant vegetation. Trop Ecol 59(2):339–350

Rupper S, Roe G, Gillespie A (2009) Spatial patterns of Holocene glacier advance and retreat in Central Asia. Quat Res 72(3):337–346

Sati SP, Ali SN, Rana N et al (2014) Timing and extent of Holocene glaciations in the monsoon dominated Dunagiri valley (Bangni glacier), Central Himalaya, India. J Asian Earth Sci 91:125–136

Scherler D, Bookhagen B, Strecker MR et al (2010) Timing and extent of late Quaternary glaciation in the western Himalaya constrained by 10Be moraine dating in Garhwal. India. Quat Sci Rev 29(7–8):815–831

Sharma MC, Owen LA (1996) Quaternary glacial history of NW Garhwal, central Himalayas. Quat Sci Rev 15:335–365
Sharma S, Chand P, Bisht P et al (2016) Factors responsible for driving the glaciation in the Sarchu Plain, eastern Zanskar Himalaya, during the late Quaternary. J Quat Sci 31(5):495–511
Sharma S, Shukla AD (2018) Factors governing the pattern of glacier advances since the Last Glacial Maxima in the transitional climate zone of the Southern Zanskar Ranges, NW Himalaya. Quat Sci Rev 201:223–240
Shi Y (2002) Characteristics of late Quaternary monsoonal glaciation on the Tibetan Plateau and in East Asia. Quat Int 97:79–91
Shukla T, Mehta M, Jaiswal MK et al (2018) Late Quaternary glaciation history of monsoon-dominated Dingad basin, central Himalaya, India. Quat Sci Rev 181:43–64
Singh DS, Mishra A (2002) Role of tributary glaciers on landscape modification in the Gangotri glacier area, Garhwal Himalaya, India. Curr Sci 82:567–571
Singh P, Haritashya UK, Kumar N (2008) Modeling and estimation of different components of streamflow for Gangotri Glacier basin, Himalayas Glacier Gangotri. Himalaya. Hydrol Sci J53(2):309–322
Singh P, Haritashya UK, Ramasastri KS et al (2005) Prevailing weather conditions during summer seasons around Gangotri glacier. Curr Sci 88(5):753–760
Srivastava D (2012) Status Report on Gangotri Glacier, science and engineering research board, department of science and technology, New Delhi. Himalayan Glaciol Tech Rep 3:102
Srivastava P, Agnihotri R, Sharma D et al (2017) 8000-year monsoonal record from Himalaya revealing reinforcement of tropical and global climate systems since mid-Holocene. Sci Rep 7(1):1–10
Tewari AP (1972) Study of the Gangotri glacier, Uttarkashi district, (Central Himalaya), Uttar Pradesh. Misc Pub Geol Surv India 15:69–74
Thomas ER, Wolff EW, Mulvaney R et al (2007) The 8.2 ka event from Greenland ice cores. Quat Sci Reviews 26(1–2):70–81
Yang B, Bräuning A, Dong Z et al (2008) Late Holocene monsoonal temperate glacier fluctuations on the Tibetan Plateau. Glob Planet Chang 60:126–140
Yang W, Yao T, Guo X et al (2013) Mass balance of a maritime glacier on the southeast Tibetan Plateau and its climatic sensitivity. J Geophys Res: Atmos 118(17):9579–9594
Zhu H, Xu P, Shao X et al (2013) Little Ice Age glacier fluctuations reconstructed for the southeastern Tibetan Plateau using tree rings. Quat Int 283:134–138

Chapter 7
Flood Hazards in the Jhelum River Are Mainly Controlled by the Piggyback Thrusting of the Kashmir Basin and Less so by Factors like Climate Change and Urbanization

A. A. Shah, Navakanesh M. Batmanathan, Asiya Qadir, and Rajesh Kumar

Abstract Historical data suggest that flood hazards in the Kashmir basin, located in the Indian portion of the NW Himalaya, are not new, and some of the major flood hazards have turned into disasters in the past, which questions the studies that exclusively relate the flood hazards in the Kashmir basin to factors like urbanization and climate change. The past studies also suggest that one of the grossly overlooked factors is the role of the geological and structural setting of the piggyback Kashmir basin in flood hazards. Therefore, the present work was aimed to review the previous works on the structural setting of the basin with the main focus on the flood river hazards in the Jhelum River. We have used 30-m spatial resolution shuttle radar topography, bedrock geology, drainage pattern, topographic profile, and knick point data to investigate the formation and development of the river. These data are related to the historical and previously published data on floods to understand the dominant factor that controls the flood hazards in the Kashmir basin. The present study suggests that structurally piggyback basin development is the single-most-important cause of drainage development, asymmetrical nature of Jhelum watershed, tilting, and formation of the Jhelum River, which are directly controlling the behavior of flooding and flood hazards in the basin. Although the climate- and urbanization-based arguments are important aspects that do contribute toward the flood hazards, however, the historical and contemporary data on the magnitude and intensity of flood hazards in the Kashmir basin in the combination with our new data suggest

A. A. Shah (✉)
Department of Physical and Geological Sciences, Universiti of Brunei, Darussalam, Brunei

N. M. Batmanathan
Southeast Asia Disaster Prevention Research Initiative, Universiti Kebangsaan Malaysia, Bandar Baru Bangi, Malaysia

A. Qadir
Srinagar, India

R. Kumar
Department of Geography, School of Human Sciences, Sikkim University, Tadong, Gangtok, Sikkim, India

S. Rani and R. Kumar (eds.), *Climate Change*, Springer Climate,
https://doi.org/10.1007/978-3-030-92782-0_7

that flood hazards have occurred in the past when these factors were not part of the standard scientific discourse on floods.

Keywords Kashmir basin · Jhelum River · Floods · SRTM DEM · Knick points · Longitudinal profile

Introduction

The occurrence of various types of floods (e.g., related to earthquake, landslide, and precipitation) in intermontane basins would be expected because such structural settings offer ideal conditions for inundation, which could be easily exploited by the prolonged or intense spells of precipitation (Shah 2016). Kashmir and Leh basins are the two intermontane basins that exemplify such structural settings in the NW Himalaya, which has made it extremely prone to floods hazards, and such data are documented in historical as well as in the sediment records (Rajatarangini 1149; Lawrence 1895; Bilham et al. 2010; Bilham and Bali 2013; Meraj et al. 2015; Shah 2015; Shah 2016; Shah et al. 2018; Shah et al. 2020a, b). And importantly, the history of flood hazards in these regions has categorically demonstrated that major floods have occurred much before the imprints of climate change have been observed on the Earth, and that includes the Himalayan regions. Therefore, the structural, tectonics, and geological framework of these basins provides an ideal opportunity to map and understand the fundamental aspects of drainage development, tilting, basin subsidence, and sediment flow regimes, and such elements have not been discussed in any detail in the previously published works, except the seminal works of Burbank and Johnson (1982, 1983) and recent works by Shah (2013, 2016). Here, we extend our previous works (Shah 2015, 2016, 2018) by using the 30-m spatial resolution shuttle radar topography to map the drainage pattern and to study the longitudinal river and topographic profiles. The drainage pattern map was overlaid on the geological map to understand the relationship between them. The longitudinal profile of the Jhelum River was used to map the presence of major knick points, which could be related to active faults or lithological contacts. The present study indicates that drainage development in the NW Himalayan region is largely a reflection of bedrock geology and faulting. The topographic profiles across the Peshawar and Kashmir basins indicate prominent tilting toward the ~NE, which suggests structural control at the depth where reverse fault ramp and produces the piggyback geometries. Importantly, the watershed map shows that Jhelum River receives most of its discharge from the eastern portions, and that makes it a classic example of an asymmetrical watershed in the region, which is strongly controlled by the active Jhelum fault. The three prominent knick points are mapped on the longitudinal profile of the Jhelum River, and these lie on the Jhelum fault system thereby suggesting tectonic origin.

Study Area

The present study has been conducted in the Kashmir basin which is a piggyback basin (Fig. 7.1) that has formed and evolved during the recent phase of the ongoing collisional tectonics between the lithospheric plates of India and Eurasia (Burbank and Johnson 1983; Shah 2013; Shah et al. 2020a, b). The Jhelum River is the only river that drains water out of the Kashmir basin, and structurally, it seems quite remarkable because the river is aggrading and could not possibly cut the mountainous terrain

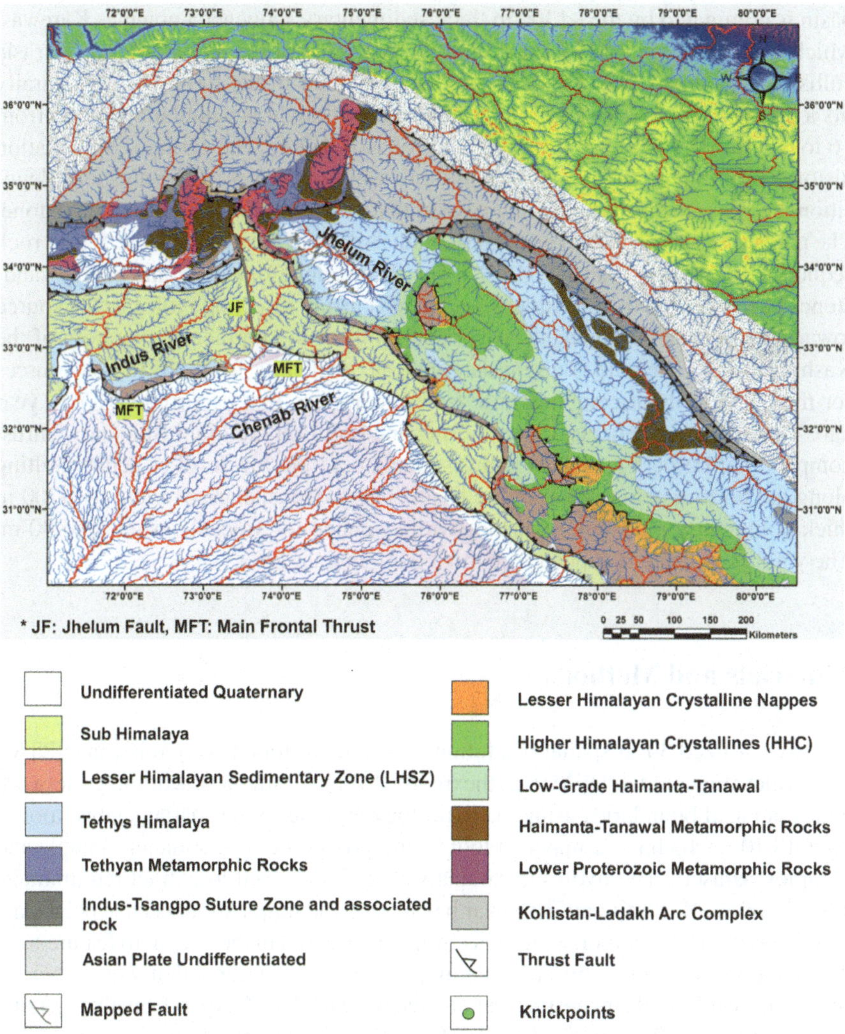

Fig. 7.1 Geological and tectonic setting of the NW Himalaya. The drainage pattern is a reflection of bedrock geology (DiPietro and Pogue 2004), and major streams are highlighted in red

in the Baramulla region where it exits the basin (Fig. 7.1). The geological setting shows that the river cuts through Plio-Pleistocene to Holocene sediments throughout its length until it reaches the exit point where it pierces through competent basement rocks that are composed of sedimentary, metamorphic, and igneous rocks. Shah and Malik (2017) mapped the eastern extension of the Tunda-Balakot–Bagh fault system and suggested that the Jhelum River follows the fault. We have reviewed the literature on the tectonics and sedimentary history of the Kashmir basin and found (Burbank and Johnson 1982, 1983) the original seminal works with details on the sedimentology of the Karewas sediments. Therefore, we have used these data to summarize the geology of sedimentary cover sequence of the Kashmir basin. The basin is dominated by the ~1300 m thick sedimentary sequence known as Karewas, which was deposited ~4 Ma ago during the last stage of the ongoing India–Eurasia collisional tectonics. The deposits have accumulated in the basin that structurally fits a piggyback tectonic setting, and the rate of sediment accumulation varied from 16 to 64 cm per 1000 year (Burbank and Johnson 1983). The earlier sedimentation history is dominated by lacustrine conditions, which indicates a calm and quiet depositional environment in the Kashmir basin that resulted in >200 m thick mudstone. The mudstone was deposited on the basement rocks that are related to the Tethys rock sequence and Panjal Traps (Fig. 7.1). The fluvial conglomerates and coarse sandstones have overlaid the mudstones, and paleo current direction indicates the source toward the northeast, which suggests rapid uplift of the northeastern margin of the Kashmir basin. The paleo current data indicate switching from NE to SW sources for the ~300 m think conglomeratic facies sediments that occurred at 1.7 m year ago. This transition reflects active faulting and uplift on the Main Boundary Thrust complex that borders it to the southwest and a possible low level of active faulting along the northeastern margin of the basin. These deposits are overlain by >700 m thick dominantly mudstone sequence with evidence for volcanic ashes at ~900 m. The whole sequence is capped by a thin conglomerate layer.

Materials and Methods

We have utilized 30-m spatial resolution shuttle radar topography mission (SRTM) digital elevation model (DEM) (Earthexplorer 2021) to compute the drainage network and watershed boundaries using the hydrology module of the spatial analyst tool of ArcGIS 10.3 which is a competent tool to analyze, process, and manage simple and complex datasets. The hydrological analysis tool was used to extract the drainage systems of the Kashmir and Peshawar basins, and the major streams have been classified via stream order tool, which was further processed in the stream to feature tool. These tools enable us to differentiate major and minor streams that were superimposed on the hill shade image that was created from the DEM. Apart from that, a transverse profile from Peshawar to Kashmir basin (W–E orientation) was generated using interpolate line tools. The changes in topographic expression, slope development, basin length were determined and used as indicators of the tectonic complexity of the region. The incorporation of various parameters (attributes such as fill, flow direction, flow accumulation) helps us to find more comprehensive and reliable datasets, especially the longitudinal river profile and knickpoints. These data are used to interpret the influence of active tectonics on the basin formation, flooding, and drainage patterns and subsequently compared with the previous data on flood hazards in the Jammu and Kashmir region (e.g., Hatwar et al. 2005; Bhat and Romshoo 2009; Yadav and Bhan 2010; Nandargi and Dhar 2011; Romshoo and Rashid 2014; Mishra 2015; Meraj et al. 2015; Rather et al. 2016; Kumar and Acharya 2016; Yadav et al. 2017).

Results and Discussion

Structural Control on the Formation of Jhelum River

The present structural and morphological setup of the Jhelum River shows that it originates in the southeastern portion of the Kashmir basin and flows toward the northwest, which is parallel to the ~NW–SE trend of the basin. Instead of continuing as usual, it abruptly turns ~west near the Baramulla region (Figs. 7.2 and 7.3), which is related to the active fault system (Shah 2016). The source of water in the river is a combination of spring water, precipitation, and glaciers. Therefore, it is not true that the river solely originates in Verinag spring, which lies in the southeast. If the

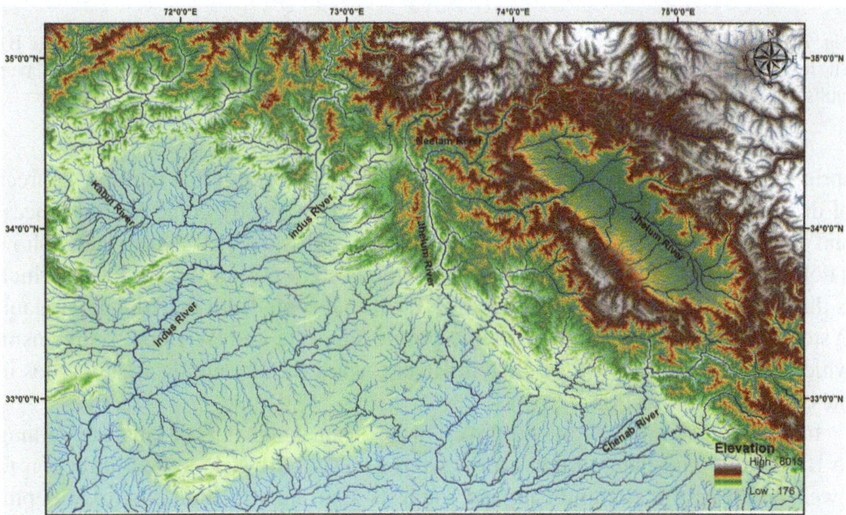

Fig. 7.2 Oval-shaped Kashmir basin resembles a centipede with a trunk stream represented by the Jhelum River, which is the lifeline for the Kashmir valley. The drainage pattern is a reflection of bedrock geology

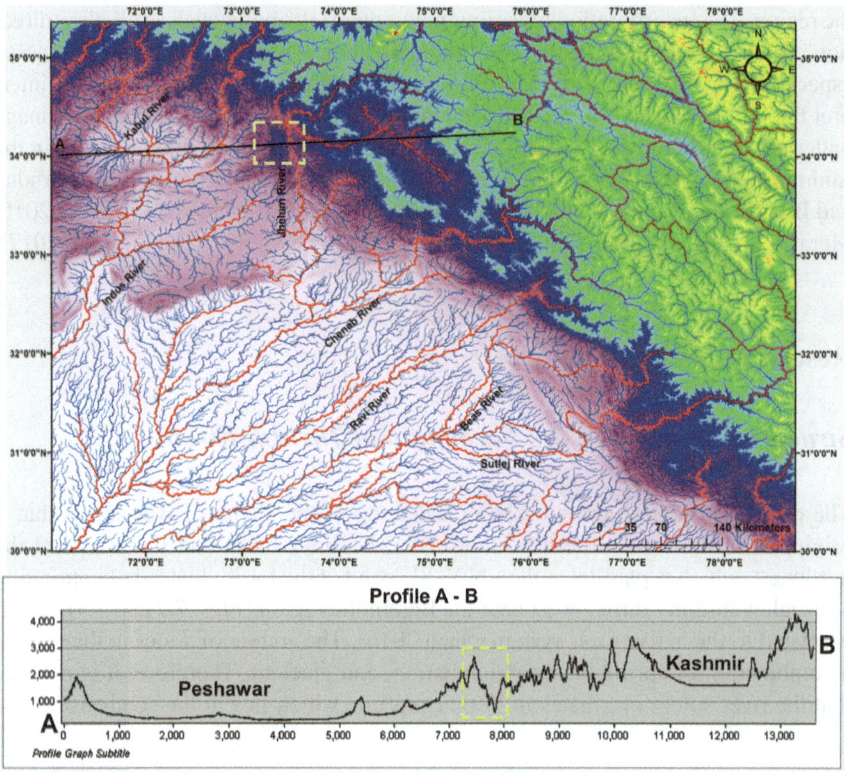

Fig. 7.3 Topography across the Peshawar and Kashmir intermontane basins is shown (A and B). The higher topographic position of the Kashmir basin is tectonically controlled by major Himalayan faults that shoulder it at the south (see Fig. 7.1 for faults)

spring vanishes tomorrow, the river will still exist because it has multiple sources of drainage systems that are a feeder to it, and these mostly come from northeast and southwest (Figs. 7.2 and 7.3). The drainage and watershed maps clearly show a dominance of drainage networks in the southwestern portions of the basin, which is directly controlled by the observed ~NE tilt that roots from the Himalayan fault systems (Shah 2013). The topographic profile suggests a ~NW slope of the basin, which controls the flow of the Jhelum River toward the Baramulla area that lies in the downstream direction.

Interestingly, the river flows northwest in the basin, which is >100 km from Verinag to Baramulla, and then abruptly turns from its usual downslope flow direction to ~west-southwest. It is significant and suggests a strong structural control at depth. Once the river exits the basin, it flows toward the Muzaffarabad region in Pakistan and follows the Balakot–Bagh fault that extends toward the southeast because the Tunda fault system possibly terminates under the Kashmir basin (Shah and Malik 2017). The fault system hosted the 7.6 magnitude earthquake in 2005 that caused a loss of ~80,000 people (Avouac et al. 2006). The Neelum/Kishanganga River joins the Jhelum River at Muzaffarabad and makes a sharp turn of almost 90° at Tunda and changes the flow direction toward the south. The major abrupt turning and flowing of the river toward the south is because of the interaction between the Tunda fault and the Jhelum fault (Fig. 7.1) that have formed a triangular ridge (Shah et al. 2020a; b). The fault interaction could also be a major reason for the abrupt stoppage of the fault rupture propagation associated with the 2005 Kashmir event (Shah et al. 2020a; b). The Jhelum fault has truncated the Tunda fault zone, and such a structural setting could potentially form a structural discontinuity that would stop the fault propagation, and instead cause fault segmentation. The ~N–S trend of the Jhelum fault makes it easier for the Jhelum River to just follow the structures until it reaches the frontal mountain ranges. It reaches Mangla dam and flows south. It develops a prominent anastomosing channel pattern that is observed for the first time after the river leaves the dam and flows southeast for >20 km before it turns southwest. The turning of the river seems to be controlled by the > 40 km long and ~NE–SW trending doubly plunging anticline that bounds it at the south-southwest and forces it to flow parallel to the trend of the fold. The river flows southwest for >150 km before turning south and finally merges with the Chenab River (Figs. 7.3 and 7.4).

Geological and Geomorphic Setting of the Jhelum River

The geological map shows that the Jhelum River follows the Kashmir basin that is filled with Plio-Pleistocene and Holocene sediments and slices through a variety of bedrock sections mostly when it exits the basin at Baramulla (Fig. 7.1). It pierces through a variety of bedrock lithologies until it merges with the Chenab River (Figs. 7.1, 7.2, 7.3, 7.4 and 7.5). The abrupt turning of the river at Baramulla coincides with the previously mapped active ~NW dipping Tunda-Balakot–Bagh fault system (Fig. 7.1), which seems to have restricted its flow regime to remain in a dominantly degradational profile. The Jhelum River then follows the ~N–S trace of the Jhelum fault system (Shah et al. 2020a; b) until it reaches the frontal portions. The remarkable structural discontinuities mark the track of the Jhelum River, and it is marked by several topographic breaks that are mapped as ruptured ridge crests, triangular facets, and truncated lithologies (Shah and Malik 2017; Shah et al. 2020a, b). The triangular facets show characteristic youthful geomorphic features, which are suggestive of active faulting. It indicates that out-of-sequence faults are potential candidates for hosting major earthquakes, which by extension means that drainage blockage, migration, reversals, and re-emergences are common occurrences in these regions. These changes are tectonics and would be expected even in the absence of anthropogenic impacts of climate change and urbanization on flood hazards, etc.

The eastern extension of the Tunda fault system (Shah and Malik 2017) in the Kashmir basin is untraceable. However, the trace of the active Kashmir basin fault (Shah 2013, 2016) is truncated by the Tunda fault, which indicates that the fault is younger than the Kashmir basin fault. It could mean that the Tunda fault system is

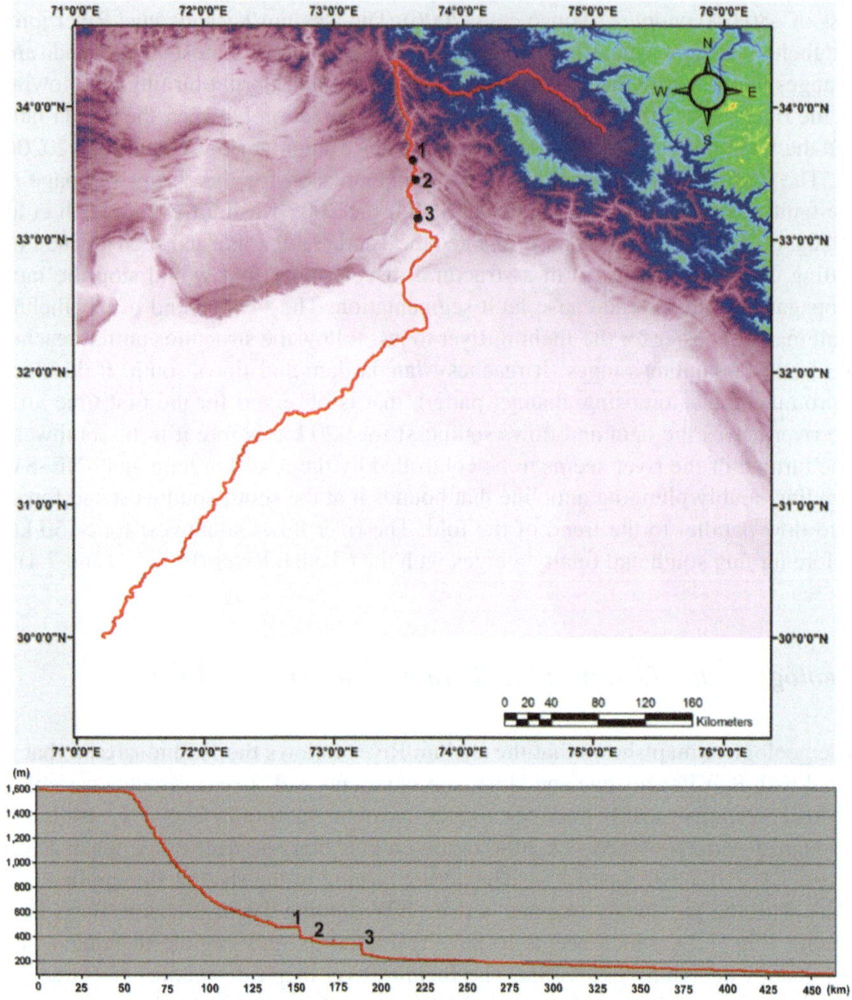

Fig. 7.4 Jhelum River is shown in red, and the longitudinal profile with knick points is shown below. The three major knick points are prominently developed along the track of the Jhelum River that follows the trace of the Jhelum fault

buried under the sediments of the Kashmir basin and could extend across the basin and in the Kashmir Tethys and Panjal Trap basement rocks.

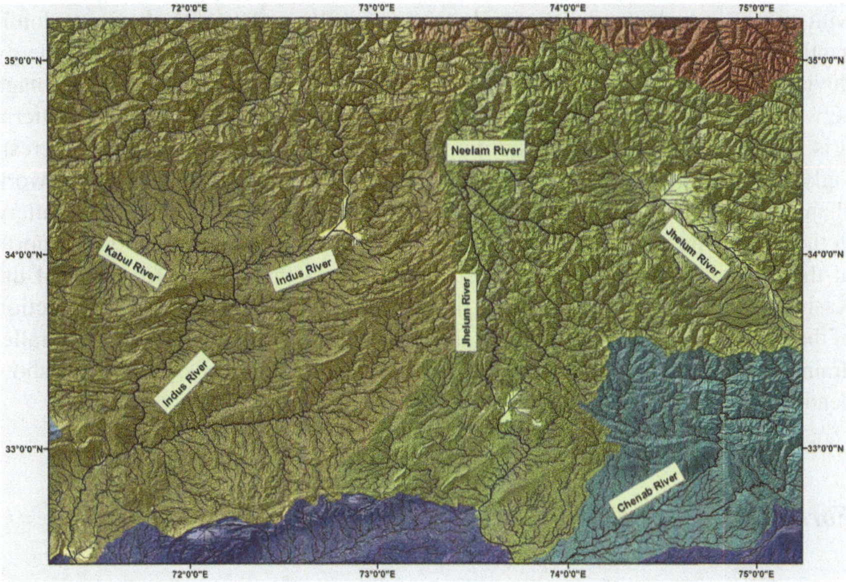

Fig. 7.5 Topography and drainage pattern clearly show an asymmetrical pattern of the drainage network in the Jhelum catchment area, which is a reflection of bedrock geology and tectonics

Drainage Network, Pattern, and Topography

The stunning topographic expression of the Kashmir region is primarily dictated by the isolated nature of the large oval-shaped piggyback basin, the Kashmir basin, in the NW Himalaya (Fig. 7.5). It sits on top of the major Himalayan reverse fault systems and runs ~parallel to the ~NW–SE trend of the faults. The topographic profile is drawn across the Kashmir and Peshawar basins (Fig. 7.2). It reveals the spectacular geomorphic expression of these tectonic landforms. The Kashmir basin stands at an average elevation of >1500 m above mean sea level (AMSL), which is a very distinctive feature when compared with the incredibly subdued topography of the nearby Peshawar basin that has an average elevation of <500 m AMSL. Importantly, the sharp elevation difference marks a transition that coincides with the emergence of the left-lateral strike-slip fault system (Shah et al. 2020a; b), the Jhelum fault.

The drainage pattern within the Kashmir basin is more like a centipede pattern with a major trunk stream of the Jhelum River being fed by a large number of tributaries that surround it. Since the basin is elongated in ~NW–SE direction, therefore, the majority of the feeder tributaries are draining from the NE and SW portions (Figs. 7.2 and 7.5). These feeders are dominantly coming from the SW portions of the basin because of the uplift related to the faults in the piggyback Kashmir basin (Burbank 1985; Bilham and Bali 2013; Shah 2016). This has created subdued basin morphology toward the northeast portions, which are relatively submerged with mostly aggradational streams. However, the SW portions of the basin are dominantly degradational and, therefore, highly susceptible to erosion with high-energy fluvial conditions that will mainly transport gravels as bed load. The overall drainage network pattern of the tributaries in the basin resembles a parallel drainage pattern. It is mainly because the streams drain through the Quaternary sediments. Interestingly, the Kashmir region marks the position where the regional drainage network changes pattern with dominantly dendritic pattern in the north and parallel pattern in the south. The major river streams are flowing toward ~NW in the regions north of the Kashmir basin and ~southwest in the regions that are located south of the Kashmir basin. These changes in the drainage developments are largely a reflection of the bedrock geology. The sub-Himalayan sequences are dominated by the parallel drainage patterns, and lesser, Tethys, and Higher Himalayan rock sequences show dendritic patterns with examples of the fault- and fold-controlled channels.

Formation of the Kashmir Basin and the Drainage Reversal

The catchment area of the Jhelum River is asymmetrical (Fig. 7.5) with a major contribution from a vast network of tributaries feeding the northeastern portions of the catchment region. The asymmetrical behavior is attributed to the existence of the Jhelum fault, which forms the truck stream of the largely dendritic drainage networks. This suggests that the drainage development before the existence of the Kashmir basin was different, and the role of faulting is central in the formation and development of drainage in the region. The drainage before the existence of the Kashmir basin would have ~southwest flow direction of the streams, as is even today in the entire eastern portion of the catchment. However, once the Kashmir basin was established, the drainage network has to reverse the flow direction at the southeast portions of the Kashmir basin. The geomorphic evidence for it has been shown (Shah and Malik 2017). This has implications for the paleo water level, sediment supply, and flow regime in the Jhelum River, and how it has changed subsequently once the Kashmir basin was established. Therefore, it controls the water and sediment discharge budget in the river, which is also directly linked with tectonic development that also controls and influences the climate of the region.

Longitudinal Profile and Knick Points

The longitudinal profile of the Jhelum River reveals that the river flows at ~1600 m AMSL for a distance of ~50 km and without changing in the downstream direction. However, the abrupt drop in elevation is marked after 50 km, which consistently drops until it reaches the first knick point (Fig. 7.4). The other two similar points are marked in the downstream direction, and all are numbered as 1–3 (Fig. 7.4). The first knick point shows ~100 m of abrupt vertical drop in the downstream direction that coincides with the emergence of the Jhelum fault system (Fig. 7.1). The other knick point is located a few meters downstream from the first one, and the third knick point resembles the first one. We have interpreted that both could be related to the Jhelum fault. The river flows further downstream and maintains a low altitude with <200 m AMSL.

Kashmir Piggyback Basin Development, Active Faulting, and Jhelum River

The basic understanding of the causes of flooding in the Kashmir region cannot be initiated without a solid understanding of the formation of the basin and in particular the development of drainage systems with a major focus on the Jhelum River. The existing scientific literature lacks data on the structural development of the JR except for the seminal works of Burbank and Johnson (1982, 1983 and their other works on Kashmir) and later Shah (2015, 2016). Therefore, herein, we demonstrate that the formation of a drainage system in the Kashmir basin is directly influenced by the tectonic forces. The basin was formed ~4 Ma ago in response to the ongoing active tectonic collision between the lithospheric plates of India and Eurasia, which laid the foundation of the basin (Burbank and Johnson 1983). The Jhelum River has established the trunk stream status later than the tributaries, which are older and dominantly feeding it from the east and west. The often-quoted reference that Jhelum has originated in Verinag is not true. The geological origin of the Jhelum River is in response to the uplift that has tilted it toward the ~NE, and that is why, the river is more toward the structurally inherent tilt direction. The ~NW flow direction of the river in the basin is governed by the overall elevation difference within the basin (Figs. 7.1, 7.2, 7.3, 7.4 and 7.5). The source water in the Jhelum River is a combination of spring water, precipitation, and glaciers. Therefore, it is not true that the river solely originates in Verinag spring, which lies at the easternmost portion. If that spring vanishes, the river would sustain because it has multiple sources of drainage systems that feed it and mostly from northeast and southwest.

Burbank and Johnson (1982, 1983) have argued that the Jhelum River originated some 4 Ma ago when the basin was created by the reverse faulting in the southern portion that is related to the ongoing active tectonic convergence between India and Eurasia. The sedimentary history of the basin shows a thick blanket of Plio-Pleistocene to Holocene sediments that are largely composed of sandstone and mudstone, which indicates prolonged lacustrine conditions have existed within the basin. The Karewas is remnants of this history and has been explored since the late 1900s, but what has remained unexplored is the transition from a largely lacustrine environment to the formation of the Jhelum River as it is today. The data presented above demonstrate the asymmetrical nature of the river catchment region that has a large network of tributaries that feed the trunk stream of the Jhelum River, which originate in the southeastern portions of the Kashmir region. What makes the river special is that it originates at an average altitude of ~1600 m and keeps that elevation until it exits the basin. There are few changes in the elevation and tilt at which it flows in the Kashmir basin, which suggests largely aggradation conditions and, hence, depositional flow regime with bedload dominated by silt and sand. However, a sudden drop in elevation is observed in the downstream portions at a distance of >50 km on the longitudinal profile (Fig. 7.4). This suggests degradational conditions and erosion-dominated streams with high stream power to pick gravels and pierce through the hard and competent lithologies. The geological map supports such a link and shows that the river flows through the bedrock sequence that is mainly composed of metamorphic, igneous, and sedimentary units (Fig. 7.1). The important observation about the Jhelum River is the fact that it exits from the Kashmir basin where it sheds its bed loads as sand and silt but gets rejuvenated once it leaves the basin which cannot be possible without the role of major fault(s). The reason for it is simple: how could a river that is unable to erode and has lost its power in the low-lying region of the Kashmir basin cuts through the competent basement rocks (Fig. 7.1). Therefore, the existence of a fault or lithological discontinuity is the most plausible explanation for the exit of the Jhelum River at the Baramulla region, which otherwise would have been impossible. Shah and Malik (2017) mapped the eastern extension of the Tunda-Balakot–Bagh fault system that was the source of the moment magnitude (Mw) 7.6 Muzaffarabad earthquake of 2005 (Avouac et al. 2006; Pathier et al. 2006). This fault has placed older rocks onto the dominantly sub-Himalaya units, for example, the Murree Formation (Searle et al. 1996) on the Pakistan side. This suggests that the River Jhelum has followed a pre-existing fault (Tunda fault), which before the formation of the Kashmir basin would have been active. The formation of the Kashmir basin has occurred 4 Ma ago, which indicates that the Jhelum River within the Kashmir basin would have formed during and after the development of the basin. Since the development of the basin was mainly caused by the Himalayan thrust system that shoulders the basin, the drainage development would be dominated by the streams that flow from the southwestern portions because of the tilt and elevation. This would mean that the present configuration of the Jhelum River was largely shaped after the basin development, which was later filled with more sediments that largely derived from the SW portions because of the dense network of the tributaries that are linked to the uplift through reverse faulting at the south and southwest.

Tectonic Subsidence, Flooding in the Kashmir Basin

The Kashmir basin is a product of tectonics, and more specifically, it was formed by the Himalayan fault systems that shoulder it from the south. This has also formed the Jhelum River, which is the only stream that drains water out of the basin, and therefore, it is the lifeline for Kashmir. Historically, many floods have occurred in the region, and some were very destructive (e.g., Rajatarangin 1149; Lawrence 1895; Bilham et al. 2010, 2013; Shah 2015; Ballesteros-Cánovas et al. 2020). Importantly, the historical data show that flood hazards have occurred in the Kashmir region when the climate change discourse was not even born (Oreskes 2004; Shah 2015) and urbanization would have been almost non-existent (Shah 2015, 2016). This takes us to an important aspect of how to understand, untangle, and uncover the peculiarities of the scientific histology of flood hazards and disasters in the basin (Shah 2016). This could only be possible if we understand the formation and evolution of the region and in particular the Kashmir basin. The initiation of India–Eurasia collisional tectonics has formed the geological, structural, and topographic systems that we see today. It has also evolved river systems, modified climate, and impacted the sediment supply that enters the trunk streams and eventually to sedimentary sinks. Therefore, the understanding of the tectonic evolution of a region is a fundamental step toward climatic evolution, and to forecast the coming events (Singh et al. 2016, 2017). If the tectonic evolution of a region is misunderstood because of various reasons, then it will dramatically impact the climate discourse and, therefore, the science on floods, etc. This applies to the Kashmir region where previous works have attributed recent floods to various factors, such as loss of wetlands, deforestation, population growth, climate change (Romshoo and Rashid 2014; Rather et al. 2016), changes in the land use pattern (Bhat and Romshoo 2009), the geography of the region, siltation in the Jhelum River, increased precipitations (e.g., Hatwar et al. 2005; Yadav and Bhan 2010; Nandargi and Dhar 2011; Mishra 2015; Kumar and Acharya 2016; Yadav et al. 2017), and unplanned urbanization (Meraj et al. 2015). However, there are many and obvious scientific concerns that question the linking of a particular flood hazard(s) in Kashmir to the above-mentioned factors. Some of these are raised in Shah (2015, 2016) and Shah et al. (2018) where it is shown that ignoring the tectonic framework of the Kashmir basin could dramatically impact any flood hazard study in the region. Tectonics governs the climate of the region to a large extent, and therefore, it is virtually impossible to link the causes of floods in Kashmir to various factors without providing scientifically valid reasoning with a backbone of the tectonic setup of a region. For example, the often-discussed urbanization and climate arguments are invalid if the used data have a temporal range of fewer than 30 years. The geological and tectonic development of the Kashmir basin demonstrates that tectonic forces have played a major role in the formation of the basin, and climate is also greatly influenced by the orographic setup (Mishra 2015), which is directly governed by mountain building that is linked with the regional tectonic convergence between the lithospheric plates of India and Eurasia (Shah et al. 2020a; b). Therefore, both the tectonics and climate are contributing toward the flooding problems in the region, apart from the anthropogenic factors such as the construction of embankments, urbanization, etc. The lack of understanding will lead to severe problems in the interpretation of data, and that is evident when we relook at the controversial scientific argument for the cause of the 2014 flooding in the Jhelum River. The reasons have migrated from reckless urbanization to precipitation to climate change. For example, (Meraj et al. 2015) writes "In light of the research findings from this research, it is suggested that the reckless and unplanned urbanization of the floodplains and conversion of wetlands in the Jhelum basin need to be stopped forthwith. This practice is *the single most important reason responsible for the enhanced extreme flooding event of September 2014 in Jhelum.*" And in the subsequent publications, this claim has been transported into different factors for example: (Romshoo et al. 2018) "*Though, the primary trigger for the extreme flood event was the high-intensity and widespread rainfall observed in the entire catchment……but the flooding in Jhelum is significantly influenced by the geomorphic set up of the basin exacerbated by the anthropogenic drivers.*" Such conclusions were previously addressed in detail by Shah (2013) where it was shown that the tectonic tilt of the basin has created two major tectono-geomorphic features in the basin where NE portions are prone to floods, which is exactly what was witnessed during the 2014 floods. Shah (2016) further showed that tectonics is the major factor of floods in the basin and he writes "*…the Kashmir basin is mainly shaped by an active interaction between the tectonics and climate. The historical and geological record (shown above) demonstrates that climate change and unplanned urbanization are NOT the major reasons for flooding; thus, it is an over-exaggeration to say that unplanned urbanization is the single-most-important factor in the flooding of Jhelum in 2014. It is clear from the study of the geology, topography, and geomorphology that flooding in the Kashmir basin is controlled by an interaction between tectonics and climate.*"

Significantly, a new study (Ballesteros-Cánovas et al. 2020) has categorically questioned the climate and urbanization centric arguments for the occurrence of floods in Kashmir by demonstrating that that extreme flood events in Kashmir have a prolonged history, and they state "*we use a millennium-long record of past floods in Kashmir based on historical and tree-ring records to assess the probability of 2014-like flood events in the region. Flood chronology (635 CE–nowadays) provides key insights into the recurrence of flood disasters and propels understanding of flood variability in this region over the last millennium, showing enhanced activity during the Little Ice Age. We find that high-impact floods have frequently disrupted the Kashmir valley in the past.*" The commentary by Shah (2016) on the cause of the 2014 floods by Meraj et al. (2015) has raised some of the important concerns about the science of flooding in Kashmir, and some of those are now partially answered in Ballesteros-Cánovas et al. (2020). It suggests that urbanization was not the single-most-important reason for flooding in the Kashmir basin. Instead, the actual cause of the flooding is directly related to the ongoing tectonic activity. The tectonic landforms are controlled by the faults, which have caused tilting and enhanced aggradation in the Jhelum River and, therefore, made the basin prone to more flood hazards (Shah 2016 and the present study). The historical flood disasters complement the tectonic causes (Ballesteros-Cánovas et al. 2020) and, therefore, offer new insight into the science of flood hazards in the Kashmir region. It cautions that to create a scientifically robust and viable dataset, one must not indulge in exaggeration, manipulation, knee jerk reaction, and mixing of unrelated claims to prove a point. There is an immediate need to comprehensively work on the generation of scientifically rigorous details on the science of floods.

Conclusions

The flood hazards in the Kashmir region are not new, which is suggested by the historical data. It has, therefore, questioned the scientific vigor of most of the post-2014 flood-related research studies on the causes of flood hazards in the Kashmir basin. This is primarily because most of the studies have related the causes to urbanization and climate change. However, the work presented here shows that flood hazards in the piggyback Kashmir basin are genetically related to the structural setup of the basin that is controlled by the major Himalayan faults systems that mostly shoulder the basin at the south and southwest. Some of these faults have caused the tilting of the basin, which has led to aggradation in the Jhelum River and, hence, made it more prone to flood hazards. The river is largely controlled by faults because it follows the Tunda-Balakot–Bagh and Jhelum fault systems, which have caused it to change from a largely aggradational river profile in the basin to degradational outside the basin. The topographic profile across the Peshawar and Kashmir basins indicates that the Jhelum River flows at a higher elevation than the river systems in the Peshawar basin. The little changes in the elevation and tilt toward the ~NW direction in the basin are a reflection of the aggradational behavior of the Jhelum River while in the basin. The abrupt change in the tilt in the downstream direction makes it largely degradational when it enters the Tunda-Balakot–Bagh fault systems. Therefore, the data presented here demonstrate that the tectonic setting is the main contributor toward the formation, geometry, and configuration of the Kashmir basin and, therefore, the drainage formation, development, and evolution, which are directly controlled by the major Himalayan fault systems. Although the climate- and urbanization-based arguments are important components and contribute toward the flood hazards in a region, however, the historical and contemporary data on the magnitude and intensity of flood hazards in the Kashmir basin suggest that floods hazards have occurred in the past when these factors were not part of the standard scientific discourse on floods.

References

Avouac JP, Ayoub F, Leprince S (2006) The 2005, Mw 7.6 Kashmir earthquake: sub-pixel correlation of ASTER images and seismic waveforms analysis. Earth Planet Sci Lett 249:514–528

Ballesteros-Cánovas JA, Koul T, Bashir et al (2020) Recent flood hazards in Kashmir put into context with millennium-long historical and tree-ring records. Sci Total Environ 722:1–9

Bhat AS, Romshoo SA (2009) Assessing the impact of geomorphology and land cover on a surface runoff of a micro-watershed. J Himalayan Eco Sustain Devel 4:67–76

Bilham R, Bali BS, Bhat MI, Hough S (2010) Historical earthquakes in Srinagar, Kashmir: clues from the Shiva temple at Pandrethan. Geol Soc Am Spec Pap 471:107–117

Bilham R, Bali BS (2013) A ninth century earthquake-induced landslide and flood in the Kashmir Valley, and earthquake damage to Kashmir's medieval temples. Bull Earthquake Eng 12(1):1–31

Burbank DW (1985) The age of the Karewas, Kashmir as determined from fission-track dating and magnetostratigraphies. In: Agrawal DP (eds) Climate and Geology of Kashmir and Central Asia: The Last 4 Million Years, Today and Tomorrow's Printers, New Delhi, pp 19–26

Burbank DW, Johnson GD (1983) The late Cenozoic chronologic and stratigraphic development of the Kashmir intermontane basin, northwestern Himalaya. Palaeogeogr Palaeoclimatol Palaeoecol 43(3–4):205–235

Burbank DW, Johonson GD (1982) Intermountane basin development in the past 4 Myr in the north west-east Himalaya. Nature 298:432–436

DiPietro JA, Pogue KR (2004) Tectonostratigraphic subdivisions of the Himalaya: a view from the west. Tectonics 23:1–20

EarthExplorer (2021) SRTM 1 arc-secon global. https://earthexplorer.usgs.gov/. Accessed 29 May 2021

Hatwar HR, Yadav BP, Rama Rao YV (2005) Prediction of western disturbances and associated weather over Western Himalayas. Curr Sci 88(6):913–920

Kumar R, Acharya P (2016) Flood hazard and risk assessment of 2014 floods in Kashmir Valley: a space-based multisensor approach. Nat Hazards 84:437–464

Lawrence WR (1895) The valley of Kashmir. Oxford University Press, UK

Meraj G, Romshoo SA, Yousuf AR et al (2015) Assessing the influence of watershed characteristics on the flood vulnerability of Jhelum basin in Kashmir Himalaya. Nat Hazards 77:153–175

Mishra AK (2015) A study on the occurrence of flood events over Jammu and Kashmir during September 2014 using satellite remote sensing. Nat Hazards 78(2):1463–1467

Nandargi S, Dhar ON (2011) Extreme rainfall events over the Himalayas between 1871 and 2007. Hydrol Sci J 56:930–945

Oreskes N (2004) Beyond the ivory tower: the scientific consensus on climate change. Science 306:1686

Pathier EF, Wright TJ, Walker R et al (2006) Displacement field and slip distribution of the 2005 Kashmir earthquake from SAR imagery. Geophys Res Lett 33(20):1–5

Rajatarangini Kalhanas, 1149, n.d Translated by Dutt (1879) Kings of Káshmír: Being a Translation of the Sanskrita Work Rájatarangginí of Kahlana Pandita. In: Ballesteros-Cánovas A, Koul T, Bashir A, Maria J et al (2020) Recent flood hazards in Kashmir put into context with millennium-long historical and tree-ring records Sci Total Environ 722:137875

Rather MI, Rashid I, Shahi N et al (2016) Massive land system changes impact water quality of the Jhelum River in Kashmir Himalaya. Environ Monit Assess 188(3):185

Romshoo SA, Altaf S, Rashid I et al (2018) Climatic, geomorphic and anthropogenic drivers of the 2014 extreme flooding in the Jhelum basin of Kashmir, India. Geomat Nat Haz Risk 9:224–248

Romshoo SA, Rashid I (2014) Assessing the impacts of changing land cover and climate on Hokersar wetland in Indian Himalayas. Arab J Geosci 7(1):143–160

Searle MP, Spencer, DA, Peter JT (1996) Himalayan Tectonics (Geological Society Special Publication No. 74). Mt Res Dev 14(4):436

Shah AA (2013) Earthquake geology of Kashmir Basin and its implications for future large earthquakes. Int J Earth Sci 102:957–1966

Shah AA (2015) Assessing the influence of watershed characteristics on the flood vulnerability of Jhelum Basin in Kashmir Himalaya by Gowhar et al., 2015. Nat Hazards 77(3):2139–2143

Shah AA (2016) The Kashmir Basin fault and its influence on the fluvial flooding in the Kashmir Basin, NW Himalaya. Geol Soc Am special Papers 520

Shah AA, Abd Manan N, Aliudin NAB et al (2020) Formation, rotation, and present-day configuration of Kashmir and Peshawar Basins in NW Himalaya. Front Earth Sci 8:1–21

Shah AA, Khwaja S, Shah BA et al (2018) Living With Earthquake and Flood Hazards in Jammu and Kashmir. NW Himalaya Front Earth Sci 6:179

Shah AA, Malik JM (2017) Four major unknown active faults identified, using satellite data, in India and Pakistan portions of NW Himalaya. Nat Hazards 88:1845–1865

Shah RA, Achyuthan H, Lone AM et al (2020b) Holocene palaeoenvironmental records from the high-altitude Wular Lake, Western Himalayas. Holocene 30(5):733–743

Singh AK, Jaiswal MK, Pattanaik JK et al (2016) Luminescence chronology of alluvial fan in North Bengal, India: implications to tectonics and climate. Geochronometria 43(1):102–112
Singh AK, Pattanaik JK, Jaiswal MK et al (2017) Late Quaternary evolution of Tista River terraces in Darjeeling-Sikkim-Tibet wedge: implications to climate and tectonics. Quat Int 443(10):132–142
Yadav BP, Bhan SC (2010) Meteorological factors associated with July 2005 floods in river Jhelum. Mausam 61(1):39–46
Yadav BP, Kumar N, Lotus S (2017) Synoptic and climatological aspects of extreme rainfall over western Himalays towards end of 2014 southwest monsoon season. Mausam 68(4):597–606

Chapter 8
Climate-Induced Glacier Retreats and Associated Hazards: Need for Robust Glaciers and Glacial Lake Management Policy in Sikkim Himalaya, India

R. K. Sharma, Rajesh Kumar, Pranay Pradhan, and Arpan Sharma

Abstract Himalayan glaciers, the freshwater tower of South and East Asia, are strongly affected by climate change impacts. The rapid melting of glaciers due to climate warming has resulted in the formation and expansion of many glacial lakes in the Indian Himalayan Region (IHR), which has posed the threat of glacial lake outburst floods. Sikkim Himalaya, an important part of Eastern IHR, possesses 84 glaciers and 14 potentially dangerous glacial lakes. However, the in situ-based study on the glaciers and glacial lakes dynamics is limited in the region due to the minimal capacity of human resources and remote and rugged terrain of glaciers. More importantly, there is a lack of glacier protection laws and glacial lake management plans to cope with climate-induced effects and hazards. The present study critically reviews glaciological researches and examines the scientific gaps and challenges controlling the research and management of glaciers and glacial lakes in Sikkim Himalaya. Further, the scientific suggestions and recommendations placed in the present study can be vital elements to frame effective climate change policy required for investigation, adaptation, and mitigations against adverse climate change impacts.

Keywords Indian Himalayan Region · Glacial lake outburst floods · Climate change vulnerability · Glacier Protection Laws · Glacial Lake Management Plan · Climate science and media reporting · Sikkim Himalaya

R. K. Sharma (✉) · R. Kumar
Department of Environmental Science, School of Basic Sciences and Research, Sharda University, Greater Noida, Uttar Pradesh, India

R. Kumar
Department of Environmental Science, School of Earth Sciences, Central University of Rajasthan, Ajmer, Rajasthan, India

R. K. Sharma · P. Pradhan
Sikkim State Council of Science and Technology, Vigyan Bhawan, Deorali, Gangtok, Sikkim, India

A. Sharma
Department of Computer Applications, Sikkim Manipal Institute of Technology, Majitar, Sikkim, India

S. Rani and R. Kumar (eds.), *Climate Change*, Springer Climate,
https://doi.org/10.1007/978-3-030-92782-0_8

Introduction

The Hindu Kush Himalaya (HKH) region, also mentioned as the "third pole," is the most significant mountain ecosystem on the earth and a critical geo-ecological asset of South Asia (Bolch et al. 2019). The glacier of HKH covers ~59 × 10^3 km^2, out of the world's total of ~540 × 10^3 km^2 mountain glacier area (Dyurgerov and Meier 1997, 2005). Further, the Himalaya and the trans-Himalaya regions consist of nearly 50% of all glaciers outside the polar region (Mayewski and Jeschke 1979). Sangewar and Shukla (2009) provided the glacier inventory and reported the 9575 mountain glaciers in the Indian Himalayan Region (IHR). Further, in the IHR, many authors have reported a total of ~10,000 glaciers, excluding the ~2000 glaciers in Nepalese and Bhutanese Himalaya (Raina and Srivastava 2008; Bajracharya and Shrestha 2011; Bajracharya et al. 2015). However, the glacier distributions in the Himalayas are controlled by altitude, orientation, slope, and climate zone in which they are located. The precipitation's contributions to the glacier mass budget in the Central, Western, and Eastern Himalaya are distinctly variable. As per the study conducted by Ageta and Pokhral (1999), monsoonal rain contributes approximately 80% of the mass balance inputs in the Eastern Himalaya. In contrast, monsoonal rain adds about 15% of the mass balance flux in the Central Himalaya. However, in the Western Himalaya, winter precipitation controls 85% of the mass balance influx. Therefore, there is a diverse range of climatic variability in Eastern, Central, and Western Himalayan regions. About ~80% supply of the freshwater on earth originates from the mountain regions (Barry and Chorley 1998), where glaciers, snow-clad regions, and glacial lakes are significant freshwater sources. Due to climate change, the melting glaciers will adversely affect the water supply in the most critical hugely populated region like Indus and Brahmaputra basins (Barnett et al. 2005).

Generally, the Himalayan region is witnessing a warmer climate. The Intergovernmental Panel on Climate Change (IPCC) (2014) reported that the warming trend in the Himalaya is more than the global averages supported by the evidence-based scientific reports (Shrestha et al. 1999; Ali et al. 2018). The retreat of glaciers in Himalaya and other parts of the world due to climate warming is evident since the beginning of the last century (Dyurgerov and Meier 1997; Paul et al. 2007; Bolch et al. 2012; Kääb et al. 2012; Gardelle et al. 2013; Shea et al. 2015a); however, the glaciers in the Karakoram have reported as surging (Copland et al. 2011; Bhambri et al. 2013; Quincey et al. 2015).

On the other hand, climate-induced glacier's retreat is usually associated with the formation and expansion of glacial lakes in the Himalayan belts (Costa and Schuster 1988; Richardson and Reynolds 2000; TanDong et al. 2010; Raj 2010; Petrakov et al. 2011; Raj et al. 2012; Sharma et al. 2018). It is reported that the expansion of glacial lakes can pose a possible risk of glacial lake outburst floods (GLOFs). GLOF is a destructive geomorphological hazard associated with the disastrous release of impounded water from the glacial lake (Richardson and Reynolds 2000; Clague and Evans 2000; Benn et al. 2012; Westoby et al. 2014; Worni et al. 2014). The high-speed water release due to the glacial lake outbursts can erode and carry enormous sediments downstream (Breien et al. 2008). It is one of the most destructive hazards that can cause risk to human life and property (Raj et al. 2013).

Despite such importance, the studies on glacier retreat and glacial lakes in Eastern Himalaya, particularly in Sikkim Himalaya, are very few. On the other hand, there are reports that the overall annual temperature rise is 0.01–0.04 °C (Sharma et al. 2009), and the rate of warming is the highest during the winter season in the Eastern Himalayan region (Chettri et al. 2010). As a result, Sikkim Himalaya is reported to have a substantial increase in temperature from the past and warmer winter due to the impact of climate change (Sharma and Shrestha 2016). A systematic mapping of glaciers in Sikkim Himalaya started way back in 2000 through remote sensing satellite images that record ~84 Himalayan glaciers located primarily in west and north districts (Bahuguna et al. 2001). However, there are no reports of in situ-based monitoring of glaciers in Sikkim Himalaya till 2013. Few glaciological interventions have been carried out in the northern part of Sikkim way back in the 1980s and 1990s. However, the policy and plans related to glacier management and mitigation of GLOFs in the region are scarce. Further, there are no systematic approaches to the management of freshwater in the area. Therefore, the present chapter critically examines and deals with the following scientific facts and policy gaps in the Sikkim Himalayan region:

- Qualitative analysis of the glaciological studies in the higher catchment of Sikkim Himalaya.
- Critical reviews on the probability of GLOFs.
- Climate-induced vulnerability and associated adaptation and mitigation programs.
- Observed climate and people perceptions on climate change.
- Scientific gaps and challenges in the glaciology and glacier-related hazards.
- Policy recommendations on climate change and glaciers.

Study Area

Sikkim (a state), situated in the northeastern part of India, is selected for the present study. It is considered as a biodiversity hotspot in the Eastern Himalaya. It shares 52.03% of the aerial extent of Eastern Himalaya in India with other northeastern states of India and Darjeeling Hills of the state of West Bengal (Tse-ring et al. 2010). It covers a total area of 7096 km^2 and is one of the highest national parks in the world; i.e., Khangchendzonga National Park is located in Sikkim Himalaya that covers about 35% of the total geographical area. Sikkim has four districts (Sikkim Primary Census Abstract 2011); however, it shares the borders with China in the north and east districts, Bhutan in the east, Nepal in the west, and West Bengal in the south district, making it a geographically important area. The Tista River, flowing in the north–south direction, is the main river of the state and is joined by the Rangit River in the downstream areas (Krishna 2005). As per the latest population census, Sikkim's total population is 610,577, which consists of 456,999 (~75%) people residing in rural areas (Sikkim Primary Census Abstract 2011). Sikkim, being a high earthquake-prone environment (Nath et al. 2005), lies in Zone IV of the seismic zonation map of India. However, many developmental actions have been carried out in Sikkim, but there is no scientific account that directly shows the quantum of climate change impacts and vulnerability in the region (Sharma and Shrestha 2016). Though small in areal extent, Sikkim Himalaya covers the various eco-zones from subtropical, temperate, and subalpine to alpine due to sharp elevation gradients of 300–8586 m above mean sea level (AMSL) (Krishna 2005; Basnet et al. 2013). According to the Forest Survey of India (FSI) Report (2017), 3344 km^2 of the area is covered by forests that comprise 47.13% of the total geographical area of Sikkim. As per the latest glacier inventory of the Geological Survey of India (GSI), there are 449 glaciers in Sikkim Himalaya covering an area of 705.54 km^2 distributed in different sub-basins, viz., East Rathong (36 glaciers; 58.44 km^2), Talung (61 glaciers; 142.90 km^2), Changme Khangpu (102 glaciers; 144.35 km^2), and Zemu (250 glaciers; 359.85 km^2) (Raina and Srivastava 2014). The location map of Sikkim, together with the elevation gradients and drainage system, is shown in Fig. 8.1.

Status of Glaciological Studies in Sikkim Himalaya

The first in situ-based study in the Sikkim Himalayan glacier was carried out at Changme Khangpu Glacier on different study components like radiometric chronology, glacio-hydrology, suspended sediment, and accumulation rate of the glacier during the 1980 and 1990s in the North Sikkim Himalaya (Shukla et al. 1983; Nijampurkar et al. 1985; Puri 1999). However, after almost three decades, Debnath et al. (2019) studied the time series satellite data from 1975 to 2016 and reported that ~20% of glacier area (18.56 $\pm$ 2.61 km^2) had lost since 1975 at an average rate of -0.453 ± 0.001 km^2 a^{-1} with a higher rate of area shrinkage (-0.665 $\pm$ 0.243 km^2 a^{-1}) during the recent decades (2001–2016). The largest glacier of Sikkim is Zemu Glacier, which is situated below Mount Kanchenjunga (8586 m AMSL). As reported, the glacier was first mapped during an expedition of German researchers (Finsterwalder 1935). After many decades, Basnet et al. (2013) reported that the Zemu Glacier shows no substantial changes in the terminus from 1990 to 2010. Their assumption was based on its large size, steep slope, and thick debris cover at the glacier's tongue (Venkatesh et al. 2011). However, they did not study the vertical mass loss/mass balance of the glacier. A recent geodetic study carried out by Rashid and Majeed (2020) found that the Zemu Glacier has lost 6.78 $\pm$ 2.05 Gt of mass at a rate of 84.8 Mt a^{-1} between 1931 and 2012 with an accelerated mass loss rate of 276.5 Mt a^{-1} during 2000 and 2012. They also recorded the terminus retreat of ~797 m ($\pm$19.7 m) during 1931 and 2018 at a rate of 9.1 m a^{-1} and projected the formation of a potential future proglacial lake in the area. Therefore, systematic mapping and in-depth understanding of the glacier–climate relationship are the need of the present time.

A joint program of the Space Applications Centre-Indian Space Research Organisation (SAC-ISRO) and State Remote Sensing Applications Centre (SRSAC) of

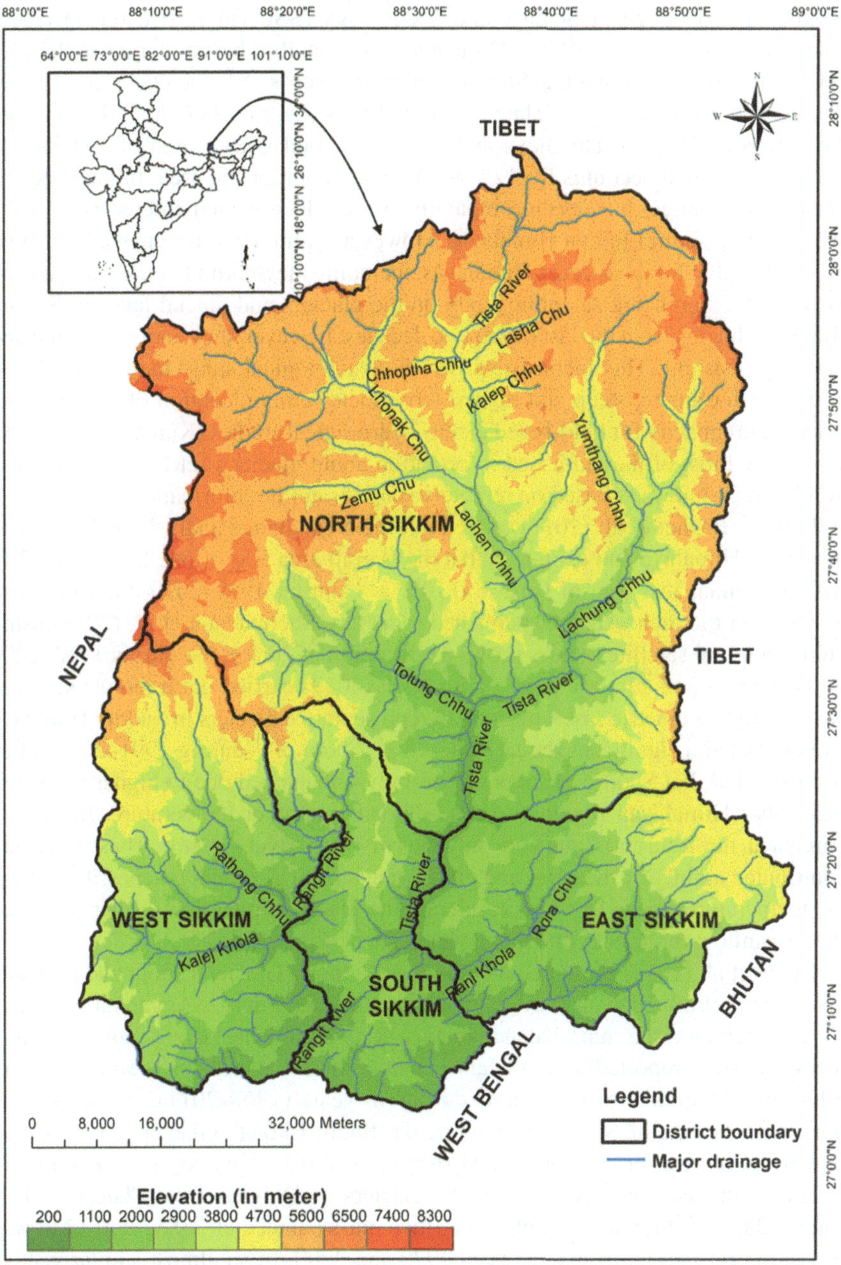

Fig. 8.1 Location and extent of Sikkim state of India showing the elevation gradients (as per ASTER DEM) and drainage system

Sikkim State Council of Science and Technology (SSCST) carried out extensive glacier inventory for the Sikkim Himalaya using satellite-based images in the year 2001. The inventory shows that Sikkim Himalaya records ~84 glaciers located mostly in west and north districts (Bahuguna et al. 2001). Aggrawal et al. (2017) mapped the glaciers in Sikkim Himalaya and reported a total glacier area of 883 km^2 in the year 2013 that accounts for 12% of the total geographical area. In the IHR, the glacier area changes have been comparatively well documented and recorded in the Western and Central Indian Himalayas. However, limited and fragmented information is available for the Eastern Himalayas, including the Sikkim Himalayan glaciers. Most of the significant scientific study on the glaciers and glacial lakes in Sikkim Himalaya begins from the last one to two decades; however, they are fragmented and are not consistent. This inconsistency in the glaciological study may be attributed to the harsh climate, steep and rugged terrain and remote location of glaciers, and inadequate capacity for field-based glacio-hydrological studies (Kumar et al. 2020b). Therefore, most of the published information about glacier's retreat and mass loss in Sikkim Himalaya is based on remotely sensed satellite information.

Further, Basnet et al. (2013) carried out the area change analysis of 39 glaciers in Sikkim Himalaya and estimated that the overall area loss of 6.9 ± 1.5 km^2 (3%) with an annual loss rate of $0.16 \pm 0.10\%$ a^{-1} during 1989–2010 due to climate warming in the area. Another published report by Racoviteanu et al. (2015) using multi-temporal satellite images shows deglaciation of $20.1 \pm 8\%$ during 1962–2000 in the Sikkim and adjacent areas. Garg et al. (2019) carried out a remote sensing-based multi-parametric assessment of 23 representative glaciers in Sikkim Himalaya and estimated a significant retreat of 17.78 ± 2.06 m a^{-1} during 1991–2015. They further stated that Sikkim glaciers are strongly imbalanced due to higher mass loss than other Himalayan regions because of rising summer temperature. In Sikkim Himalaya, the Tista and Rangit are only the two major river systems. These rivers are contributed mainly by the snow and glacial melts from the upper reaches. Therefore, the temperature rise is likely to accelerate the glacier retreat that further increases the streamflow in the Sikkim Himalayan glaciers.

Luitel et al. (2012) estimated that the East Rathong Glacier's snout had retreated 460 m from 1980–2012 at an average retreat rate of 13.9 m a^{-1}. Aggarwal and Tayal (2015) attempted the mass balance of East Rathong Glacier using remote sensing methods. They reported a significant negative mass balance for the glacier, with a loss of ~11 m w.e. or 0.047 km^3 during 48 years (1963–2011). In the extreme northwestern side of Sikkim Himalaya, the South Lhonak Glacier is reported to retreat by 1.9 km during 1962–2008 (Raj et al. 2013). Very recently, Zhou et al. (2018) estimated the mass loss for the glaciers of Qinghai-Tibet Plateau and its surroundings and found the highest negative mass balance of -0.32 ± 0.12 m w.e. a^{-1} in the Sikkim Himalaya between 1970 and 2000. In an almost similar year of observations, Bhattacharya et al. (2018) reported that the glaciers in the northwestern part of Sikkim Himalaya and surrounding areas (including Nepal) were losing mass moderately during 1975–1999 (0.10 ± 0.11 m w.e. a^{-1}). However, they observed accelerated mass loss of -0.48 ± 0.51 m w.e. a^{-1} during recent years (1999–2006) with an overall mass budget -0.19 ± 0.16 m w.e. a^{-1} during the entire study period

(1975–2006). This difference in the mass loss may be subject to various possible reasons, viz., differences in spatial resolution of digital elevation model (DEM), human errors (including mapping, processing, and pixel errors), and errors involved in the area's delineation. Therefore, these variations in the results using satellite images and DEM further require field-based validation. However, it is evident from the above-published reports that glaciers in Sikkim Himalaya have been facing the consequences of climate change as in other parts of the world. Therefore, it is expected that future projections on the climate and its impact on freshwater resources would be very crucial for building adaptation and mitigation plans to combat climate change.

In the Himalaya, in situ-based mapping and monitoring of every glacier are not possible due to the harsh environment and steep terrain. However, there are many benchmark glaciers in the Himalayas where in situ-based records of mass loss, snout retreat, and glacier runoff are vital to know the flow forecasting and establish climate–glacier relationships. In Sikkim Himalaya, the in situ-based study on glacier dynamics is very minimal. Kumar et al. (2020b) carried out the first in situ-based study on glacier's melt runoff process and hydro-meteorological characteristics of East Rathong Glacier. They concluded that there is a significant hydro-meteorological relationship of meltwater discharge with the temperature, rainfall, and relative humidity in the East Rathong Glacier catchment. Recently, Sharma et al. (2021) also studied the in situ-based suspended sediment dynamics for three consecutive melt seasons (2013–2015) and reported substantial meteorological controls on the sediment transport process in the East Rathong meltwater stream. However, the data on the contribution of glacier's melts and sediment transport to the streamflow during the melting (ablation) season in the Eastern Himalaya are very few that necessitate an in-depth investigation of different glacier's catchments to understand insights of glacio-hydrological processes in the Sikkim Himalaya. The major glaciers in Sikkim Himalaya are shown in Fig. 8.2.

Potentially Vulnerable Glacial Lakes and the Threat of GLOFs

In Sikkim Himalaya, melting glaciers, the formation of proglacial lakes, and the growth of existing glacial lakes are well-known due to climate warming (Basnet et al. 2013; Sharma et al. 2018). Besides, the accelerated rate of glaciers retreat from the last two decades leads to the continuous expansion of glacial lakes. A total of 320 glacial lakes have been reported in the Sikkim Himalaya (Raj et al. 2013). Among them, 14 glacial lakes are potentially vulnerable to lake outbursts (Mool and Bajracharya 2003; Raj et al. 2012). The major glacial lakes in the Sikkim Himalaya are shown in Fig. 8.2. Raj et al. (2013) identified 85 new glacial lakes in their study compared to the previous inventory of glacial lakes prepared by the International Center for Integrated Mountain Development (ICIMOD) (Campbell 2005). Worni et al. (2013) carried out the risk assessment of Shako Cho Glacial Lake

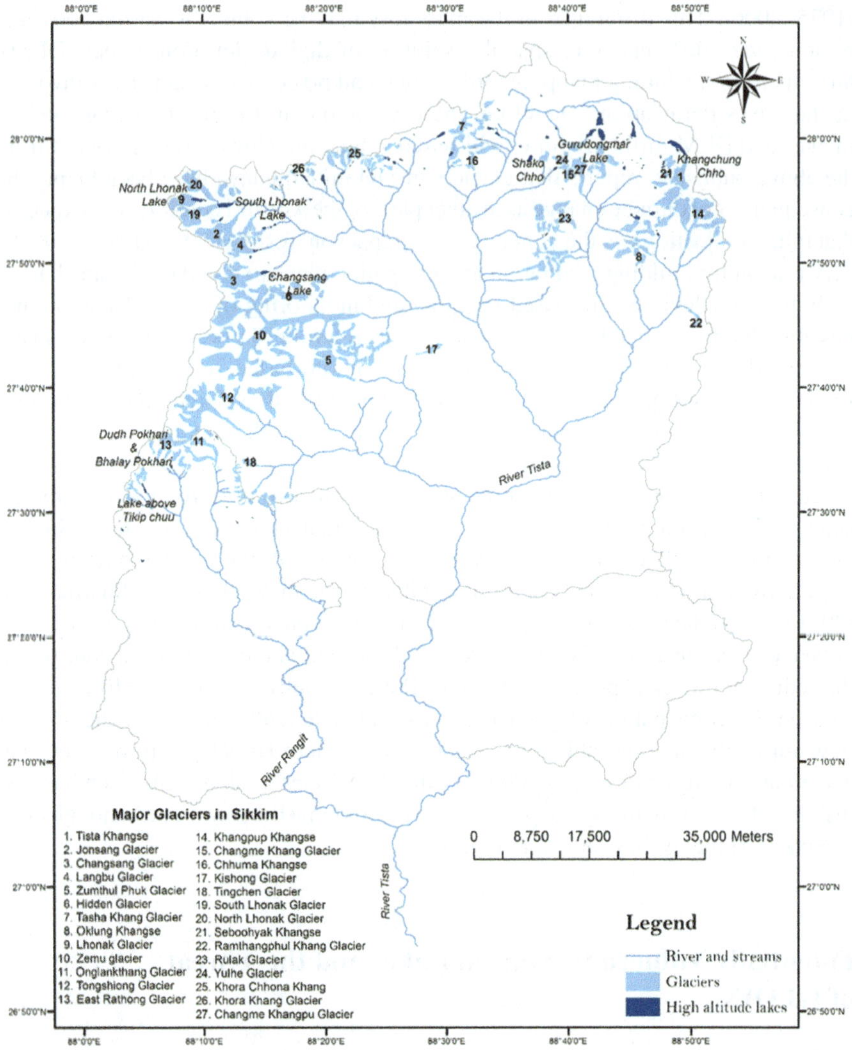

Fig. 8.2 Map showing the major glaciers and glacial lakes in Sikkim Himalaya. The number on the map indicates the names of glaciers. The names of the glacial lakes are shown on the map itself

in North Sikkim. They have advocated that the lake is highly vulnerable to glacial lake outbursts.

Similarly, Raj et al. (2013) studied the hazard assessment of South Lhonak Lake (SLL) located at the extreme northwest part of Sikkim Himalaya. They also suggested a high outburst probability of SLL. Sharma et al. (2018) carried out the remote sensing- and field-based assessments of SLL and calculated the lake's volume. They have estimated a lake volume of 65.81 million m^3 with a maximum depth of 131 m.

Fig. 8.3 Scientific team from Sikkim State Climate Change Cell carrying out the bathymetric survey of South Lhonak Lake in 2014. Massive calving of ice in the lake can be seen in the photographs

Their study was the first record of lake bathymetry carried out through a field-based survey in Sikkim Himalaya (Fig. 8.3). They further suggested many mitigation measures that need to be taken up in the SLL. In the remote sensing-based assessment of glacial lakes, Begam and Sen (2019) also identified SLL to have a potentially very high risk of generating GLOF events anytime in the future. Therefore, in Sikkim Himalaya, SLL is considered a highly vulnerable lake with a high probability of outbursts. Sattar et al. (2019) carried out hydrodynamic breach modeling of SLL. They stated that the flood wave could reach the Lachen town located at a distance of 46 km downstream from the lake, at 3 h and 38 min after the breach's initiation, if in case of lake bursts. It would have a peak flood of 3928.16 m^3s^{-1} and a maximum flow velocity of 13.6 ms^{-1}. Similarly, after reaching the Chungthang town located at a distance of 62.35 km from South Lhonak Lake, the flood wave can potentially inundate settlements on the side of the flow channel. The model estimates a peak flood of 3828.08 m^3s^{-1} which is achieved after 4 h of the dam breach as per their simulation.

Shukla et al. (2018) mentioned that glacier melting could lead to the formation of new lakes and expansion of existing lakes in the Sikkim that increased the potential of glacial lake outbursts floods and suggested that persistent attention is required at the earliest level. Further, there are 14 potentially vulnerable glacial lakes in Sikkim Himalaya in terms of glacial lake outbursts. Therefore, climate change impacts are expected to be tangible in other glacial lakes in terms of their expansion of size and volume, if the temperature warming is continuous in the Sikkim Himalayan region.

Observed Climate and People Perceptions

Altitudinal variation with high mountains across Sikkim is the driving factor controlling the climate and weather patterns. Sikkim Himalaya faces the unprecedented consequences of climate change; however, the published records on climatic change and variability are few and fragmented. According to Sharma et al. (2009), the overall increase in temperature is 0.01–0.04 °C annually in the Eastern Himalayan region.

Further, Chettri et al. (2010) found that warming rates are the highest in winter in the Eastern Himalayan region. Additionally, Gupta and Das (2006) also reported an increase in mean minimum temperature and climate warming in Sikkim. Telwala et al. (2013) analyzed the historical and recent temperature records. They noted that the warmest and coldest months' mean temperature had been increased by 0.76 ± 0.25 and 3.65 ± 2 °C, respectively, in Lachen and Lhonak Valleys of North Sikkim Himalaya.

The observed meteorological records in Sikkim indicate no significant change in maximum temperature, but there is an increase in minimum temperature by 2.5 °C during 1957–2009 (SAPCC-Sikkim 2014). The observed records show that the total rainfall has been increased by 250 mm between the periods of 1983–2009. Sharma and Shrestha (2016) analyzed the instrumental meteorological data (1978–2009) of Tadong and Gangtok stations in Sikkim and reported that annual mean minimum temperature and mean minimum winter temperature (DJF) have been significantly increased from the past. More recently, Kumar et al. (2020a) also carried out the meteorological analysis for the Gangtok and Tadong meteorological stations (1961–2017) in Sikkim and found positive increasing trends in minimum temperature. They have noticed accelerated climate warming during the last two decades with an increased probability of extreme temperatures. Therefore, the recent climate warming has had a substantial impact on many developmental sectors like agriculture, forests, water, disaster, etc. People have perceived reduced frost and snowfall, increased hailstorms, warmer winter, and intense rainfall in recent decades (SAPCC-Sikkim 2014). Nowadays, information about regional climate change is growing in demand for an in-depth understanding of climate change impacts on different ecosystems at a regional scale (Liu and Chen 2000). Therefore, community-level perceptions on climate change are equally crucial to developed village-level plans and policies. Sharma and Shrestha (2016) carried out semi-structured interviews on climate change perceptions in 12 villages in Sikkim Himalaya, covering 3–4 villages in a district. They reported a significant increase in temperature and decreased rainfall compared to a decade earlier, warmer winter, water springs that are drying up, and the changes perceived in spring-water recharge phenomenon (locally known as *Mul Phutnu*). People also perceived the spring season changes, low crop yields, and mosquitoes' incidences during winter as per their study. The local knowledge-based information on climate can help formulate effective policies and adaptive strategies to cope with climate change (Chaudhary and Bawa 2011).

Climate Change Vulnerability, Adaptation, and Mitigation Programs

As per the IPCC Report (2014), the impacts of climate-related extremes like floods, droughts, heatwaves, cyclones, and wildfires have significant vulnerability and exposure to some ecosystems and many human systems to current climate variability. The

Himalayan countries in the HKH region face problems and challenges in meeting the demand for food, water, and energy due to fast socio-economic development, a rise in population, and urbanization (Mukherji et al. 2015). Therefore, it is relatively challenging to minimize and manage climate change impacts and risks through adaptation and mitigation measures. Thus, assessing the risks and vulnerability of a system is essential to prioritize the adaptations and mitigation plans to minimize the climate change impacts. In an early assessment carried out to prepare the climate change action plan in Sikkim, the south district of Sikkim was found highly vulnerable to climate change impacts as most of the district's subtropical areas come under drought-prone zones (Tambe et al. 2011; SAPCC-Sikkim 2014). In a recent vulnerability assessment carried out in IHR using a common framework, the east district of Sikkim was a highly vulnerable district due to its high population density and a high degree of slope (DST-NMSHE2018–19). The methodology framework of vulnerability assessment was based on the guidelines of the Fifth Assessment Report of the IPCC (IPCC 2014).

Owning to the significant adaptation measures for climate change in Sikkim, the *Dhara Vikas* (springshed development programme), initiated in the year 2008, is an innovative program to revive and maintain drying springs, streams, and lakes (Bhutia 2019). The program has emerged as a robust climate change adaptation strategy for Sikkim's drought-prone districts (west and south). It aimed to enhance water availability in drought-prone areas and was undertaken through the Mahatma Gandhi National Rural Employment Guarantee Act (MGNREGA) to convergence various line departments and non-government organizations (NGOs). The program has succeeded in reviving 55 springs in villages like Kaluk, Rhenock, Ravangla, Sumbuk, Jorethang, Namthang and has recharged 1035 million liters of groundwater annually in the last four years (Bhutia 2019). Sikkim implements the water sector recommendations of the State Action Plan on Climate Change (SAPCC) utilizing the funds from the MGNREGA. However, the other sectoral recommendations of the climate change action plan have not been implemented effectively (Jogesh and Dubash 2015).

In terms of some considerable scientific and mitigation measures in Sikkim, Sikkim State Climate Change Cell (SSCCC) has been actively engaged in climate change-based scientific interventions. SSCCC was established in October 2014 under Sikkim State Council of Science and Technology, an autonomous organization of the State Science and Technology Department, Government of Sikkim. The SSCCC is operational in a project mode under the National Mission for Sustaining the Himalayan Ecosystem (NMSHE), supported by the Department of Science and Technology (DST), Government of India (https://sikkim.gov.in/departments/department-of-science-technology/sikkim-state-climate-change-cell). The SSCCC carried out the first bathymetric survey of potentially vulnerable SLL located in Sikkim Himalaya to know the storage volume of the lake (Sharma et al. 2018). Eventually, GLOF mitigation measures like siphoning of impounded water and debris clearance near the lake outlet were carried out in SLL in collaboration with the Land Revenue and Disaster Management Department (LRDMD), Government of Sikkim. SECMOL-Ladakh technically supported the mitigation measures in SLL.

Fig. 8.4 GLOF mitigation (siphoning of impounded water) at South Lhonak Lake in 2016

Three (3) number of 8-inch diameter high-density polyethylene (HDPE) pipes have been connected to siphon the impounded water to minimize the threat of GLOF. About 150–180 lit/sec of water has been siphoned with three sets of pipes from SLL (Fig. 8.4).

In the IHR, the NMSHE and the National Mission on Strategic Knowledge for Climate Change (NMSKCC) under the eight National missions of the National Action Plan on Climate Change (NAPCC) have been coordinated by the Ministry of Science and Technology, Govt. of India from the year 2010. These missions mainly emphasize building capacities, making institutional networks, researching strategic knowledge, and recognizing the knowledge gaps on climate change in the IHR (DST 2010a; DST 2010b). Therefore, identifying central and state-specific climate-related developmental plans and programs is crucial for operationalizing short-term climate change adaptation activities in Sikkim. At the national level, the National Adaptation Fund on Climate Change (NAFCC), established in 2015, has been supporting many states and union territories that are mainly exposed to adverse impacts of climate change in the sectors like agriculture, animal husbandry, water, forestry, tourism. The projects under NAFCC prioritize the requirements that build climate resilience in the sectoral areas identified under the SAPCC. In Sikkim, NAFCC supports the project entitled "Addressing Climate Change vulnerability of Water Sector at Gram Panchayat Level in drought-prone areas of Sikkim," which has been executed by the Rural Development Department, Government of Sikkim (https://www.nabard.org/content.aspx?id=585).

Gaps and Challenges for Climate and Glaciers in Sikkim Himalaya

Sikkim Himalaya is one of the biological hotspots having varied geographical settings and elevation gradients from 300 to 8586 m AMSL. These diverse geographical areas have distinct microclimatic variations. Therefore, climate warming impacts would be very different in every sector like water, agriculture, forests, health. Consequently, it is important to identify the gaps and challenges that need to be addressed to develop adaptation and mitigation plans to deal with the changing climate. Some of the critical gaps and challenges that require the attention in Sikkim Himalaya are mentioned below:

Limited Instrumental Observatories

Sikkim Himalaya is recognized as a part of India's biodiversity hotspots; however, it is one of the most data-deficit regions of Eastern IHR. There are only two reliable instrumental meteorological observatories located in the Gangtok and Tadong. Broadly, the scarcity of long-term instrumental meteorological datasets in the Himalayas is a major problem as it has impacted the scientific findings (Kumar 2020b). High-altitude meteorological datasets are very minimal in Sikkim that limit the development of climate–glacier relationships and projections. Recently, a few in situ instrumental measurements on glaciohydro-meteorological observations have been made in the East Rathong Glacier (Kumar et al. 2020b; Sharma et al. 2021). Further, there are limited meteorological observatories in the lower reaches of Sikkim.

Institutional Capacity Building

Institutional capacity building is the base for creating a skilled workforce for livelihood upliftments, planning and formulating the developmental plans and policies, and any research and developmental activities. At the state level, the State Institute of Capacity Building (SICB) has been actively working to create a highly skilled workforce empowered with enhanced skills, knowledge and prepared themselves for livelihood upliftments and increased participation of youth, women, and others disadvantaged sections (https://sdedsikkim.in/SICB.aspx#). As a mandate of the NMSHE, DST, the Government of India is supporting the Sikkim for institutional capacity building on climate change adaptation and planning (DST 2010a) in collaboration with the technical partners like the Swiss Agency for Development and Cooperation (SDC), Indian Himalayas Climate Change Adaptation Programme (IHCAP). Further, the German Agency for International Cooperation (GIZ) has been an active technical state partner for Sikkim right from the formulation of the state action plan on climate change (SAPCC-Sikkim 2014). However, capacity building of the nodal officials of the department and sectoral officials is crucial since they can provide vital inputs to climate change plans and policies at the local level.

Absence of Glacier Protection Laws (GPLs) and Glacial Lake Management Plan

It is a widespread fact that climate warming in the Himalayan region is more than the global average. The temperature rise has impacted the glaciers to shrink, and the formation and expansion of proglacial lakes in the Himalayas are evident due to melting glaciers. Similar observations have been made for the Sikkim Himalaya where temperature rise has impacted the glaciers and glacial lakes (Basnet et al. 2013; Sharma and Shrestha 2016; Sharma et al. 2018; Sattar et al. 2019; Kumar et al. 2020b). As compared to Western and Central IHR, the field-based study on glaciers and glacial lakes in Sikkim Himalaya is minimal. Any scientific assessment of glaciers and glacial lakes may highlight the key management plans to help the policy decision. Kumar et al. (2020b) study the hydrometeorology of the East Rathong Glacier in Sikkim Himalaya and have shown the linkages of climate change programs that help build capacities, institutional networks and identify the knowledge gaps in the IHR. Sikkim Himalaya is considered a data-deficit zone on account of the limited study of the freshwater resources in Eastern IHR. As Sikkim caters to many glaciers and snow-dominated zones, Sharma et al. (2021) suggested the decision-makers about the necessity of a dedicated "Regional Centre for Glaciology" for a detailed understanding of the impacts on glacier resources and future changes due to climate change. Through such intervention, they have expected improved glaciological research facilities in the Eastern IHR. Further, such development can be highly beneficial to the policymakers for formulating plans and policies on the management of Himalayan glaciers and potentially vulnerable glacial lakes.

In countries like Argentina and Chile, Glacier Protection Laws (GPLs) have been adopted to protect the cryosphere from harmful activities since the mining activities have damaged glaciers (Anacona et al. 2018). However, by prohibiting activities on glaciated areas, GPLs may limit or even prevent the timely execution of mitigation measures on glacial-related hazards. Therefore, it has overlooked the effect of climate change on the glacial environment that needs to be managed and mitigated. In Sikkim Himalaya, the majority of people are dependent on tourism as it is fast evolving as one of the most critical drivers of the state economy (STP 2018). Adventure tourism (trekking, paragliding, mountaineering, river rafting, etc.) has been substantially increased nowadays in Sikkim. Many communities in the high-altitude areas are yak and sheepherders, and they are also directly or indirectly related to tourism for their livelihoods. Therefore, to protect the glacial environment, it is recommended to have community-friendly regulations or laws and sustainable environmental perspectives. Before enacting the laws for glacier protection or any rules about glaciers and other freshwater resources, it is recommended to have clear communication to the locals and generate climate change awareness regarding the importance of glaciers and glacial hazards to avoid the social pressure in hazard mitigation (Anacona et al. 2018).

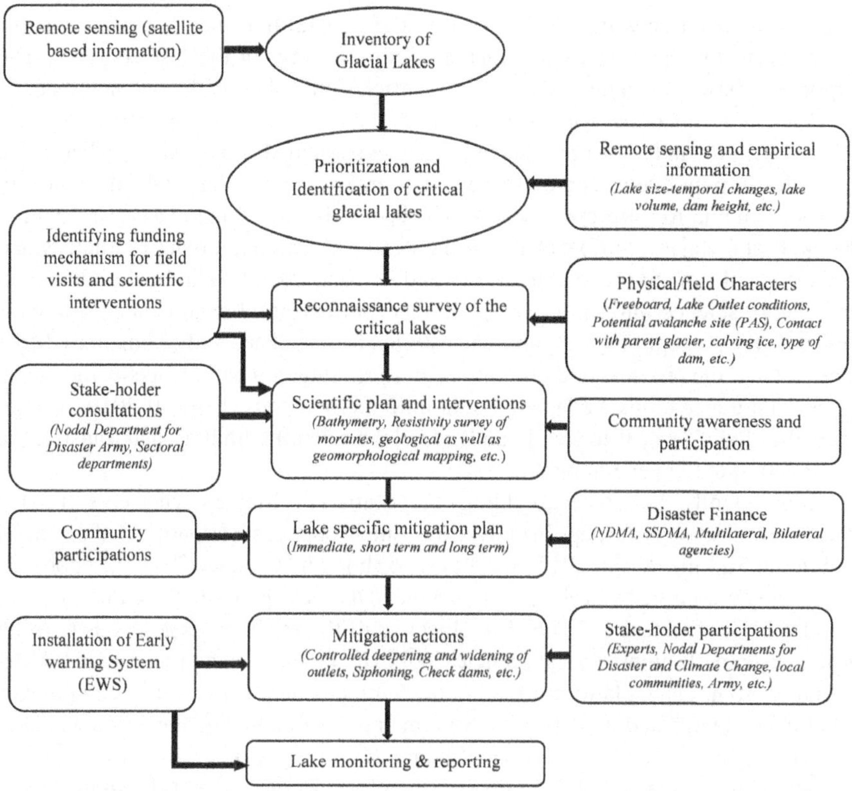

Fig. 8.5 Conceptual representation of glacial lake management plan for Sikkim Himalaya

Further, many studies confirm 14 potentially vulnerable glacial lakes in Sikkim Himalaya that can bring risk to downstream communities and infrastructures, including hydropower plants. However, there is no glacial lake management/mitigation plans exist in Sikkim Himalaya to date. Therefore, to mitigate the threats of GLOFs in Sikkim Himalaya, a *Glacial Lake Management Plan* is urgently required with a robust policy for glacial lakes. Sensitization and awareness to the downstream communities regarding the potential GLOFs can help in the prevention of disasters. Further, installing an early warning system (EWS) in the vulnerable glacial lakes is an essential recommendation to tackle the worst hits of floods (Fig. 8.5).

Climate Science and Media Reporting: Bridging the Gaps

Around the globe, media plays a crucial role in disseminating information and opinions through newspapers, magazines, television, radio, and the Internet. Therefore, media can play a vital role in climate change awareness and target a vast audience, even from social networks. Keller et al. (2019) found that India's climate change coverage in the newspapers has increased significantly in the last 20 years. The report was based on an analysis of 18,224 articles published in the two newspapers.

However, climate change coverage is minimal in the local newspapers in Sikkim. There is significantly less presence of local television media in Sikkim, probably due to the lesser population. Therefore, scientists and researchers may take the platform of social media to raise awareness about recent climate change issues in Sikkim. Boykoff and Rajan (2007) stated that the dearth of training among journalists and a shortage of time to investigate a story and its background could be detrimental in translating science into information. Further, portraying climate change problems based on the perceptions with fabricated stories creates chaos to the audiences. More importantly, the progress and difficulties in translating climate science by the media greatly influence climate adaptation and mitigation actions (Boykoff 2010). Therefore, the media's role is to clearly understand the scientific findings and translate the information to the general public.

Consequently, the capacity building of journalists in climate science is essential in the region to address the gaps in delivering climate change information to the general public. In Sikkim, during 2017, a media workshop on "Climate Change Reporting" was jointly organized by the IHCAP under the SDC, Centre for Media Studies (CMS), and DST, Government of India. The objective of the media workshop was to bring scientists/experts, international agencies, policymakers, and media persons in Sikkim together on the same platform for an in-depth understanding of various components of climate change and its impact on Sikkim and the Indian Himalaya as a whole (a report—media workshop on Climate Change Reporting 2017). Climate change is a global phenomenon with local impacts; therefore, local media still required more capacity building and specialization in climate science reporting.

Need for Maintaining the Science and Policy Linkages

The scientific analysis required the knowledge of observations and interpretations of the results. However, sometimes making plans and policies does not include evidence-based scientific findings in the decision-making process due to various reasons. Pullin et al. (2009) have concluded their study with the following observation: "*There is a mismatch between broad holistic questions typically posed in policy formation and narrow reductionist questions that are susceptible to the scientific method. This inhibits the two-way flow of information at the science-policy interface and weakens the impact of applied ecology on environmental policy.*" Therefore, it is crucial to make plans and policies supported by scientific analysis and evidence-based scientific observations. In the present era of evidence-based practices, climate policy decisions need to be developed using sectoral priorities supported by scientific evidence. In a nutshell, recent policy decisions may involve the engagement of mainly three major stakeholders—the government (decision maker), the sectoral bureaucrats, and the scientists. Further, policymaking is a complex process, where evidence-based scientific findings act as a vital element along with social, ethical, cultural, legal, economic, and political aspects that are often considered (Montuschi 2017).

Conclusions

In Eastern Himalaya, as discussed earlier, the studies on climate change impacts on glaciers and glacial lakes using field-based observations are scarce, resulting in a limited understanding of glacial melt processes and associated glacial hazards. In Sikkim Himalaya, the little field-based research about glacier retreat and lake formation warrants policymakers' urgent attention to formulate an effective plan for capacity building to study the glaciers and glacial lakes. Further, due to the risk imposed by harsh weather conditions in the glacier, the field-based observations on glaciers and glacial lakes in Sikkim Himalaya are very few and far. Besides, there are insufficient meteorological observatories to developed climate–glacier relationships and projections. Therefore, the following suggestions have been proposed to have wholesome climate change understanding and effective management of glaciers and glacial lakes in Sikkim Himalaya:

- Capacity building for the local researchers on the dynamics of the glaciers and glacial lakes.
- Increase the number of meteorological observatories for in situ-/field-based monitoring of the glacial lakes.
- Effective mitigation interventions in potentially vulnerable lakes (siphoning of impounded water and installation of early warning system).
- Build linkages with the nationally coordinated missions like NMSHE, NMSKCC.
- Capacity building of the media persons for the appropriate translation of climate science to the general masses.
- Sensitization and awareness about the importance of glaciers, climate risks, and vulnerability, and glacial hazards
- Linkages need to be built between scientists, bureaucrats, and policymakers (science policy linkages).
- A dedicated center for climate change adaptation, management, and mitigations.

Studies also show that many glacial lakes are continuously expanding in size and volume. Therefore, both satellite-based monitoring and field-based monitoring of the glacial lakes for assessing hazard potential are needed. Besides, future research lies in the impact assessment and risk modeling for glacial lakes management. Likewise, the high outburst-sensitive lakes need in-depth investigations of the glaciological and geomorphological features for designing sustainable mitigation measures. Also, it is suggested to have continuous monitoring for moderate and low outburst-sensitive lakes. A detailed and comprehensive understanding is necessary to know the lake's formation and evolution since many glacial lakes are still continuously expanding in their size in Sikkim Himalaya.

Acknowledgements The authors acknowledge the Department of Science and Technology (DST), Government of India, and Land Revenue and Disaster Management Department (LRDMD), Government of Sikkim (GoS), for supporting the climate change activities and its mitigation measures in Sikkim. The authors would like to thank the DST, Vigyan Bhawan-Sikkim, and Research and Technology Development Centre (RTDC), Sharda University, Uttar Pradesh, for providing the infrastructural support. The overall support from Shri D.G. Shrestha, Director (Remote Sensing and Climate Change Division), and Dr. N.P. Sharma, Asst. Scientific Officer (RS), is deeply acknowledged.

References

A Report-Media Workshop on Climate Change Reporting (2017) Jointly organized by Indian Himalayas Climate Adaptation Programme (IHCAP) under Swiss Agency for Development and Cooperation (SDC), Centre for Media Studies (CMS) and Department of Science and Technology (NMSHE), Gangtok, Mar 25–27, 2017

Aggarwal S, Rai SC, Thakur PK, Emmer A (2017) Inventory and recently increasing GLOF susceptibility of glacial lakes in Sikkim, Eastern Himalaya. Geomorphology 295:39–54. https://doi.org/10.1016/j.geomorph.2017.06.014

Ageta Y, Pokhral AP (1999) Characteristics of mass balance components of summer accumulation type glacier in the Nepal Himalaya. Seppyo 45:81–105

Agrawal A, Tayal S (2015) Mass balance reconstruction since 1963 and mass balance model for East Rathong glacier, Eastern Himalaya, using remote sensing methods. Geogr Ann Series Phys Geogr 97:695–707. https://doi.org/10.1111/geoa.12109

Ali SN, Dubey J, Ghosh R, Quamar MF, Sharma A, Morthekai P, Dimri AP, Shekhar M, Arif M, Agrawal S (2018) High frequency abrupt shifts in the Indian summer monsoon since Younger Dryas in the Himalaya. Sci Rep 8:9287. https://doi.org/10.1038/s41598-018-27597-6

Anacona PI, Kinney J, Schaefer M, Harrison S, Wilson R, Segovia A, Mazzorana B, Guerra F, Farı´as D, Reynolds JM, Glasser NF (2018) Glacier protection laws: potential conflicts in managing glacial hazards and adapting to climate change. Ambio 47:835–845https://doi.org/10.1007/s13280-018-1043-x

Bahuguna IM, Kulkarni AV, Arrawatia ML, Shrestha DG (2001) Glacier Atlas of Tista Basin-Sikkim Himalaya. Technical Report No. SAC/RESA/MWRG-GLI/SN/16/2001. Space Applications Centre, Ahmedabad and State Remote Sensing Applications Centre, SSCS&T, Sikkim

Bajracharya SR, Maharjan SB, Shrestha F, Guo W, Liu S, Immerzeel W, Shrestha B (2015) The glaciers of the Hindu Kush Himalayas: current status and observed changes from the 1980s to 2010. Int J Water Resour D 31(2):161–173

Bajracharya SR, Shrestha B (2011) The status of glaciers in the Hindu Kush–Himalayan region. International Centre for Integrated Mountain Development, ICIMOD, Kathmandu

Barnett TP, Adam JC, Lettenmaier DP (2005) Potential impacts of warming climate on water availability in snow-dominated regions. Nature 438:303–309. https://doi.org/10.1038/nature04141

Barry RG, Chorley RJ (1998) Atmosphere, weather and climate. Routledge Basin at the Himalayan Scale. Effect of scale on interpretation and management of the sediment and water quality. Int Assoc Hydrol Sci 226:17

Basnett S, Kulkarni AV, Bolch T (2013) The influence of debris cover and glacial lakes on the recession of glaciers in Sikkim Himalaya, India. J Glaciol 59(218):1035–1046. shttps://doi.org/10.3189/2013jog12j184

Begam S, Sen D (2019) Mapping of moraine dammed glacial lakes and assessment of their areal changes in the central and eastern Himalayas using satellite data. J Mt Sci 16(1):77–94. https://doi.org/10.1007/s11629-018-5023-1

Benn DI, Bolch T, Hands K, Gulley J, Luckman A, Nicholson LI, Quincey DJ, Thompson S, Toumi R, Wiseman S (2012) Response of debris-covered glaciers in the Mount Everest region to recent warming and implications for outburst flood hazards. Earth Sci Rev 114:156–174

Bhambri R, Bolch T, Kawishwar P, Dobhal DP, Srivastava D, Pratap B (2013) Heterogeneity in glacier response in the upper Shyok valley, northeast Karakoram. Cryosphere 7:1385–1398. https://doi.org/10.5194/tc-7-1385-2013

Bhattacharya A, Ghosh S, Mukherjee K (2018) Multi-decadal Mass Budget and Area Change of some Eastern Himalayan Glaciers (Nepal-Sikkim) using Remote sensing Techniques. In: 4th International conference on recent advances in information technology (RAIT-2018). 978-1-5386-3039-6/18/$31.00©2018 IEEE.

Bhutia UD (2019) Dhara Vikas- Reviving the Springs of Sikkim- Policy brief. (Edited by Ms. Neha Bharati, TERI). Published by Integrated Mountain Initiative, F-5, Ground Floor, Kailash Colony, New Delhi–110 048, India

Bolch T, Kulkarni AV, Kääb A, Huggel C, Paul F, Cogley JG, Frey H, Kargel JS, Fujita K, Scheel M, Bajracharya S, Stoffel M (2012) The state and fate of Himalayan glaciers. Science. https://doi.org/10.1126/science.1215828

Bolch T, Shea JM, Liu S, Azam FM, Gao Y, Gruber S, Immerzeel WW, Kulkarni A, Li H, Tahir AA, Zhang G (2019) Status and change of the cryosphere in the Extended Hindu Kush Himalaya Region. In: The Hindu Kush Himalaya Assessment, Springer, Cham, pp 209–255

Boykoff M (2010) Indian media representations of climate change in a threatened journalistic ecosystem. Clim Change 99:17–25. https://doi.org/10.1007/s10584-010-9807-8

Boykoff MT, Rajan, SR (2007) Signals and noise. Mass-media coverage of climate change in the USA and the UK. EMBO Rep 8(3):207–211. https://doi.org/10.1038/sj.embor.7400924

Breien H, Blasio FV De, Elverhoi A, Hoeg K (2008) Erosion and morphology of a debris flow caused by a glacial lake outburst flood, Western Norway. Landslides, pp 271–280. https://doi.org/10.1007/s10346-008-0118-3

Campbell JG (2005) Inventory of glaciers and glacial lake and the identification of potential glacial lake outburst floods (GLOFs) affected by global warming in the Mountains of India, Pakistan and China/Tibet Autonomous Region. Final report for APN project 2004-03-CMY-Campbell, International Centre for Integrated Mountain Development (ICIMOD) and Asia-Pacific Network for Global Change Research (APN).

Chaudhary P, Bawa KS (2011) Local perceptions of climate change validated by scientific evidence in the Himalayas. Biol Lett. https://doi.org/10.1098/rsbl.2011.0269

Chettri N, Sharma E, Shakya B, Thapa R, Bajracharya B, Uddin K, Oli KP, Choudhury D (2010) Biodiversity in the eastern Himalayas: status, trends and vulnerability to climate change. Climate change impact and vulnerability in the eastern Himalayas-Technical report 2 (p. 7). Kathmandu: ICIMOD

Clague JJ, Evans SG (2000) A review of catastrophic drainage of moraine-dammed lakes in British Columbia. Quaternary Sci Rev 19:1763–1783

Copland L, Sylvestre T, Bishop MP, Shroder JF, Seong YB, Owen LA, Bush A, Kamp U (2011) Expanded and Recently Increased Glacier Surging in the Karakoram. Arct Alp Res 43(4):503–516. https://doi.org/10.1657/1938-4246-43.4.503

Costa JE, Schuster RL (1988) The formation and failure of natural dams. Geol Soc Am Bull 100:1054–1068

Debnath M, Sharma MC, Syiemlieh, HJ (2019) Glacier Dynamics in Changme Khangpu Basin, Sikkim Himalaya, India, between 1975 and 2016. Geosci 9(6):259. https://doi.org/10.3390/geosciences9060259

DST (2010a) National Mission for Sustaining the Himalayan Ecosystem Under National Action Plan on Climate Change-Mission Document. Department of Science & Technology (DST), Ministry of Science & Technology, Govt of India, New Delhi. Available at: http://dst.gov.in/sites/default/files/NMSHE_June_2010.pdf. Accessed 20 May 2020

DST (2010b) National Mission on Strategic Knowledge for Climate Change (2010) Under National Action Plan on Climate Change-Mission Document. Department of Science & Technology (DST), Ministry of Science & Technology, New Delhi. Available at: http://dst.gov.in/sites/default/files/NMSKCC_July_2010.pdf). Accessed 20 May 2020

DST-NMSHE (2018–2019) Climate vulnerability assessment for the Indian Himalayan region using a common framework. Report submitted by Indian Institute of Technology-Guwahati & Indian Institute of Technology-Mandi. In collaboration with Indian Institute of Science-Bengaluru under the project 'Capacity Building on Climate Change Vulnerability Assessment in the States of Indian Himalayan Region'

Dyurgerov MB, Meier MF (1997) Year to year fluctuations of global mass balance of small glaciers and their contribution to sea level. Arct Alp Res 29(4):392–402

Dyurgerov MB, Meier MF (2005) Glaciers and the changing earth system: a 2004 snapshot. Occasional Paper, vol 58

Finsterwalder R (1935) On the map of the Zemu Glacier. Himalayan J 7:125–138

FSI (2017) Forest Cover. In: India State of forest report, forest survey of India, Ministry of Environment, Forest & Climate Change (MoEFCC)

Gardelle J, Berthier E, Arnaud Y, Kääb A (2013) Region-wide glacier mass balances over the Pamir-Karakoram-Himalaya during 1999–2011. Cryosphere. https://doi.org/10.5194/tc-7-1263-2013

Garg PK, Shukla A, Jasrotia AS (2019) On the strongly imbalanced State of glaciers in the Sikkim, eastern Himalaya, India. Sci Total Environ. https://doi.org/10.1016/j.scitotenv.2019.07.086

Gupta RC, Das M (2006) Recent trends in temperature (cold wave conditions) and rainfall over Gangtok city. Lett Editor Mausam 57(3):511–512

IPCC (2014) Summary for policymakers. Climate Change 2014: impacts, adaptation, and vulnerability. part a: global and sectoral aspects. In: Field CB, Barros VR, Dokken DJ, Mach KJ, Mastrandrea MD, Bilir TE, Chatterjee M, Ebi KL, Estrada YO, Genova RC, Girma B, Kissel ES, Levy AN, MacCracken S, Mastrandrea PR, White LL (eds) Contribution of working group II to the fifth assessment report of the intergovernmental panel on climate change. Cambridge University Press, Cambridge, United Kingdom and New York, NY, USA, pp 1–32

Jogesh A, Dubash NK (2015) State-led experimentation or centrally-motivated replication? A study of state action plans on climate change in India. J Integr Environ Sci 12(4):47–266. https://doi.org/10.1080/1943815X.2015.1077869

Kääb A, Berthier E, Nuth C, Gardelle J, Arnaud Y (2012) Contrasting patterns of early twenty-first-century glacier mass change in the Himalayas. Nature. https://doi.org/10.1038/nature11324

Keller TR, Hase V, Thaker J, Mahl D, Schäfer MS (2019) News media coverage of climate change in India 1997–2016: using automated content analysis to assess themes and topics. Environ Commun 17. https://doi.org/10.1080/17524032.2019.1643383

Krishna AP (2005) Snow and glacier cover assessment in the high mountains of Sikkim Himalaya. Hydrol Process 19(12):2375–2383

Kumar P, Sharma MC, Saini R, Singh GK (2020) Climatic variability at Gangtok and Tadong weather observatories in Sikkim, India, during 1961–2017. Sci Rep 10:15177. https://doi.org/10.1038/s41598-020-71163-y

Kumar R, Sharma RK, Pradhan P, Sharma N, Shrestha DG (2020a) Melt runoff characteristics and hydro-meteorological assessment of East Rathong Glacier in Sikkim Himalaya, India. Earth Systems and Environment. https://doi.org/10.1007/s41748-020-00168-4

Liu X, Chen B (2000) Climate warming in the Tibetan plateau during recent decades. Int J Climatol 20:1729–1742

Luitel, KK, Shrestha DG, Sharma NP, Sharma RK (2012) Impact of Climate change on East Rathong Glacier in Rangit Basin, West Sikkim. In Climate change in Sikkim-Patterns, impacts and initiatives. Information and Public Department, Government of Sikkim, Gangtok

Mayewski PA, Jeschke PA (1979) Himalayan and Trans-Himalayan Glacier Fluctuations since AD 1812. Arctic and Alpine Research 11(3):267–87. Accessed 24 Mar 2021. https://doi.org/10.2307/1550417

Montuschi E (2017) Using science, making policy: what should we worry about? Euro Jnl Phil Sci 7:57–78. https://doi.org/10.1007/s13194-016-0143-3

Mool PK, Bajracharya SR (2003) Tista Basin, Sikkim Himalaya inventory of glaciers, glacial lakes, and the identification of potential glacial lake outburst floods (GLOFs) affected by global warming in the mountains of Himalayan region. ICIMOD, Kathmandu, p 145

Mukherji A, Molden D, Nepal S, Rasul G, Wagnon P (2015) Himalayan waters at the crossroads: issues and challenges. Int J Water Resour Dev 31(2):151–160. https://doi.org/10.1080/07900627.2015.1040871

Nath SK, Vyas M, Pal I, Sengupta P (2005). A seismic hazard scenario in the Sikkim Himalaya from seismotectonics, spectral amplification, source parameterization, and spectral attenuation laws using strong motion seismometry. J Geophys Res 110 (B01301). https://doi.org/10.1029/2004JB003199

Nijampurkar VN, Bhandari N, Borole DV, Bhattacharya U (1985) Radiometric chronology of Changme Khangpu Glacier Sikkim. J Glaciol 31(107):28–33

Paul F, Kaab A, Haeberli W (2007) Recent glacier changes in the Alps observed by satellite: consequence for future monitoring strategies. Glob Planet Change 56:111–122. https://doi.org/10.1016/j.gloplacha.2006.07.007

Petrakov DA, Tutubalina OV, Aleinikov AA, Chernomorets SS, Evans G, Kidyaeva, VM, Krylenko IN, Norin SV, Shakhmina S, Seynova IB (2011) Monitoring of Bhaskara Glacier lakes (Central Caucasus, Russia) and modeling of their potential outburst. Nat Hazards. https://doi.org/10.1007/s11069-011-9983-5

Pullin AS, Knight TM, Watkinson AR (2009) Linking reductionist science and holistic policy using systematic reviews: unpacking environmental policy questions to construct an evidence-based framework. J Appl Ecol 46(5):970–975. https://doi.org/10.1111/j.1365-2664.2009.01704.x

Puri VMK (1999) Glaciohydrological and suspended sediment load studies in the meltwater channel of Changme Khangpu glacier. In: Symposium on snow, ice and glaciers-Himalayan prospective, Lucknow, India, vol 1

Quincey DJ, Glasser NF, Cook SJ, Luckman A (2015) Heterogeneity in Karakoram glacier surges. J Geophys Res Earth Surf 120:1288–1300. https://doi.org/10.1002/2015JF003515

Racoviteanu AE, Arnaud Y, Williams MW, Manley WF (2015) Spatial patterns in glacier characteristics and area changes from 1962 to 2006 in the Kanchenjunga-Sikkim area, eastern Himalaya. The Cryosphere 9:505–523. https://doi.org/10.5194/tc-9-505-2015

Raina VK, Srivastava D (2008) Glacier Atlas of India. Geological Society of India (GSI), Bangalore, p 315

Raina VK, Srivastava D (2014) Glacier Atlas of India. GSI Publications

Raj KBG (2010) Remote sensing based hazard assessment of glacial lakes: a case study from Zanskar basin, Jammu and Kashmir. Geom Nat Hazards Risk 1:339–347

Raj KBG, Kumar KV, Remya SN (2012) Remote sensing based inventory of glacial lakes in Sikkim Himalaya: semi-automated approach using satellite data. Geom Nat Hazards Risk. https://doi.org/10.1080/19475705.2012.707153

Raj KBG, Remya SN, Kumar KV (2013) Remote sensing-based hazard assessment of glacial lakes in Sikkim Himalaya. Curr Sci 104(3):359–364

Rashid I, Majeed U (2020) Retreat and geodetic mass changes of Zemu Glacier, Sikkim Himalaya, India, between 1931 and 2018. Reg Environ Change 20(4). https://doi.org/10.1007/s10113-020-01717-3

Richardson SD, Reynolds JM (2000) An overview of glacial hazards in the Himalayas. Quat Int 65:31–47

Sangewar CV, Shukla SP (2009) Inventory of the Himalayan glaciers. Geological Survey of India, Special Publication No. 34, ISSN 0254–0436, Kolkata

SAPCC-Sikkim (2014) The Sikkim State action plan on climate change report. Published by the Government of Sikkim, Science and Technology Department

Sattar A, Goswami A, Kulkarni AV (2019) Hydrodynamic moraine-breach modeling and outburst flood routing-A hazard assessment of the South Lhonak lake, Sikkim. Sci Total Environ 668:362–378. https://doi.org/10.1016/j.scitotenv.2019.02.388

Sharma E, Chettri N, Tse-ring K, Shrestha AB, Jing F, Mool P, Eriksson M (2009) Climate change impacts and vulnerability in the eastern Himalayas. ICIMOD, Kathmandu, pp 5–6

Sharma RK, Kumar R, Shrestha DG (2021)Suspended sediment dynamics and associated hydro-meteorological interrelations in East Rathong Glacier, Eastern Himalaya, India. Mater Today Proc. https://doi.org/10.1016/j.matpr.2021.01.082

Sharma RK, Pradhan P, Sharma NP, Shrestha DG (2018) Remote sensing and in situ-based assessment of rapidly growing South Lhonak glacial lake in eastern Himalaya, India. Nat Hazards 93:393–409. https://doi.org/10.1007/s11069-018-3305-0

Sharma RK, Shrestha DG (2016) Climate perceptions of local communities validated through scientific signals in Sikkim Himalaya. India. Environ Monit Assess 188:578. https://doi.org/10.1007/s10661-016-5582-y

Shea JM, Immerzeel WW, Wagnon P, Vincent C, Bajracharya S (2015a) Modelling glacier change in the Everest region, Nepal Himalaya. Cryosphere. https://doi.org/10.5194/tc-9-1105-2015

Shrestha AB, Wake CP, Mayewski PA, Dibb JE (1999) Maximum temperature trends in the Himalaya and its vicinity: An analysis based on temperature records from Nepal for the period 1971-94. J Clim 12(9):2775–2786. https://doi.org/10.1175/1520-0442(1999)012<2775:MTTITH>2.0.CO;2

Shukla A, Garg PK, Srivastava A (2018) Evolution of glacial and high-altitude lakes in the Sikkim, Eastern Himalaya Over the Past Four Decades (1975–2017). Front Environ Sci 6:81. https://doi.org/10.3389.fenvs.2018.00081

Shukla PN, Bhandari N, Nijampurkar VN, Rao DK, Puri VMK, Sharma S (1983) Ice accumulation rate in Changme-Khangpu glacier, Sikkim. Proc Indian Acad Sci (earth Planet Sci) 92(3):255–260

Sikkim Primary Census Abstract (2011) Census of India. Directorate of Census Operations, Sikkim

STP (2018) State Tourism Policy-2018. Department of tourism and civil aviation, Government of Sikkim, Gangtok, Sikkim

Tambe S, Arrawatia ML, Bhutia NT, Swaroop B (2011) Rapid, cost-effective and high resolution assessment of climate-related vulnerability of rural communities of Sikkim Himalaya. India Curr Sci 101(2):165–173

TanDong Y, ZhiGuo L, Wei Y, XueJun G, LiPing Z, ShiChang K, YanHong W, WuSheng Y (2010) Glacial distribution and mass balance in the Yarlung Zangbo River and its influence on lakes. Chinese Sci Bull 55(20):2072–2078. https://doi.org/10.1007/s11434-010-3213-5

Telwala Y, Brook BW, Manish K, Pandit MK (2013) Climate-induced elevational range shifts and increase in plant species richness in a Himalayan biodiversity epicentre. PloSOne. https://doi.org/10.1371/journal.pone.0057103

Tse-ring K, Sharma E, Chettri N, Shrestha A (2010) Climate change vulnerability of mountain ecosystems in the eastern Himalayas; climate change impact and vulnerability in the eastern Himalayas-synthesis report (p. 6). Kathmandu: ICIMOD

Venkatesh TN, Kulkarni AV, Srinivasan J (2011) Relative effect of slope and equilibrium line altitude on the retreat of Himalayan glaciers. Cryos Discuss 5(5):2571–2604. https://doi.org/10.5194/tcd-52571-2011

Westoby MJ, Glasser NF, Brasington J, Hambrey MJ, Quincey DJ, Reynolds JM (2014) Modelling outburst floods from moraine-dammed glacial lakes. Earth-Sci Rev 134:137–159. https://doi.org/10.1016/j.earscirev.2014.03.009

Worni R, Huggel C, Clague JJ, Schaub Y, Stoffel M (2014) Coupling glacial lake impact, dam breach, and flood processes: a modeling perspective. Geomorphology 224:161–176

Worni R, Huggel C, Stoffel M (2013) Glacial lakes in the Indian Himalayas-From an area-wide glacial lake inventory to onsite and modeling based risk assessment of critical glacial lakes. Sci Total Environ. https://doi.org/10.1016/j.scitotenv.2012.11.043

Zhou Y, Li Z, Li J, Zhao, R, Ding X (2018) Glacier mass balance in the Qinghai–Tibet Plateau and its surroundings from the mid-1970s to 2000 based on Hexagon KH-9 and SRTM DEMs. Remote Sens Environ 210:96–112. https://doi.org/10.1016/j.rse.2018.03.02

Chapter 9

Trends in Observed Discharge (1963–2012) at Rampur Hydrological Station of Satluj River Basin, Western Indian Himalaya

Riyaz Ahmad Mir and Sanjay K. Jain

Abstract Trend analysis of river discharge of Himalayan basins is essential for water resource management and policymaking because of emerging climate change. In this study, trends in the observed discharge (Rampur hydrological station −1021 m AMSL) of the Satluj River in the Western Indian Himalaya (WIH) over 5 decades (1963–2012) were evaluated. The data were classified into average annual discharge (AAD), minimum annual discharge (MnAD), maximum annual discharge (MxAD), average winter discharge (AWD), average pre-monsoon discharge (APrD), average monsoon discharge (AMD), average post-monsoon discharge (APoD), average accumulation discharge (AAcD), and average ablation discharge (AAbD) as per the prevailing climatic conditions of the basin. The Sen's slope and Mann–Kendall tests were used for identifying the direction, magnitude (Qi), and significance (Zs) of trends. The findings indicated declining trends in annual and seasonal discharge during the last 5 decades (50 years). The trends were statistically significant (95% confidence) for AAD, MxAD, AMD, APoD, and AAbD. The discharge is declining at a rate of 0.60 m^3 s^{-1} $year^{-1}$ (AAD), 1.98 m^3 s^{-1} $year^{-1}$ (MxAD), 1.78 m^3 s^{-1} $year^{-1}$ (AMD), 0.56m^3 s^{-1} $year^{-1}$ (APoD), and 1.0 m^3 s^{-1} $year^{-1}$ (AAbD). The highest and lowest decline is observed in MxAD and APoD during the study period. The AAD is strongly correlated (R^2) and controlled by the AAbD (0.99), AMD (0.91), APoD (0.58), APrD (0.50), AAcD (0.46), and AWD (0.31). Nevertheless, there exists a sequential pattern in the decline of annual and seasonal discharge in terms of R^2, Qi, and Zs values. The declining discharge at the station may have serious implications for water resource management and policymaking in the area.

Keywords Trend analysis · Discharge · Rampur station · Satluj River · Western Indian Himalaya

R. A. Mir (✉)
Department of Earth Sciences, Indian Institute of Technology, Roorkee, Uttarakhand, India

Geological Survey of India, Srinagar, Jammu and Kashmir (UT), India

S. K. Jain
Water Resource System Division, National Institute of Hydrology, Roorkee, Uttarakhand, India

S. Rani and R. Kumar (eds.), *Climate Change*, Springer Climate,
https://doi.org/10.1007/978-3-030-92782-0_9

Introduction

In the Himalayan region, analyzing the general trends in time series data of river flow and discharge is very essential to understand, quantify, and assess the effects of ongoing climate change on the water resources, and to evaluate the general policy for its management (Kundzewicz et al. 2014; Coulthard et al. 2005; Vliet et al. 2012; Abghari et al. 2013; Afshar et al. 2016). The river discharge production is known to be dependent on several factors including both natural factors such as precipitation, base-flow contribution, evaporation, glaciers and snow cover resources as well as other related anthropogenic factors (Kundzewicz et al. 2014; Mir et al. 2015, 2017, 2018) with global warming as the predominant factor (Mir et al. 2017; Saadi et al. 2017; Mir et al. 2018). The global warming effect due to the huge and enhanced emission of greenhouse gases has led to a significant increase in the atmospheric temperature and erratic behavior in precipitation in the world which in turn has undesirably affected the global hydrologic cycle (Bates et al. 2008; Biemans et al. 2013; Abdi et al. 2016; Lobanova et al. 2018; Ahmed et al. 2018).

The river discharge responds very sensitively to the variations in precipitation (Biemans et al 2013; Lobanova et al. 2018). All over the world and at the Indian level as well, the changes in discharge have been recorded over the several decades in response to climate change (Mimikou and Fotopoulos 2005; Xu ct al. 2011; Vicuña et al. 2011). In the Indian region, for example, Bouwer et al. (2006) studied the discharge pattern in the Krishna River basin and reported an annual runoff varying at 14–34 mm from 1901 to 2000 as a result of climatic changes. Bhutiyani et al. (2008) reported significantly declining trends in average annual and monsoon discharge of Beas River and insignificant declining trends for the Ravi River from 1947 to 2000. But, they recorded significantly increasing trends in winter discharge of the Chenab River during the same period. Nune et al. (2012) showed a large insignificant decline in stream flow of the Himayat Sagar catchment in Southern India. Panda et al. (2013) have investigated the stream flow trends in the Mahanadi River basin in Eastern India. Abeysingha et al. (2016) have also recorded a gradual declining trend of annual, monsoonal, and winter seasonal stream flows of the Gomati River, India. But, it is notable to mention that there are still very sparse studies on trend analysis of river discharge in India attributed mainly to the non-availability of long-term datasets.

All over the world, studies have suggested an increase in the atmospheric water vapor content, retreat of glaciers, changes in biodiversity, and extensive droughts (Jain et al. 2010b; Mir et al. 2018; Stagge et al. 2015; Jain and Mir 2019; Mir 2021). These changing patterns of the natural factors have a direct influence on the river water quantity as well as its quality. Nevertheless, the construction of large hydraulic structures such as dams and weirs also cause a significant variation and changes in the river discharge patterns (Shiru et al. 2018; Qutbudin et al. 2019). Thus, under this situation, the sustainable management of water resources in the Himalayan region is more challenging than ever (Hirsch et al. 2015; Guo et al. 2018). Furthermore, to assess more precisely the climatically induced hydrological changes, a detailed analysis of the data from near-pristine drainage basins at a small scale than regional scale is required. The near-pristine basins are generally not affected largely by human activities like urbanization, deforestation, reservoirs, changes in drainage systems, water abstraction, and river engineering (Kundzewicz et al. 2014), while the studies on small-scale basins lead to perseveration of spatial information on the variable (Rai et al. 2010).

Several parametric and non-parametric models and approaches have been employed for the assessment of river discharge trends (Mir et al. 2015a, b, c; Tosunoglu and Kisi 2017). But, the Mann–Kendall (MK) (Mann 1945; Kendall 1975) and Sen's slope tests (Sen 2011) are widely used for trend analysis (Tabari et al. 2015; Demir and Kisi 2016). These methods have numerous advantages such as handling of missing data, the requirement of only a few assumptions, and independence of the data distribution (Kisi 2015; Öztopal and Sen 2017; Wu and Qian 2017). Moreover, the influence of autocorrelation, if any, in the dataset is generally removed through pre-whitening techniques (Su et al. 2006; Mir et al. 2015a; Wang et al. 2017; Sanikhani et al. 2018; Malik and Kumar 2019).

Considering the importance of trend analysis of river discharge, the main objectives of the present study are to evaluate any variability and presence of possible trends in the observed discharge at Rampur hydrological station in the Satluj basin located in the Western Indian Himalaya (WIH) during the period from 1963–2012. The study would be useful for water resource management in the study area.

Study Area

The present study has been done in the Satluj River basin of WIH. The Satluj river originates at an elevation of more than >4500 m above mean sea level (AMSL) from the Tibetan Plateau in the southern area of Mount Kailash. Here, in this area, it has a very low discharge before it enters into India in Himachal Pradesh. The Satluj basin gains river discharge in Himachal Pradesh in spring season after joining Spiti tributary whose catchment receives tremendous snowfall in winters owing to the western weather disturbances (Singh and Kumar 1997). In India, the Satluj basin encompasses 22,305 km^2 area with elevation varying between 500 and 7000 m AMSL. It covers a length of near fifteen hundred (1500) kilometers up to Asia's second-largest dam, Bhakra of Himachal Pradesh (Fig. 9.1). In addition, the basin is known to be covered by 65% of regular snow in winters and 12% of permanent glaciers (Singh and Bengtsson 2004), receiving its discharge mainly due to snowmelt from upper reaches during spring season (Jain et al. 1998; Singh and Jain 2002).

The study area witnesses a tropical and warm temperate climates in lower parts (receiving only rain) (Mir et al. 2015a, a, c), a cold temperate climate in middle parts (receiving both rain/snow), very cold in the upper parts, and permafrost in the uppermost parts. The average annual rainfall in the Outer, Middle, and Greater Himalayan ranges of the basin, respectively, is about 1300, 700, and 200 mm (Singh and Kumar 1997).

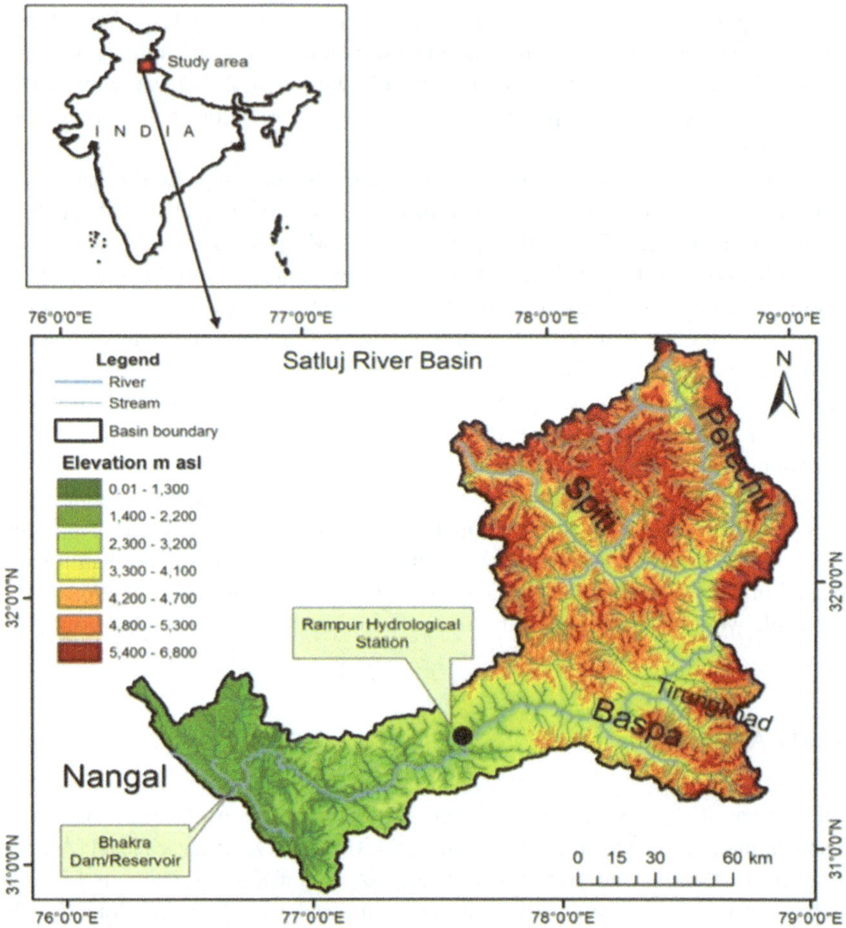

Fig. 9.1 Location and extent of the Satluj River basin with Rampur hydrological station. ASTER DEM in the background shows the elevation of the area

People in this basin are dependent on agriculture/farming income and, therefore, cultivate almost all kinds of crops at various altitudes. More than 75% of population receives water through pipelines and the rest directly from seasonal streams and natural sources (Asian Development Bank ADB 2011). About 5.8 and 17.6% utilize water from perennial/seasonal streams and natural sources, respectively (ADB 2011). The area is represented by high elevations, structural hills, elevated terrain, and deep valleys. The area during recent decades has witnessed a continuous recession/loss of glaciers due to warming climate (Philip and Sah 2004; Berthier et al. 2007; Kulkarni et al. 2007; Bolch et al. 2012; Mir et al. 2013, 2014, 2015a, b, c, 2017).

Hydroelectric Potential

The study area has a very vast potential for hydroelectricity generation owing to the promising precipitation and river flow throughout the year. It is supported by various hydroelectric projects at various places whose dams efficiently hold excess water and prevent the downstream areas from floods. The prominent projects in the basin include Nathpa Jhakri Hydro-electric Project (NJHEP), 1500 MW (WAPCOS 2007), the Rampur Hydroelectric Project (412 MW), Luhri Hydro-electric Project (700 MW) located in Shimla, Kullu, and Mandi districts of Himachal Pradesh (WAPCOS 2007), Khab Dam Project (1020 MW), located in district Kinnaur, Baspa Hydro-electric Project (300 MW), Dhamwari Sunda Hydro-electric Project (70 MW), Karcham Wangtoo Hydro-electric Project (1000 MW) (Economic Survey 2007). Hence, it is essential to understand the emerging changes in river discharge which is directly related to these energy resources and its management in the area.

Materials and Methods

The daily river discharge data (m^3 s^{-1}) of the Rampur hydrological station located at an elevation of 1021 m AMSL (Fig. 9.1) of Satluj River were obtained from the Bhakra Beas Management Board (BBMB) of India for a period from 1963–2012 (50 years). The quality of the data was assessed and checked statistically as well as visually. First, the monthly average discharge data were calculated for each month from January to December for each year. From the monthly average data of each year, the minimum, maximum, mean, and total (sum) annual discharge data were calculated.

Further analysis and processing of daily river discharge data were done as per the procedure discussed elsewhere in Mir et al. 2015a a, c, 2017. The months including November, December, January, and February (NDJF) were classified as winter season, March, April, and May (MAM) as pre-monsoon season, July and August (JA) as monsoon season, and the September (SO) as post-monsoon season. From October to March (ONDJFM), an accumulation period was defined, and from April to September (AMJJAS), an ablation season was defined to reflect and understand the trends in discharge. During the winter season, minimum river flow is observed (Jain et al. 2010a). In addition, the different periods classified in the investigation include the average annual discharge (AAD), minimum annual discharge (MnAD), maximum annual discharge (MxAD), average winter discharge (AWD), average pre-monsoon discharge (APrD), average monsoon discharge (AMD), average post-monsoon discharge (APoD), average accumulation discharge (AAcD), and average ablation discharge (AAbD) respectively. The trends were estimated from the descriptive statistical parameters employing simple linear regression analysis, Mann–Kendall test, and Sen's slope estimator test (Mann 1945; Kendall 1975).

Descriptive and Bivariate Statistics

The statistical parameters of the investigation (i.e., mean, standard deviation, and minimum, maximum, median, skewness, kurtosis, and coefficient of variation (CV) of discharge) were computed and presented in (Table 9.1). In addition, the bivariate statistical analysis (correlation) between the annual average discharge and other components of discharge was also determined and presented in this study.

Trend Analysis

The slope of the trends obtained from ordinary least square (OLS) indicate the mean temporal change of the studied variable. Its positive and negative values, respectively, indicate increasing and decreasing trends of the slope. The significance of the trends was achieved and understood using Mann–Kendall test (Mann 1945; Kendall 1975). To avoid autocorrelation, a pre-whitening procedure was performed. Subsequently, the Mann–Kendall test was applied on the "pre-whitened" series obtained as $(x_2 - r_1x_1, x_3 - r_1x_2, \ldots.., x_\mathrm{n} - r_1x_\mathrm{n} - 1)$ where $x_1, x_2, x_3 \ldots x_\mathrm{n}$ are the data points of the series.

The Mann–Kendall test is defined as

$$S = \sum_{k=1}^{N-1} \sum_{j=k+1}^{N} \mathrm{sgn}(x_j - x_k) \tag{9.1}$$

where N is number of data points, and the value of $(X_j - X_k)$ is computed as

$$\mathrm{sgn}\left(x_j - x_k\right) = \left\{ \begin{array}{ll} 1 & \mathrm{if}\, x_j - x_k > 1 \\ 0 & \mathrm{if}\, x_j - x_k = 1 \\ -1 & \mathrm{if}\, x_j - x_k < 0 \end{array} \right\} \tag{9.2}$$

This statistic represents the numbers of positive differences minus the number of negative differences for all the differences considered. The variance of the statistic S is calculated as

$$\mathrm{Var}(S) = \frac{\left[N(N-1)\,(2N+5) - \sum_{k=1}^{n} \mathrm{tk}(\mathrm{tk}-1)(2\mathrm{tk}+5)\right]}{18} \tag{9.3}$$

where N is the total number of data points in the data series, n and tk are the number of tied groups in the data series and numbers of tied groups in the kth tied group, respectively. Mann–Kendall "Z-statistic" is computed as

Table 9.1 Summary statistics (min, max, mean, St. dev, CV, skewness, and kurtosis), magnitude (Qi), and significance of trends (Zs) of average discharge during the classified periods obtained using Mann–Kendall (MK) and Sen's slope at Rampur stations, Satluj River

S. no	Periods	Classes	Id	Min	Max	Mean	St. dev	CV	Skewness	Kurtosis	Qi	Zs	P
1	Annual	Average annual discharge	AAD	44.18	123.66	83.89	15.13	5.46	0.13	0.32	0.60	**−2.75**	0.01
2		Minimum annual discharge	MnAD	500.08	1305.03	943.37	207.76	91.51	−0.004	−1.00	0.15	−1.01	0.31
3		Maximum annual discharge	MxAD	204.08	483.02	334.82	60.27	21.70	0.046	−0.23	1.98	**−2.89**	0.004
4	Winter season	Average winter discharge	AWD	76.81	133.93	101.05	13.87	3.81	0.16	−0.76	0.18	−0.95	0.34
5	Pre-monsoon season	Average pre-monsoon discharge	APrD	119.27	403.78	205.04	57.80	32.59	0.81	1.38	0.54	−0.48	**0.63**
6	Monsoon season	Average monsoon discharge	AMD	447.60	1128.07	804.54	169.25	71.21	0.014	−0.75	1.78	**−3.25**	0.001
7	Post–monsoon season	Average post-monsoon discharge	APoD	197.36	423.26	292.47	53.87	19.85	0.44	0.17	0.56	**−2.25**	0.02
8	Accumulation period	Average accumulation discharge	AAcD	86.73	145.15	114.05	15.17	4.04	0.017	−0.84	0.16	−1.68	0.09

(continued)

Table 9.1 (continued)

S. no	Periods	Classes	Id	Min	Max	Mean	St. dev	CV	Skewness	Kurtosis	Qi	Zs	P
9	Ablation period	Average ablation discharge	AAbD	316.78	841.33	555.59	110.74	44.15	0.14	−0.16	1.00	**−2.94**	0.003

Note The—sign indicates declining trends, whereas the + sign (if any) indicates increasing trends. The bold font Z-values indicate significant trends at 95% confidence level. The bold font *p*-values indicate the existence of no trend

$$Z = \begin{cases} \frac{S-1}{\sqrt{\mathrm{Var}(S)}} & \text{if } S > 0 \\ & \text{if } S = 0 \\ \frac{S+1}{\sqrt{\mathrm{Var}(S)}} & \text{if } S < 0 \end{cases} \tag{9.4}$$

Based on the Z-statistics (Zs) value, the null hypothesis of no trend is accepted or rejected at a 95% confidence interval. A Zs value of 1.96 is considered for the determination of the significance of the trend at a 95% confidence level. The Zs test statistic was used as a measure of the significance of the trend. Positive values of standard normal variable Zs indicated an increasing trend, whereas the negative Zs values showed decreasing trends.

The magnitude of trend in a time series was determined using a non-parametric method known as Sen's slope estimator (Sen 2011). The Sen's slope was calculated as

$$Q_i = \frac{x_j - x_k}{j - k} \; for \; i = 1, 2, \ldots., N, \tag{9.5}$$

where X_j and X_k are data values at times j and k ($j > k$), respectively. The median of these N values of Q_i is Sen's estimator of the slope. If N is odd, then Sen's estimator is computed by

$$Q_{\mathrm{med}} = Q[(N+1)/2] \tag{9.6}$$

If N is even, then Sen's estimator is computed by

$$Q_{\mathrm{med}} = \frac{1}{2}(Q[N/2]) + Q[(N+2)/2]) \tag{9.7}$$

Finally, Q_{med} is tested with a two-sided test at the 100(1 − α) % confidence interval, and the true slope may be obtained with the non-parametric test (Mir et al. 2015a).

In this work, the confidence interval was computed at two different confidence levels (α = 0.01 and α = 0.05) as follows

$$C_\alpha = Z_{1-\alpha/2}\sqrt{Var(S)} \tag{9.8}$$

where Var (S) has been defined in Eqs. (9.4) and (9.3) and $Z_{1-\alpha/2}$ is obtained from the standard normal distribution.

Then, $M_1 = (N-C_\alpha)/2$ and $M_2 = (N + C_\alpha)/2$ are computed. The lower and upper limits of the confidence interval, $Q_{\min}$ and $Q_{\max}$, are the M_1Th largest and the $(M_1 + 1)^{Th}$ largest of the N ordered slope estimates Q_i. If M_1 is not a whole number, the lower limit is interpolated. Correspondingly, if M_2 is not a whole number, the upper limit is interpolated (Mir et al. 2015a). The magnitude of slope (Qi), i.e., change per unit time was estimated using a simple non-parametric procedure developed by

Sen (2011). Moreover, for the data without any trend, the probability values were selected close to 0.5.

Results

Descriptive Statistics of River Discharge

The results indicated that the coefficient of skewness varied from −0.004 to 0.81; kurtosis varied between −1.0 and 1.38. Normally, the coefficient of skewness and kurtosis must be equal to 0 and 3 so that the time series data may be considered normally distributed. However, these statistical attributes indicated that the time series data are positively skewed except for MnAD that is negatively skewed and not normally distributed. Furthermore, the coefficient of variation was also calculated to analyze the variability of river discharge. The coefficient of variation varied between 3.8 and 91.5%. The average variation of the discharge of the Satluj River was found to be 32.7%.

Variability in Annual and Seasonal Mean Discharge

The river discharge data were analyzed as annual and seasonal, minimum, maximum, average, and sum/total for a period from 1963–2012 (50 years). The AAD varied from 204.4 to 483.0 $m^3 s^{-1}$ with a mean value of 334.8 $m^3 s^{-1}$, whereas the average total discharge (ATD) varied between 2449 $m^3 s^{-1}$ (2004) and 5796.2 $m^3 s^{-1}$ (1973) with a mean of 4017.8 $m^3 s^{-1}$. Seasonally, the results indicated that the AWD varied from 76.8 (2012) to 133.9 (1999) $m^3 s^{-1}$ with a mean value of 101.0 $m^3 s^{-1}$, whereas the average total winter discharge (ATWD) varied from 307.2 to 535.7 $m^3 s^{-1}$ with a mean value of 404.2 $m^3 s^{-1}$. Similarly, the APrD varied from 119.2 (2012) to 403.7(1973) $m^3 s^{-1}$ with a mean value of 205.0 $m^3 s^{-1}$, whereas the average total pre-monsoon discharge (ATPrD) varied from 357.8 to 1211.3 $m^3 s^{-1}$ with a mean value of 615.1 $m^3 s^{-1}$. The AMD varied from 447.6 (2004) to 1228.0(1969) $m^3 s^{-1}$ with a mean value of 805.4 $m^3 s^{-1}$, whereas the average total monsoon discharge (ATMD) varied from 1342.7 to 3384.2 $m^3 s^{-1}$ with a mean value of 2413.6 $m^3 s^{-1}$. The APoD varied from 197.3 (2004) to 423.3 (1983) $m^3 s^{-1}$ with a mean value of 292.4 $m^3 s^{-1}$, whereas the average total post-monsoon discharge (ATPoD) varied from 394.7 (2012) to 846.5 (1998) $m^3 s^{-1}$ with a mean value of 584.9 $m^3 s^{-1}$. Furthermore, the AAcD varied from 86.7 to 145.1 $m^3 s^{-1}$ with a mean value of 114.0 $m^3 s^{-1}$, whereas the average total accumulation discharge (ATAcD) varied from 520.3 (2004) to 870.8 (1973) $m^3 s^{-1}$ with a mean value of 684.3 $m^3 s^{-1}$. Similarly, the AAbD varied from 316.7 to 841.3 $m^3 s^{-1}$ with a mean value of 555.5 $m^3 s^{-1}$, whereas the average total ablation discharge (ATAbD) varied from 1900.7 to 5047.9 $m^3 s^{-1}$ with a mean value of 3333.5

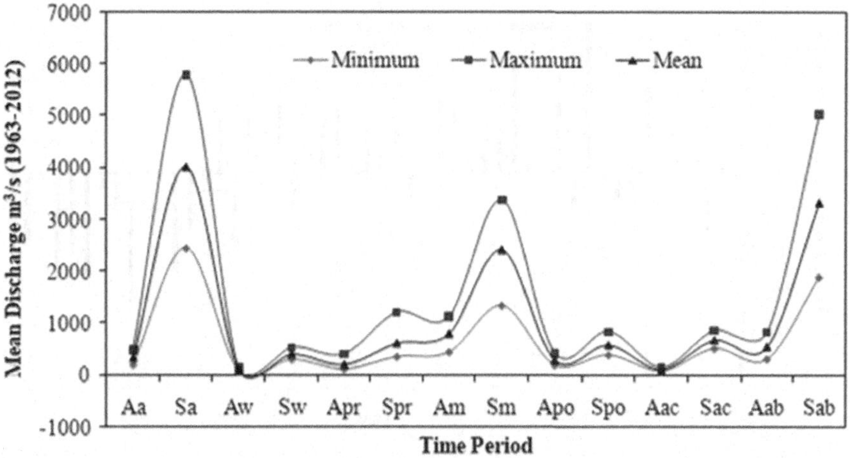

Fig. 9.2 Variation of river discharge calculated from 1963–2012 for different periods of Satluj River basin at Rampur hydrological station. (*Note* Aa—average annual, Sa—sum annual, Aw—average winter, Sw—sum winter, Apr—average pre-monsoon, Spr—sum pre-monsoon, Am—average monsoon, Sm—sum monsoon, Apo—average post-monsoon, Spo—sum post-monsoon, Aac—average accumulation, Sac—sum accumulation, Aab—average ablation, and Sab—sum ablation)

$m^3\ s^{-1}$. In the Satluj River basin, the highest discharge/flow is observed during the monsoon and ablation periods (Fig. 9.2).

Trends in Annual River Discharge

Trend analysis indicated a decreasing trend for the AAD, MnAD, and MxAD at the analyzed station. However, the R^2 values of AAD ($R^2 = 0.37$) and MxAD ($R^2 = 0.35$) are higher than the MnAD ($R^2 = 0.02$). The higher MxAD R^2 value of 0.35 is close to the R^2 value of 0.37 (AAD), and it, therefore, reflects a significant control of MxAD on the AAD data trends. Furthermore, the Z-statistics values for the AAD (p = 0.01) and MxAD ($p = 0.004$) are higher than a standard of Zs = 1.96 thereby indicating that the discharge during these two periods is declining significantly at a 95% confidence level (Table 9.1). The MxAD is observed to be declining at a higher rate of 1.98 $m^3\ s^{-1}\ year^{-1}$, whereas the AAD is declining at a lower rate of 0.60 $m^3\ s^{-1}\ year^{-1}$. On the other hand, the declining trend in MnAD are not statically significant, and the rate of decline of 0.15 $m^3\ s^{-1}\ year^{-1}$ for MnAD is also very small. The probability values (*p*-value) of 0.31 for MnAD are also very close to a standard value of 0.5 reflecting a very insignificant trend in it. Furthermore, from the linear plots (Fig. 9.3a, b, c), it is evident that the AAD, MnAD, and MxAd have been declining during the study period. Further, it is also clearly visible that the AAD has decreased from 1999 onward till 2012. The AAD and MxAD have been increasing

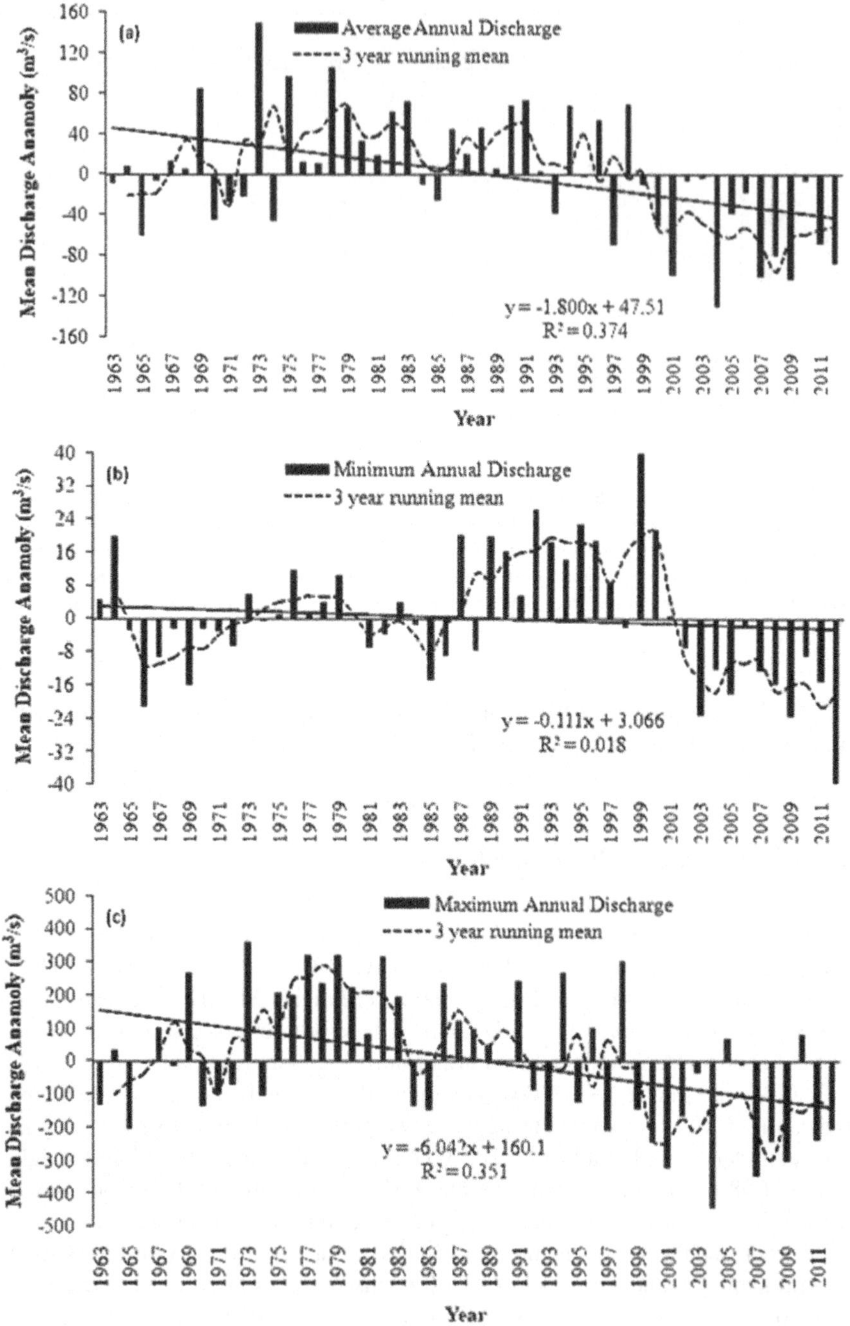

Fig. 9.3 Linear regression trends in **a** AAD, **b** MnAD, **c** MxAD of Satluj River at Rampur station. A 3-year running mean of the discharge is shown by a dotted line

until 1998 and 2002, whereas the MnAD has been decreasing until 1990 with an increase thereof until 1998. The discharge has again exhibited decreasing trend from 1999 onward to 2012.

Trends in Seasonal River Discharge

The trend analysis revealed decreasing trends in discharge during all the periods or seasons in the study area. However, it is pertinent to note that the R^2 values of AWD ($R^2 = 0.02$) and APrD ($R^2 = 0.03$) are very low as compared to correlation coefficient values of AMD ($R^2 = 0.45$) and APoD ($R^2 = 0.32$) thereby reflecting a significant decline in the river discharge during these two periods. The declining trends are very significant during these two seasons (AMD and APoD). During the monsoon period, the Z-statistic values for the AMD were found to be −3.2 ($p = 0.001$) which is very high than a standard of Zs = 1.96 thereby indicating that the discharge is declining at a 95% confidence level (Table 9.1). Similarly, during the post-monsoon period, the Z-statistics of APoD is very high −2.25 ($p = 0.02$) but lower than the AMD. Consequently, during the AMD period, the discharge is also declining at higher rates of 1.78 $m^3\ s^{-1}\ year^{-1}$ as compared to the APoD that declines at a rate of 0.56 $m^3\ s^{-1}\ year^{-1}$. The rate of decline of AWD and APrD is very low as 0.18 and 0.54 $m^3\ s^{-1}\ year^{-1}$. The rate of decline of 0.54 $m^3\ s^{-1}\ year^{-1}$ during the pre-monsoon is very close to the rate of decline of 0.56 $m^3\ s^{-1}\ year^{-1}$ during the post-monsoon season. However, the probability values (*p*-value) of 0.63 for APrD are higher than a standard value of 0.5 reflecting the existence of no and insignificant trend, respectively. Similarly, the *p*-value of 0.34 for AWD is also relatively higher and close to 0.5 reflecting the existence of no and insignificant trend. Furthermore, from the linear plots (Fig. 9.4a, b, c, d), the AWD revealed decreasing trend until 1986. From 1986, the discharge indicated the increasing trend until 2000 with a consequent decline thereof. Similarly, the APrD has been increasing till 1991 and then decreasing intermittently till 2012. Almost a similar trend is exhibited by the AMD with a significant downward trend from 2000 onward. However, the APoD pattern reveals a tendency towards an increase from 2010 onward despite the overall decreasing trend. The linear plots also clearly revealed a general decline in river discharge from 1963 to 2012.

Trends in Accumulation and Ablation River Discharge

The linear regression analysis of average annual discharge of accumulation and ablation periods revealed that the trend was feebly declining during the accumulation period, whereas the trend was declining prominently during the ablation season. The feeble trend of the accumulation period is also reflected by its very low coefficient of correlation ($R^2 = 0.09$) values than the ablation period during which it

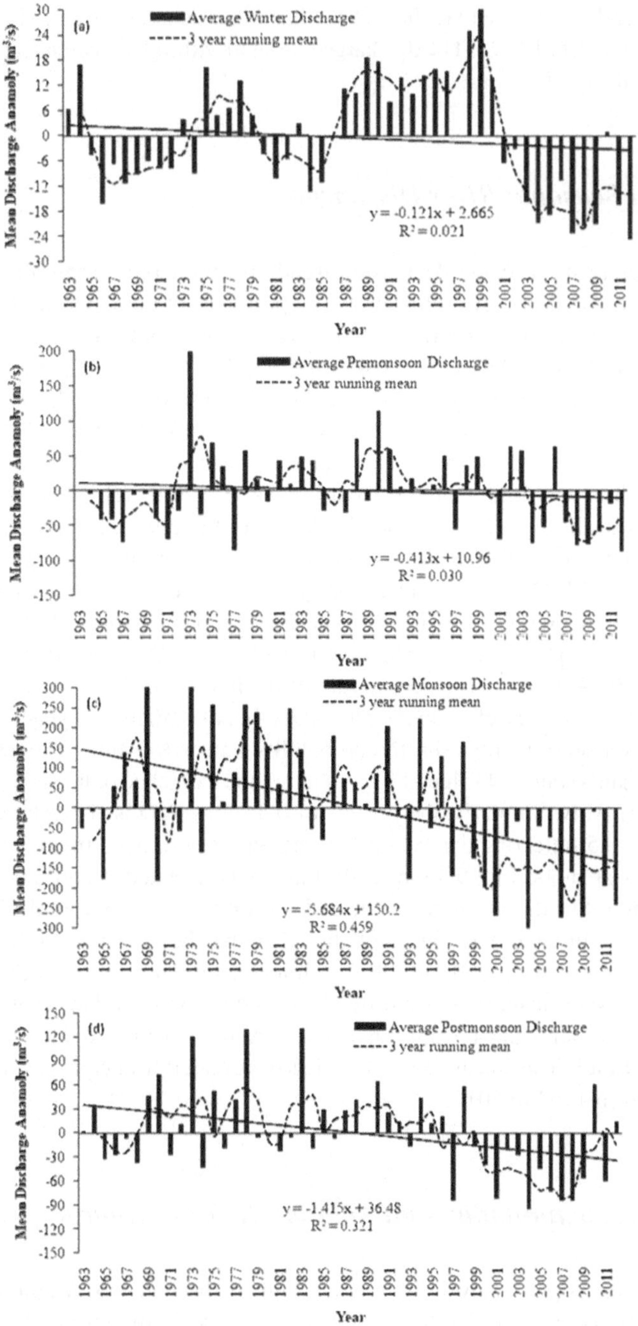

Fig. 9.4 Linear regression trends in **a** AWD, **b** AMD, **c** APrD, and **d** APoD of Satluj River at Rampur station. A 3-year running mean of the discharge is shown by a dotted line

was significantly higher ($R^2 = 0.39$). However, the Mann–Kendal test indicated that the trend in discharge during the accumulation period are not statically significant, whereas the trend was statistically significant during ablation season. But, it is very pertinent to mention here that the trend during AAcD having Z-statistics value of −1.68 are very close to the standard values of Zs = 1.96 (Table 9.1). Therefore, from this observation, it may be inferred that the declining trend during the AAcD season are also severe for the overall health of the river flow and discharge. Consequently, during the AAbD, the discharge is declining at a higher rate of 10 m^3 s^{-1} $year^{-1}$ than AAcD period during which the discharge is declining at a rate of 0.16 m^3 s^{-1} $year^{-1}$. The *p*-values of 0.003 and 0.09 are also very low for AAbD as well as AAcD, respectively. Furthermore, from the linear plots (Fig. 9.5a, b), the AAcD revealed decreasing trend until 1972. From 1972, the discharge indicated an increasing trend until 1979 with a consequent decline thereof until 1987. The discharge also showed a decline from 2001 onward till 2012 with a moderate rise during 2010. Similarly, the AAbD also exhibited decreasing trend until 1972 and a subsequent increase till 1991 with a rise during 1975 and 1985, respectively. The discharge showed a declining trend thereof till 2012 with a moderate peak during 2003. Overall, a decline in river discharge is indicated from 1963 to 2012.

Discussion

In this study, the statistical analysis of river discharge of the Satluj River basin has been carried out using parametric and non-parametric tests. For this purpose, the observed daily river discharge data of the Rampur hydrological station have been used. This station is located at a lower elevation of 1021 m AMSL in downstream areas, where the contribution from snow, glacier as well as rainfall is significant. The location is also in a pristine area and away from the urbanized areas. So, the trend analysis and signals of variation in river discharge observed here can overall represent a filtered signal of the effect of climate change in the region. Furthermore, as per the prevailing climatic conditions of the area and for better understanding of the trends and changing patterns of the river flow, the daily river discharge data were classified into different periods for the study.

Overall, the results indicated a declining trend in river discharge at this station during all the periods from 1963 to 2012 i.e., during the last 50 years. The trends were declining annually as well as seasonally in the area. The linear regression test showed a higher slope for AAD ($R^2 = 0.37$), MxAD ($R^2 = 0.35$) followed by a higher correlation coefficient value for AMD ($R^2 = 45$), APoD ($R^2 = 0.32$), and AAbD ($R^2 = 0.39$). This observation reflected a tendency towards the higher significance of declining trends during these periods. To further substantiate these results of linear regression analysis, the MK test was carried out. The MK test indicated that the declining trends were statistically very significant at 95% confidence (Zs > 1.96) for AAD ($p = 0.01$), MxAD ($p = 0.004$), AMD ($p = 0.001$), APoD ($p = 0.02$), and AAbD ($p = 0.003$), respectively (Fig. 9.6). To understand the magnitude of the

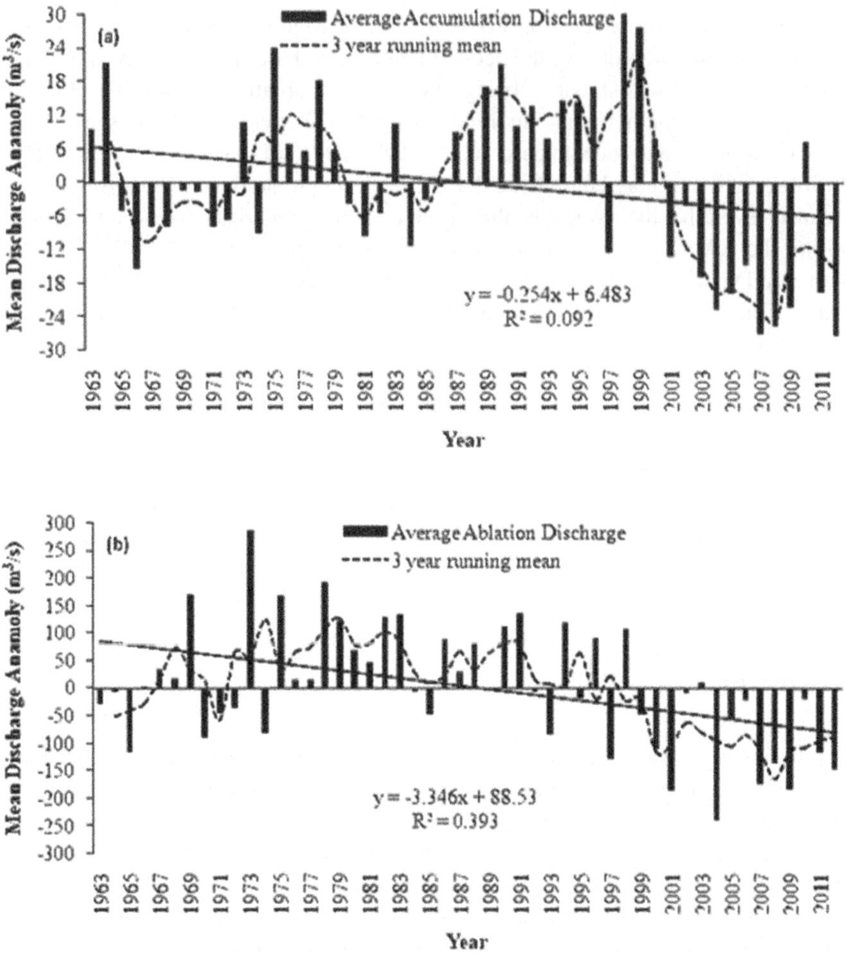

Fig. 9.5 Linear regression trends in **a** AAcD and **b** AAbD of Satluj River at Rampur station. A 3-year running mean of the discharge is shown by dotted line

trends that is change per unit per year, Sen's slope test was carried out. The AAD is declining at a rate of 0.60 m^3 s^{-1} $year^{-1}$ followed by a higher declining rate for MxAD (1.98 m^3 s^{-1} $year^{-1}$), AMD (1.78 m^3 s^{-1} $year^{-1}$), APoD (0.56 m^3 s^{-1} $year^{-1}$), and AAbD (1.0 m^3 s^{-1} $year^{-1}$). Although statically insignificant at 95% confidence level, the declining pattern of discharge for MnAD, AWD, APrD, and AAcD may be significant at a lower percentage of confidence level as well as in the near future if this declining pattern continues in the area (Fig. 9.6). The declining nature of this river discharge/flow may be directly attributed to the increased surface temperature and rapid shrinkage of glaciers over the Northwest Himalayan region as confirmed by the previous studies (Bhutiyani et al. 2009; Mir et al. 2015a, 2015b, 2017). However, the general increasing trends observed from 1970 to 2000 in this

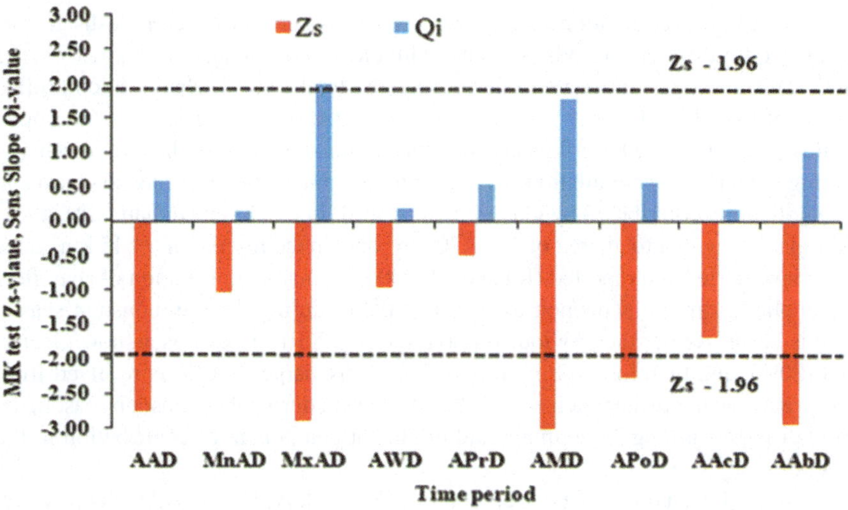

Fig. 9.6 MK test's *Z*-values and Sens slope's Qi showing the significance at 95% confidence level (Zs = 1.96) and magnitude of change per unit per year for different time intervals of river discharge of Satluj River basin at Rampur station

study during the different periods may be a direct outcome of the accelerated glacier melting in the region (Mir et al. 2017; Mir 2018; Jain and Mir 2019). Further, from the significant and continuous decline in discharge from 2000 onward to 2012, it is inferred that the volume of glacier shrinkage has probably reduced very significantly resulting in a limited contribution of glacier meltwater to the streamflow in the area. However, it is notable that a detailed investigation taken in this direction may help to understand comprehensively this response of river discharge to the glacier shrinkage in the area.

In the Himalayan region, several previous studies have also indicated a general decline in river discharge during the last 5 decades. For instance, Singh et al. (2014) have reported a declining trend of discharge of the Satluj River during summer and autumn seasons. They have reported a declining rate of 0.32 and 0.31 $m^3\ s^{-1}\ year^{-1}$ at Sunni station and a rate of decline of 0.47 and 0.34 $m^3\ s^{-1}\ year^{-1}$ at Rampur station from 1970 to 2010 which is almost similar to the findings of the declining rates of APrD and APoD of the present study. Additionally, Bhutiyani et al. (2008) have reported a significant decline in annual and monsoon discharges of the Satluj River from 1922 to 2004. Similarly, Tahir et al. (2011) have also reported a declining trend of discharge of Hunza River, Karakoram Western Himalaya. Khattak et al. (2011) have revealed declining summer trends in river discharge of the upper Indus basin in Pakistan, Western Himalaya. In the Nepal Himalayan region, the Karnali and Sapta Koshi Rivers have revealed declining trends over a period from 1962 to 2000 and 1977 to 1995 (Shrestha and Shrestha 2004). Similarly, another study on the Kosi River of Nepal Himalaya has also revealed a declining trend in annual discharge from 1947 to 1993 (Sharma et al. 2000).

Furthermore, a correlation analysis was carried out between the average discharges estimated for different periods (seasons) with the average annual discharge (Fig. 9.7 a–f). This analysis was carried out because the total river discharge is controlled and contributed by different components of the hydrological system. For example, in this river basin, the base-flow components dominantly control the river discharge during winter and accumulation seasons, whereas during the post-monsoon season, a major part is contributed by glacier melt as well. In this basin (Indian part), about 945 glaciers with a total area of 1217.70 km^2 and an ice reserve of 94.45 km^3 have been inventoried and reported (Rana et al. 2006). A major contribution of river flow is supplied from the snowmelt as well as rainfall during the pre-monsoon season in this basin. As per the previous reports, the Satluj River basin is a snow/glacier-fed (80%), and therefore, a high proportion of discharge (>60%) is received from snow/glaciers meltwater (Jain et al. 1998). However, during the monsoon season, the rainfall is performing as the main and dominant component of contribution to the river discharge.

Results indicated that the correlation of AAD with AMD and AAbD is positive, very strong, and high with R^2 values of 0.91 and 0.99 (Fig. 9.7c, f). The AAD shows a moderately strong and good correlation with APrD ($R^2 = 0.58$) and APoD ($R^2 = 0.50$) (Fig. 9.7b, d). However, the AAD shows a moderate correlation with AAcD ($R^2 = 0.46$) and weak correlation with AWD ($R^2 = 0.31$), (Fig. 9.7a, e) respectively. These results overall indicated that the annual discharge (AAD) is controlled sequentially by different components during different seasons. The controlling influence can be followed in an ascending order as AAbD > AMD > APoD > APrD > AAcD > AWD. A similar order is followed by Qi of change as AMD > AAbD > APoD > APrD > AWD > AAcD. However, the MK significance level shows minor shifts in order and follows a pattern as AMD > AAbD > APoD > AAcD > AWD > APrD. Therefore, from these observations, it can be concluded that the changes and declining nature of the discharge follow a pattern in terms of its contributions and controlling factors. The dominantly declining contributing factor will probably affect the overall nature of the annual river discharge sequentially following a descending and sequential order. However, it is important to note that the long-term decline in discharge may also be influenced by the local climate, LULC, urbanization, and other local topographic factors. These factors may have a modifying influence on the changes in river discharge.

Conclusions

The river discharge observed at the Rampur hydrological station of the Satluj River basin indicated a significant declining trend for 5 decades from 1963 to 2012 annually as well as seasonally. The trends are statistically very significant at 95% confidence for AAD, MxAD, AMD, APoD, and AAbD, respectively. However, the magnitude of decline was found to be heterogeneous and varying between a low declining rate of 0.60m^3 s^{-1} $year^{-1}$ for AAD followed by a higher declining rate for MxAD (1.98

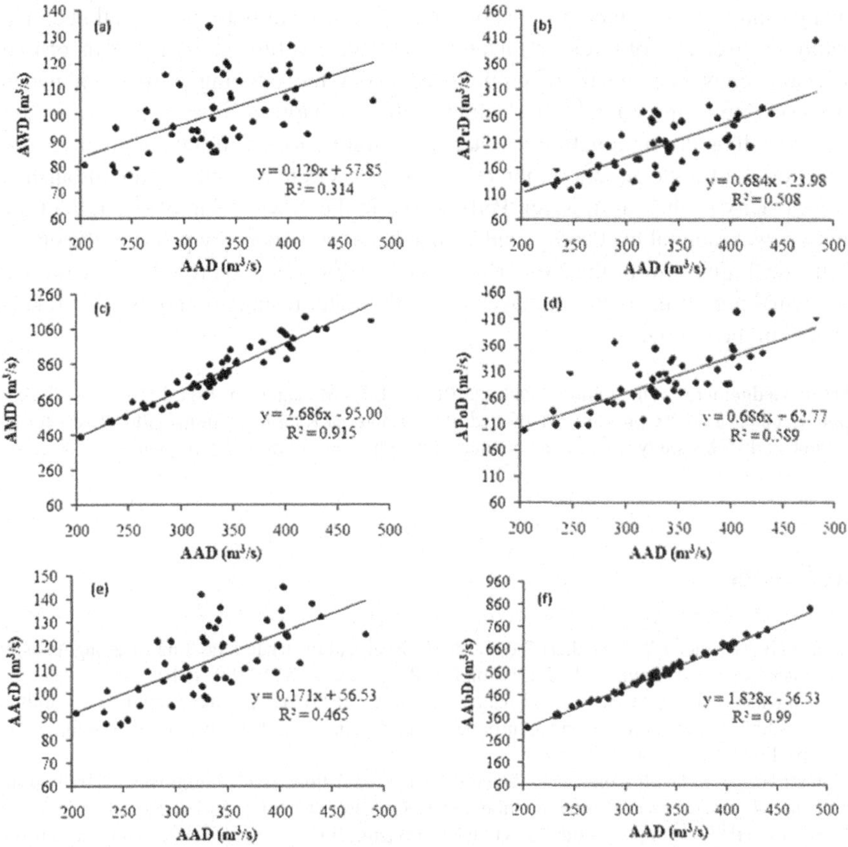

Fig. 9.7 Correlation of **a** AAD verses AWD, **b** AAD verses APrD, **c** AAD verses AMD, **d** AAD verses APoD, **e** AAD verses AAcD, and **f** AAD verses AAbD of Satluj River basin at Rampur station

$m^3\ s^{-1}\ year^{-1}$), AMD (1.78 $m^3\ s^{-1}\ year^{-1}$), APoD (0.56 $m^3\ s^{-1}\ year^{-1}$), and AAbD (1.0 $m^3\ s^{-1}\ year^{-1}$). Overall, the decline in annual discharge is a direct result of the variability and decline in discharge during different periods and seasons. The variability and decline of total annual discharge (AAD) are strongly controlled by the AAbD (0.99), AMD (0.91), APoD (0.58), APrD (0.50), AAcD (0.46), and AWD (0.31). Nevertheless, there exists a sequential pattern in the variability and decline of annual discharge and seasonal discharge in terms of R^2, Qi, and Zs values. In descending order, the sequential order based on R^2, Qi, and Zs can be identified as AbD > AMD > APoD > APrD > AAcD > AWD; AMD > AAbD > APoD > APrD > AWD > AAcD; and AMD > AAbD > APoD > AAcD > AWD > APrD, respectively, in the region.

The declining nature of river discharge can be a direct outcome of climate change in the Himalayas. The continuous glacier loss and snow depletion because of climate change and warming may also be contributing significantly to the overall declining nature of discharge of rivers in the area. Nevertheless, the declining trends of river discharge may become more pronounced under the ongoing climate change in the near future. In general, the decline in the discharge of river annually, seasonally as well as during ablation and accumulation periods will have a great adverse effect on the fragile aquatic biodiversity and ecosystem as well as natural purification capacity of the Satluj River. Particularly, in the downstream areas, agriculture, hydropower, animal husbandry, and human health are also very prone to the effects of the declining flow of the river. Thus, there is a dire need to provide strategies and policymaking for the better management of the water resources and its other related sectors in the region.

Acknowledgements The authors thank the Bhakra Beas Management Board (BBMB), Chandigarh, Punjab, India, for providing daily hydrological data of Rampur station (Satluj River) for the present study. This study was carried out during the Ph.D. program of the first author.

References

Abdi A, Hassanzadeh Y, Talatahari S et al (2016) Regional bivariate modeling of droughts using L-comoments and copulas. Stoch Environ Res Risk Assess 31(5):1199–1210

Abeysingha NS, Singh M, Sehgal V, Khanna M, Pathak H (2016) Analysis of trends in streamflow and its linkages with rainfall and anthropogenic factors in Gomti River basin of North India. The App Climatol 123(3–4):785–799

Abghari H, Tabari H, Hosseinzadeh Talaee P (2013) River flow trends in the west of Iran during the past 40years: impact of precipitation variability. Glob Planet Chang 101:52–60

ADB (2011) IWRM Scoping Study for Satluj River Basin, Himachal Pradesh improving capacity for climate change adaptation-final report prepared by Asian Development Bank, Asian Development Bank, Government of India, Government of Himachal Pradesh, pp 1–29

Afshar AA, Hasanzadeh Y, Besalatpour AA, Pourreza-Bilondi M (2016) Climate change forecasting in a mountainous data scarce watershed using CMIP5 models under representative concentration pathways. Theor Appl Climatol 129:683–699

Ahmed K, Shahid S, Nawaz N, Khan N (2018) Modeling climate change impacts on precipitation in arid regions of Pakistan: a non-local model output statistics downscaling approach. Theor Appl Climatol 1–18

Bates BC, Kundzewicz Z, Wu S, Palutikof J (2008) Climate change and water, technical paper of the intergovernmental panel on climate change. IPCC Secretariat, Geneva pp 3–4

Berthier E, Arnaud Y, Kumar R, Ahmad S, Wagnon P, Chevallier P (2007) Remote sensing estimates of glacier mass balances in the Himachal Pradesh (Western Himalaya, India). Remote Sens Environ 108(3):327–338

Bhutiyani MR, Kale VS, Pawar NJ (2008) Changing streamflow patterns in the rivers of northwestern Himalaya: implications of global warming in the 20th century. Curr Sci 95(5):618–626

Bhutiyani MR, Kal VS, Pawar NJ (2009) Climate change and the precipitation variations in the northwestern Himalaya: 1866–2006. Int J Climatol 30(4):535–548

Biemans H, Speelman LH, Ludwig F, Moors EJ, Wiltshire AJ, Kumar P, Gerten D, Kabat P (2013) Future water resources for food production in five South Asian river basins and potential for adaptation–a modeling study. Sci Total Environ 468–469:S117–S131

Bolch T, Kulkarni AK, Kääb A, Huggel C, Paul F, Cogley G, Frey H, Kargel JS, Fujita K, Scheel M, Stoffel M, Bajracharya S (2012) The state and fate of Himalayan glaciers. Science 336:310–314

Bouwer LM, Aerts JCJH, Droogers P, Dolman AJ (2006) Detecting the long-term impacts from climate variability and increasing water consumption on runoff in the Krishna River basin (India). Hydrol Earth Syst Sci 10:703–713

Coulthard TJ, Lewin J, Macklin MG (2005) Modelling differential catchment response to environmental change. Geomorphology 69(1–4):222–241

Demir V, Kisi O (2016) Comparison of Mann-Kendall and innovative trend method (Şen trend) for monthly total precipitation (Middle Black Sea Region, Turkey). In: Proceedings of the conference 3rd international Balkans conference on challenges of civil engineering, 3-BCCCE. Tirana, Albania

Economic Survey (2007) Power, available online at http://himachal.nic.in/finance/bud0203/ecosurv/eco7.htm

Guo J, Huang G, Wang X, Li Y, Yang L (2018) Future changes in precipitation extremes over China projected by a regional climate model ensemble. Atmos Environ 188:142–156

Hirsch RM, Archfield SA, Cicco LAD (2015) A bootstrap method for estimating uncertainty of water quality trends. Environ Model Softw 73:148–166

Jain SK, Goswami A, Saraf AK (2010a) Assessment of snowmelt runoff using remote sensing and effect ofclimate change on runoff. Water Res Manag 24:1763–1777

Jain S K, Keshri R, Goswami A, Sarkar A (2010b) Application of meteorological and vegetation indices for evaluation of drought impact: a case study for Rajasthan, India. Nat Hazard 54(3):643–656

Jain SK, Kumar N, Ahmed T (1998) SLURP model and GIS for estimation of runoff in a part of Satluj catchment, India. Hydrol Sci J 43(6):875–884

Jain SK, Mir RA (2019) Glacier and glacial lake classification for change detection studies using satellite data: a case study from Baspa basin, western Himalaya. Geocar Int 34(4):391–414

Kendall MG (1975) Rank correlation methods, 3rd edn. Hafner Publishing Company, New York, p 128

Khattak M, Babel M, Sharif M (2011) Hydro-meteorological trends in the upper Indus River basin in Pakistan. Climat Res 46(2):103–119

Kisi O (2015) An innovative method for trend analysis of monthly pan evaporations. J Hydrol 527:1123–1129

Kulkarni AV, Bahuguna IM, Rathore BP, Singh SK, Randhawa SS, Sood RK, Dhar S (2007) Glacial retreat in Himalaya using Indian remote sensing satellite data. Curr Sci 92(1):69–74

Kundzewicz ZW, Kanae S, Seneviratne SI et al (2014) Flood risk and climate change: global and regional perspectives. Hydrol Sci J

Lobanova A, Liersch S, Nunes JP, Didovets I, Stagl J, Huang S, Koch H, Rivas López MdR, Maule CF, Hattermann F et al (2018) Hydrological impacts of moderate and high-end climate change across European river basins. J Hydrol Reg Stud 18:15–30

Malik A, Kumar A, Guhathakurta P, Kisi O (2019) Spatial-temporal trend analysis of seasonal and annual rainfall (1966–2015) using innovative trend analysis method with significance test. Arab J Geosci 12:328

Mann HB (1945) Non-parametric tests against trend. Econometrica 13:245–259

Mimikou MA, Fotopoulos FO (2005) Regional effects of climate change on hydrology and water resources in Aliakmon River basin. Foz do Igua9u, Brazil

Mir RA (2021) Remote sensing based assessment of Glacier Resources in Parts of Ladakh Mountain Range, a Trans-Himalayan Region. In: Taloor AK et al (eds) Water, cryosphere, and climate change in the Himalayas, geography of the physical environment. https://doi.org/10.1007/978-3-030-67932-3_5

Mir RA, Jain SK, Jain SK, Thayyen R, Saraf AK (2017) Assessment of glacier changes and its controlling factors from 1976 to 2011: a case study from Baspa basin, Western Himalaya. Arct Antarc Alp Research 49(4):621–647

Mir RA (2018) Recent changes of two parts of Kolahoi Glacier and its controlling factors in Kashmir basin, western Himalaya. Remote Sen App Soc Environ 11:265–281

Mir RA, Jain SK, Saraf AK (2015a) Analysis of current trends in climatic parameters and its effect on discharge of Satluj River basin, Western Himalaya. Nat Hazard 79(1):587–619
Mir RA, Jain SK, Saraf AK, Goswami A (2013) Detection of changes in Glacier mass balance using satellite and meteorological data in Tirungkhad basin located in Western Himalaya. J Indian Soc Remote Sen 42:91–105
Mir RA, Jain SK, Saraf AK, Goswami A (2014) Glacier changes using satellite data and effect of climate in Tirungkhad basin located in western Himalaya. Geocart Int 29:293–313
Mir RA, Jain SK, Saraf AK, Goswami A (2015b) Declines in snowfall in response to temperature in Satluj basin, western Himalaya. J Earth Syst Sci 124(2):365–382
Mir RA, Jain SK, Saraf AK, Goswami A (2015c) Accuracy assessment and trend analysis of MODIS derived snow cover area in Satluj basin western Himalaya. Int J Remote Sen 36(15):3837–3858
Mir RA, Jain SK, Lohani AK, Saraf AK (2018) Glacier recession and glacial lake outburst flood studies in Zanskar basin, western Himalaya. J Hydrol 564 (2018):376–396
Nune R, George BA, Teluguntla P, Western AW (2012) Relating trends in streamflow to anthropogenic influences: a case study of Himayat Sagar catchment, India. Hydrol Earth Syst Sci Discuss 9:9295–9336
Öztopal A, Sen Z (2017) Innovative trend methodology applications to precipitation records in Turkey. Water Resour Manag 31:727–737
Panda DK, KumarA GS, Mohanty RK (2013) Stream flow trends in the Mahanadi River basin (India): linkages to tropical climate variability. J Hydrol 495:135–149
Philip G, Sah MP (2004) Mapping repeated surges and retread of glaciers using IRS-1C/1D data: a casestudy of Shaune Garang Glacier, Northwestern Himalaya. Int J Appl Earth Obs Geoinf 6(2):127–141
Qutbudin I, Shiru MS, Sharafati A, Ahmed K, Al-Ansari N, Yaseen ZM, Shahid S, Wang X (2019) Seasonal drought pattern changes due to climate variability: case study in Afghanistan. Water 11:1096
Rai RK, Upadhyay A, Ojha CSP (2010) Temporal variability of climatic parameters of Yamuna River basin, spatial analysis of persistence, trend and periodicity. Open Hydrol J 4:184–210
Rana RS, Bhagat RM, Kalia A, Sood C (2006) Inventory of two decades for glaciers and glacial lakes in Satluj River Basin of Himachal Pradesh. J Agric Phys 6(1):28–34
Saadi Z, Shahid S, Ismail T, Chung ES, Wang XJ (2017) Trends analysis of rainfall and rainfall extremes in Sarawak, Malaysia using modified Mann–Kendall test. Meteorol Atmos Phys 1–15
Sanikhani H, Kisi O, Mirabbasi R, Meshram SG (2018) Trend analysis of rainfall pattern over the central India during 1901–2010. Arab J Geosci 11:437
Sen Z (2011) Innovative trend analysis methodology. J Hydrol Eng 17:1042–1046
Sharma KP, Moore B, Vörösmarty CJ (2000) Anthropogenic, climatic and hydrological trends in the Kosi Basin, Himalaya. Clim Chang 47:141–165
Shiru MS, Shahid S, Alias N, Chung ES (2018) Trend analysis of droughts during crop growing seasons of Nigeria. Sustainability 10:871
Shrestha ML, Shrestha AB (2004) Recent trends and potential climate change impacts on Glacier Retreat/Glacier Lakes in Nepal and potential adaptation measures. In: Global forum on sustainable development. Environment policy committee, environment directorate. OECD, Paris, ENV/EPOC/GF/SD/RD(2004)6/FINAL. Paris, France: OECD, p 23
Singh D, Gupta RD, Jain SK (2014) Study of long-term trend in river discharge of Sutlej River (NW Himalayan region). Geog Environ Sus 7(3):87–96
Singh P, Bengtsson L (2004) Hydrological sensitivity of a large Himalayan basin to climate change. Hydrol Process. 18:2363–2385
Singh P, Jain SK (2002) Snow and glacier melt in the Satluj River at Bhakra dam in the Western Himalayan region. Hydrol Sci J 47:93–106
Singh P, Kumar N (1997) Impact assessment of climate change on the hydrological response of a snow and glacier melt runoff dominated Himalayan River. J Hydrol 193:316–350
Stagge JH, Kohn I, Tallaksen LM, Stahl K (2015) Modeling drought impact occurrence based on meteorological drought indices in Europe. J Hydrol 530:37–50

Su B, Jiang T, Jin W (2006) Recent trends in observed temperature and precipitation extremes in the Yangtze River basin, China. Theor Appl Climatol 83:139–151

Tabari H, Taye MT, Willems P (2015) Statistical assessment of precipitation trends in the upper Blue Nile River basin. Stoch Environ Res Risk Assess 29:1751–1761

Tahir AA, Chevallier P, Arnaud Y, Ahmad B (2011) Snow cover dynamics and hydrological regime of the Hunza River basin, Karakoram Range, Northern Pakistan. Hyd Earth Sys Sci 15(7):2275–2290

Tosunoglu F, Kisi O (2017) Trend analysis of maximum hydrologic drought variables using Mann-Kendall and Şen's innovative trend method. River Res Appl 33:597–610

Vicuña S, Garreaud RD, Mc Phee J (2011) Climate change impacts on the hydrology of a snowmelt driven basin in semiarid Chile. Clim Chang 105(3):469–88

Vliet MTHV, Yearsley JR, Franssen WHP et al (2012) Coupled daily streamflow and water temperature modelling in large river basins. Hydrol Earth Syst Sci 16:4303–4321

Wang Y, Wang D, Lewis QW, Wu J, Huang F (2017) A framework to assess the cumulative impacts of dams on hydrological regime: a case study of the Yangtze River. Hydrol Process 31:3045–3055

WAPCOS (2007) Environmental and Socio-Economic Baseline Setup, available online at http://sjvn.nic.in/Society-Environ/Chap-4%20revised_4.pdfss

Wu H, Qian H (2017) Innovative trend analysis of annual and seasonal rainfall and extreme values in Shaanxi, China, since the 1950s. Int J Climate 37:2582–2592

Xu H, Zhou B, Song Y (2011) Impacts of climate change on headstream runoff in the Tarim River Basin. Hyd Res 42(1):20–29

Chapter 10
Climate Change and Its Impact on Forest of Indian Himalayan Region: A Review

Amit Kumar Tripathi, Prem Chandra Pandey, Jyoti Kumar Sharma, Dimitris Triantakonstantis, and Prashant K. Srivastava

Abstract The Indian Himalayan Region (IHR) forms the northern boundary of India covering an area of about 5.3 lakh km^2, which constitutes 16.2% of the total geographical area. The region is remarkable for its diversity of plants and animals and provides abundant ecosystem services to people. The Indian Himalayas (IH) is richly endowed with forest resources and supports different types of forest ecosystems along with varied topography. In addition to anthropogenic factors, the forests of IH are under pressure due to changing climate. There are several pieces of evidence of significant changes in temperature, precipitation, and vegetation phenology in IHR. Climate change (CC) has a significant impact on forest ecosystems of IH which results in the upward movement of several plant species, and further changes are expected to cause the extinction of species. Due to changing climatic conditions in IHR, possibly some of the locations may become more favorable to alien invasive species causing negative impacts on native plants. Changing temperature and rainfall in the region have caused phenological changes in many economically and ecologically important plant species like Rhododendron sp., Myrica esculenta, Pinus wallichiana, etc. As a result of which these species are shifting towards higher elevations to cope up with altered climate. A significant reduction in suitable habitat and the massive decline in the population of animal species due to climate change is also predicted. The studies carried out in IHR indicated shifts in vegetation types across IHR from moderate to large scale. Higher elevation Himalayan temperate forests, subalpine forests, and alpine forests are more vulnerable to the adverse effects of climate change while Sal

A. K. Tripathi · P. C. Pandey (✉) · J. K. Sharma
School of Natural Sciences, Center for Environmental Sciences & Engineering, Shiv Nadar University, Greater Noida, Uttar Pradesh, India

D. Triantakonstantis
Department of Soil Science of Athens, Institute of Soil and Water Resources, Hellenic Agricultural Organization DIMITRA, 1 S. Venizelou str., Lycovrisi, 14123 Attiki, Greece

P. K. Srivastava
Remote Sensing Laboratory, Institute of Environment and Sustainable Development, Banaras Hindu University, Varanasi, Uttar Pradesh, India

DST-Mahamana Centre of Excellence in Climate Change Research, Banaras Hindu University, Varanasi, Uttar Pradesh, India

S. Rani and R. Kumar (eds.), *Climate Change*, Springer Climate,
https://doi.org/10.1007/978-3-030-92782-0_10

forests are predicted to shift towards the north-eastern region. Western Himalayan forests are highly susceptible to changing climate while eastern Himalayan forests are estimated to be more resilient.

Keywords Climate change impact · Indian Himalayas · Phenological changes · Habitat loss · Himalayan ecosystem

Introduction

Globally, the climate is considered one of the most important elements in controlling vegetation forms and has a major impact on the distribution, structure, and ecology of forest ecosystems (Kirschbaum et al. 1996). Climate change (CC) has caused variable (moderate to large scale) shifts in the distribution of plant species over the last few decades (Kueppers et al. 2005; Kelly and Goulden 2008; Loarie et al. 2008; Chen et al. 2011). Hence, this is reasonable to assess that climate change would alter the distribution pattern of forests (Chaturvedi et al. 2010). Climate change is projected to have a direct effect on forest ecosystems during and beyond the twenty-first century, which will cause extensive forest die-back and to some extent biodiversity (BD) loss (IPCC 2014). Climate change coupled with anthropogenic disturbances could have a synergistic impact on invasive species expansion in the future (Mungi et al. 2018).

Climate change has posed major threats to biodiversity across the globe including India and its impacts are expected to increase substantially in near future (IPCC 2014). Each species in an ecosystem has a specific niche. Climate change and biodiversity loss pose significant global threats to human well-being and may cause significant disturbance to species in different forest types that need attention (Pandey et al. 2019a; Malhi et al 2020,). Healthy ecosystems along with rich biodiversity are fundamental to the existence of life on our planet and important to enhance ecosystem productivity. BD loss due to climate change may have the potential to change the structures and alter the functions of forest ecosystems, require assessment and monitoring (Pandey et al. 2014; Pandey et al. 2019b; Anand et al 2020). According to the recent study of Trisos et al. (2020), climate change could cause sudden, potentially catastrophic biodiversity losses all over the world throughout the twenty-first century. Global warming due to the increased emission of GHG (greenhouse gases) such as CO_2, CH_4 in the environment is known to alter the structure and function of mountain ecosystems worldwide (IPCC 2014). CC is predicted to be larger for a high mountainous region like the Himalayas (Chaturvedi et al. 2010). The name Himalaya has been derived from two Sanskrit words, Hima (snow) and Alaya (abode), meaning ‘the abode of snow’. The Himalayan region is rich in biodiversity including endemic, rare, and threatened flora and fauna. The region with estimated 10,000 vascular plants, provide a home to about 300 species of mammals, 979 bird, 177 reptiles, 124 amphibians, and 269 freshwater species (Mittermeier et al. 2004). Particularly rich in biodiversity, the Himalayas, provides habitat to about one-tenth of the world’s total higher altitude flora and fauna species, and are also home to about 50% of native plant species of India (Padma 2014).

The Indian Himalayan Region (IHR) is considered an ecologically important but fragile region. This region is predominantly sensitive to climate change and slight changes in mean temperatures may result in a significant consequence on the ecosystems and it may enhance species extinction rates, especially in sensitive regions. Temperature and rainfall are two important variables of climate change but information on actual changes in the two most climatic variables is not particulate in the Himalayas (Shrestha et al. 2012). The Himalayas are one of the world's most fragile ecosystems subjected to climate change and leading to global warming rates (Shrestha et al. 2012). The varied environmental conditions such as wide variation in rainfall, weather, temperature, topography as well as soil conditions are responsible for rich floristic diversity in different ecosystems across the Indian Himalayas (IH). The IHR supports a vast variety of forest types and provides numerous ecosystem services to people. A large number of people depend on forest resources for their livelihood in the region. Further, the forest ecosystems in the IH provide shelter to numerous species of animals. Thus, this chapter presents an overview of climate change impacts on forest ecosystems in the IHR based on a literature review.

Study Area

The IHR including eastern and western Himalayas mostly consists of 11 states viz., Himachal Pradesh, Uttarakhand, Sikkim, Arunachal Pradesh, Nagaland, Manipur, Mizoram, Tripura, Meghalaya, Assam, West Bengal, and two union territories, i.e., Jammu and Kashmir and Ladakh of India (NMSHE 2015–16) (Fig. 10.1). IHR forms the northern boundary of the country encompassing the total geographical area of 5.3 lakh km^2 (−16.2%) and extends over 2500 km in length between the Indus and the Brahmaputra river systems (NMSHE 2015–16). The altitude reaches over 7000 m and its width ranges between 150 to 600 km at different places. The north-eastern region of India is considered as eastern IH while Himachal Pradesh, Uttarakhand, Jammu and Kashmir, and Ladakh are considered western IH in the present study (Fig. 10.1). In comparison to the western Himalayas, the region in the eastern Himalayas has a relatively wetter climate and records the world's highest rainfall in the Cherrapunji-Mawsynram sector of Meghalaya. Physiographically, the IH starts from the foothills in the Siwaliks and extends up to the Tibetan plateau in Trans-Himalaya. The Siwaliks, the lesser Himalaya, the Greater Himalaya, and the Trans-Himalaya are broad divisions under IHR; extending mostly continuous throughout its length (NMSHE 2015–16).

The higher altitude region in the IH is usually covered by snow throughout the year and encompasses several glaciers, prominent among them are the Siachen, the largest glacier in the IH. The region is also important for its high forest cover and enormous diversity of plants and animals. The IHR is home to nearly 51 million people and

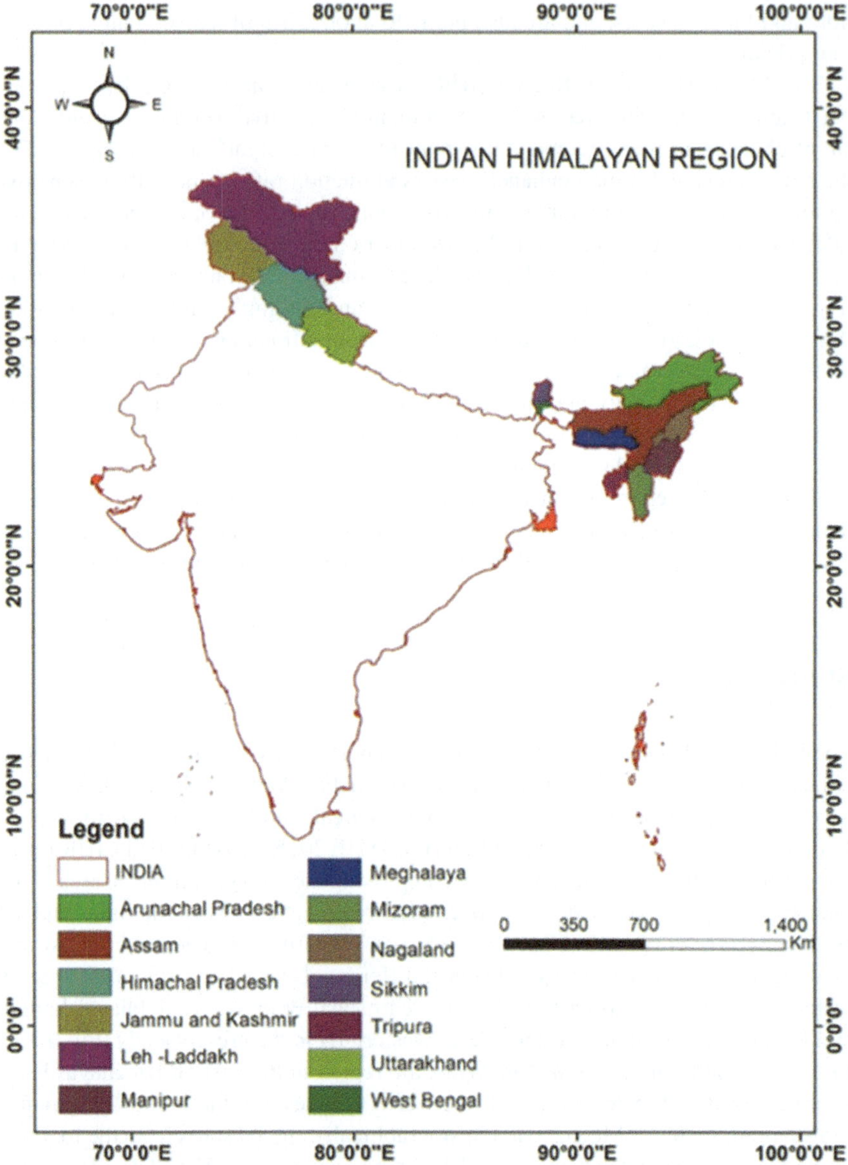

Fig. 10.1 Location and extent of IH

the majority of the population practices agriculture in various ecosystems, including species-rich forests (Tewari et al. 2017).

Materials and Methods

An extensive literature search was made to compile the climate change impacts on forest ecosystems and distribution patterns of floral and faunal species in the IHR. Published literature from 2004 to 2021 in national and international journals by different researchers on the impact of climate change in the IH over 100 years time period was consulted. The websites such as IPCC, National Oceanic and Atmospheric Administration (NOAA), National Mission for Sustaining the Himalayan Ecosystem (NMSHE), etc. were also accessed for gathering the information. The literature search was focused on the rise of temperature and rainfall alteration and climate change impact on forest ecosystems, fruit production, phenology and distribution of plants, expansion of alien invasive species, and habitat loss of animal species. The efforts were taken to access suitable reference materials among the numerous data sources.

Forest Ecosystems in IHR

India is credited to be one of the mega-biodiversity countries of the world, where forests represent over one-fifth of the geographical area. Champion and Seth (1968) divided Indian forests into 16 groups based on floristic composition and physiognomy. The IHR is part of both Himalaya and Indo-Burma hotspots of biodiversity. The parts of eastern Himalaya come under Indo-Burma hotspot and parts of western Himalayas are included in Himalaya hotspot (Mittermeier et al. 2004). The forests in the IH can be broadly categorized into tropical forests, subtropical broadleaved hill forests, temperate broadleaf forests, and subalpine forests.

Tropical forests in the IH are mainly of two types—tropical moist forests and tropical rainforests. Tropical moist forest occurs in the lower slopes of the eastern Himalayas. Sal forests dominated by *Shorea robusta*, are typically categorized as tropical moist deciduous forests. The tropical rainforests are dense and receive a high amount of rainfall and are characterized by evergreen tall trees. The understory is enriched with ferns, climbers, orchids, bamboos, etc. These forests are structurally similar to the forests of equatorial regions. The extreme spread of tropical rainforests in northern limits in the world has been recorded in Arunachal Pradesh. Recently, the existence of tropical rainforests has been discovered from Meghalaya that is the westernmost limit of the rainforests North of the Tropic of Cancer in India (Shankar and Tripathi 2017).

Subtropical broadleaved hill forests are found at lower hills and the canopy is mostly composed of coniferous and broadleaved tree species. Flora consists of a mixture of both evergreen and temperate forest species. In the eastern Himalayas, these forests are well represented in the north-eastern region of India. The deep valleys and hills support a conducive habitat for the luxuriant growth of herbaceous flora. This forest is rich in herbaceous flora and a potent source of a large number of medicinal plants. The evergreen tree members of the Lauraceae family- *Litsea*, *Machilus*, and

Phoebe species along with numerous species of orchids, ferns, bamboos, and climbers are well represented in these forests.

The temperate broadleaved forests undergo a transition into the Himalayan subtropical forests (subtropical pine and subtropical broadleaved forests) at lower elevations and subalpine conifer forests at higher elevations. In the western Himalayas, this eco-region grades into Himalayan subtropical pine forests at lower elevations while at higher elevations, it grades into Himalayan subalpine conifer forests as well Himalayan alpine shrub and meadows. The temperate broadleaf forest ecosystems are diverse and species-rich, usually with a great diversity of Oaks consisting of *Quercus, Lithocarpus* and *Castanopsis* species, *Rhododendron, Betula, Juniper, Maple, Deodar, Acer, Alnus, Litsea* and *Magnolia* species, etc. These forests support an outstanding richness of wildlife such as Gee's golden langur, Hodgson's giant flying squirrel, Namdapha flying squirrel, Brahma white-bellied rat, Himalayan serow, Assam macaque, stump-tailed macaque, red panda, etc., in the eastern Himalayas and Asiatic black bear, leopard, the Himalayan tahr, and the threatened Himalayan serow, Kashmir cave bat, etc., in the western Himalayas.

Subalpine forests are predominating conifers in higher hills (upper limit of timberline) of the IH. The forest is rich in dwarf trees and shrubs, especially species of *Rhododendrons*. Fir, spruce, etc. are also important species in this forest. Besides, epiphytic mosses, lichens, and ferns are abundant while climbers are almost absent. This forest extends its distribution from Jammu and Kashmir to Arunachal Pradesh.

Besides, grassland ecosystems are also the predominant vegetation type comprising of diverse grass species rather than trees or large shrubs in the region. The grassland ecosystems may be named savannahs, temperate grasslands, tallgrass prairies, etc. Grasslands and savannahs ecosystem are spread over plains, valleys, and hilly regions of the northeastern states of India (Champion and Seth 1968).

Results and Discussion

There are overwhelming pieces of evidence that our earth is warming. The major concern of climate change is often considered as melting glaciers. According to a climate report (NOAA 2009), the extent of snow cover, the volume of glaciers, and sea ice are decreasing worldwide. The rate of glacier meltdown is much higher in the Himalayas than elsewhere due to increasing temperature (Sharma et al. 2021). Upward shifts of plants in response to increasing temperature have already been reported from the Himalayan region (Dubey et al. 2003; Schickhoff et al. 2015). Under projected climate change during the twenty-first century, a large section of terrestrial and freshwater species are predicted to be at higher extinction risk, especially as climate change interacts with other stressors, such as habitat alteration, overexploitation, and invasive species (IPCC 2014). Globally, an increase in temperature ~~is~~ affecting ecosystems and communities. Due to climate change, plants and animal's habitats are being modified, flowering and egg-laying are being shifted, and even species are altering their home ranges (NOAA 2020). Such climate change impacts on Himalayan forest ecosystems are critical (Tewari et al. 2017). The impact of climate change in the IHR can be overviewed as follows.

Climate Change Evidence in the Indian Himalayas

The region of the IH, including both eastern and western Himalayas, has shown significant variations in temperature as summarized in Table 10.1. The increasing trend of temperature in the western Himalayas has been supported through dendrochronological studies in high-level fir (*Abies spectabilis*) that showed a mean annual temperature rise by 0.6 °C and winter temperature (December–February) by 1 °C in the twentieth century (Yadav et al. 2004). Bhutiyani et al. (2007) observed an increase of 0.16 °C per decade during the last century in the western Himalayan region. Dash et al. (2007) reported that the western part of the IH experienced a 0.9 °C warming over the last century (1901–2003) while in the case of the eastern part of the IH (northeast India) it was 1 °C. Similarly, Dimri and Dash (2012) also found a warming trend in the western Indian Himalayan Region, from 1975 to 2006 with the highest observed increase in mean maximum temperature between 1.1 and 2.5 °C. Shrestha et al. (2012) reported significant changes in temperature, rainfall, and vegetation phenology across the Himalayan regions for the periods of 25 years between 1982 and 2006 based on remotely sensed imagery and concluded the increase of the average annual mean temperature by 1.5 °C (0.06 °C per year). However, Telwala et al. (2013) covering a period of nearly 160 years reported an increase of mean temperatures in the Sikkim Himalayan region for the warmest and the coldest months by 0.76 ± 0.25 °C and 3.65 ± 2 °C, respectively during 1849–2010.

The rainfall also showed a significant variation across the Himalayas (Table 10.1). An increase in pre-monsoon (March–May) precipitation for the period of 1901–2003 was noted in the region by Guhathakurta and Rajeevan (2008). Likewise, Bhutiyani et al. (2010), based on observations at three weather stations, also reported a statistically significant decreasing trend in monsoon and average annual rainfall in the western Himalayas during 1866–2006. Dimri and Dash (2012) observed significantly decreasing winter precipitations (December-February) in the Himalayan region during 1975–2006. In addition, Shrestha et al. (2012) reported an increase in average annual precipitation by 163 mm (6.52 mm per year) in the region of the Himalayas.

Table 10.1 Rise of the temperature and changes in rainfall due to climate change in the Himalayas

SN	Region	Finding	Source
Temperature			
1	Western Himalayas	Increase in mean annual temperature by 0.6 °C and winter temperature by 1 °C in twentieth century	Yadav et al. (2004)
2	Western Himalayas	Increase of temperature 0.16 °C per decade during last century	Bhutiyani et al. (2007)
3	Western Himalayas	Average increase of 0.9 °C in temperature during 1901–2003	Dash et al. (2007)
4	Eastern Himalayas	Average increase of 1 °C in temperature during 1901–2003 in northeast India	Dash et al. (2007)
5	Western Himalayas	Increase in maximum temperature between 1.1 and 2.5 °C (1975–2006)	Dimri and Dash (2012)
6	Himalayas	Increase of the average annual mean temperature by 1.5 °C (0.06 °C per year) between 1982 and 2006	Shrestha et al. (2012)
7	Eastern Himalayas	Increase of mean temperatures for the warmest and the coldest months by 0.76 ± 0.25 °C and 3.65 ± 2 °C, respectively during the period 1849–2010 in Sikkim	Telwala et al. (2013)
Rainfall and monsoon alterations			
8	Western Himalayas	Increase in pre-monsoon (March–May) precipitation during 1901–2003	Guhathakurta and Rajeevan (2008)
9	Western Himalayas	Downward trend in monsoon and average rainfall during 1866–2006	Bhutiyani et al. (2010)
10	Western Himalayas	Decreasing winter precipitation (December-February), increase in number of warm days, decrease in number of cold days, and rising trend in number of consecutive dry days in winter during 1975–2006	Dimri and Dash (2012)
11	Himalayas	Increase in average annual precipitation by 163 mm (6.52 mm per year) for the periods of 25 years between 1982 and 2006	Shrestha et al. (2012)

Climate Change Impact on Forest Ecosystems in the Indian Himalayas

Chaturvedi et al. (2010) pointed that the northwestern part of India is projected to experience higher levels of warming and these regions will be highly vulnerable to changing climate while eastern Himalayan forests are estimated to be more resilient or least vulnerable. This conclusion was based on the changes in the area under different forest types, shifts in the boundary of forest types, and Net Primary Productivity (NPP). Furthermore, Chaturvedi et al. (2010) also estimated that mountainous forests such as Himalayan dry temperate forest, Himalayan moist temperate forests, subalpine forest, and alpine forest are more susceptible to the adverse effects of climate change due to higher elevations. Previous studies reported the significant upward shift of several plant species such as Rhododendron, Pinus at Indian Himalayan regions due to changing climate (Dubey et al. 2003; Gaira et al. 2014). Several studies have also assessed the effects of climate change in species spatial distribution (Kumar 2012), carbon stocks and several other phenological phenomenon such as early flowering and so on (Chaukiyal 2011; Gaira et al. 2014).

Climate Change Impact on Sal Forests

Shorea robusta (commonly called sal), a popular timber tree of the family Dipterocarpaceae, is a dominant tree species in sal forests in plains and lower foothills of the Himalayas. Its natural range lies between 20–32°N lat. and 75–95°E long., where the distribution of sal is usually controlled by climate (Chitale and Behera 2012). Through the species distribution model, Chitale and Behera (2012) identified moisture as the main player to influence the distribution of sal to shift towards northern and eastern India, with more than 90% certainty due to changing climate. The easternmost limit of *S. robusta* distribution recorded from Khasi hill sal forest ecosystem in Meghalaya where mixed dominance of a conifer species, i.e., *Pinus kesiya* was remarkable (Tripathi and Shankar 2014). However, on the south of Meghalaya, sal forests recorded in Tripura where this ecosystem is shifting towards relatively wetter and plain areas due to warming (Majumdar et al. 2014). Sal forests are shifting towards the north-eastern region from a northwestern direction due to higher water-holding capacity, greater moisture content, and other favored edaphic factors with changing climate (Chitale and Behera 2012). Climate change studies also suggest that the north-eastern region of India is much wetter compared to the rest of the area in-country (Ravindranath et al. 2006).

Deshingkar et al. (1997) discussed potential climate change impacts on forests in the northern state of Himachal Pradesh. Furthermore, Ravindranath et al. (2006) in a study concluded that 77 and 68% of the forested grids in India are likely to experience a shift in forest types for climate change under A2 and B2 scenarios, respectively. The subalpine conifer forest zones are estimated to shift upward by over 400 m during 2000–2050 (Zomer et al. 2014).

Climate Change Impact on the Orchard

It is anticipated that the rise in temperature leads to the change in snowfall patterns in different ranges in the Himalayas. Climate change has affected both the apple cultivation and tourism industries in Himachal Pradesh and most apple farmers in Himachal Pradesh. Shetty (2018) reported a decline in apple production and delayed harvest, blaming reduced snowfall due to warmer temperatures.

Climate Change Impact on Phenology and Distribution of Plant Species

Globally, there is evidence of climate change's impact on phenology (flowering cycle) and persistence of species (Gaira et al. 2014). The changes in two important climatic factors, i.e., temperature and rainfall revealed changes in phenology that ultimately influence the natural vegetation of a place (Shrestha et al. 2012) as shown in Table 10.2.

Myrica esculenta is an important ethnomedicinal plant belonging to the family Myricaceae and most commonly found in the subtropical western and eastern Himalayas. Unseasonal (early) flowering in *M. esculenta* in the western Himalayas was one of the impacts of rising temperature due to climate change (Chaukiyal 2011). Similarly, *M. esculenta* is highly impacted by climate change in Nepal where early fruiting was recorded in the plant due to higher temperature (Alamgir et al. 2014).

Rhododendron species found in the IH, are an important, ethnomedicinal, and primitive group of flowering plants known for their considerable ecological and economic importance. The cold, moist slopes and deep valleys provide a conducive habitat for the luxuriant growth of *Rhododendron* species. Arunachal Pradesh (119 taxa) and Sikkim (42 taxa) are home to the highest number of the *Rhododendron* species in India (Mao et al. 2017). Kumar (2012) predicted the shrinkage of Rhododendrons habitat in Sikkim under future climate change scenarios. Though home to 97% of the *Rhododendron* taxa in the country, Rhododendrons in forest ecosystems in the eastern Himalayas are under threat due to indiscriminate felling and loss of habitat (Mao et al. 2017). The phenology (flowering cycle) in *Rhododendron* species in lower elevations starts in February and continues till April while in the case of the higher elevations, it appears in late May and continues till June. During studying the phenology of Rhododendrons, Mao et al. (2017) revealed that *Rhododendrons* are prominent indicators of climate change. In a study conducted in the Sikkim Himalayan region, Telwala et al. (2013) reported a mean upward shift of endemic plants to 27.53 ± 22.04 m/decade between 1849–2010.

Based on Generalized Additive Model (GAM) using real-time field observations (2009–2011) and herbarium records (1893–2003), Gaira et al. (2014) predicted 88–97 days early flowering over the last 100 years in *Rhododendron arboreum* (locally called Buransh) in Uttarakhand. Authors further depicted that annual mean maximum

Table 10.2 Early phenology, distribution of plant species, a decline of fruit production, and habitat loss due to climate change in the Himalayas

S No	Study area	Finding	Source
Early phenology			
1	Western Himalayas	Unseasonal (early) flowering in Myrica esculenta due to rise in temperature	Chaukiyal (2011)
2	Indian Himalayas	88–97 days early flowering over the last 100 years in Rhododendron arboreum due to annual mean maximum temperature	Gaira et al. (2014)
Distribution of plant species			
1	Eastern Himalayas	Shrinkage of habitat of Rhododendron spp.	Kumar (2012)
2	Eastern Himalayas	Upward shift of endemic plants	Telwala et al. (2013)
3	Western Himalayas	Upward shift of Rhododendron arboreum	Gaira et al. (2014)
4	Western Himalayas	Upward shift of Pinus wallichiana of 19 and 14 m per decade	Dubey et al. (2003)
Fruit production			
3	Western Himalayas	Decline in production and delayed harvest in Apple due to rise in temperature	Shetty (2018)
Habitat loss (Animals)			
4	Himalayas	Massive decline of about 73% of the Himalayan Brown Bear habitat by the year 2050 due to rise in temperature	Mukherjee et al. (2021)
5	Himalayas	Habitat loss of the cold-water fish species (Snow trout) of Himalayan rivers by 16% in 2050 and nearly 27% by 2070 due to climate change	Sharma et al. (2021)

temperature is accountable for shifts in phenology in *R. arboreum* and predicted shift of *R. arboreum* towards higher elevations to cope with climate change. *R. arboreum* is a culturally and ecologically important tree and supports the local economy in Uttarakhand, where local people prepare juice/squash, jam, chutney, and even traditional medicine from its flowers. In addition, under future climate scenarios, the habitat of *Betula utilis* will shift northward throughout the Himalayan region as predicted by Schickhoff et al. (2015). One of the studies carried out in the western Himalayas has recorded an upward shift of Himalayan pine (*Pinus wallichiana*), i.e., 1.9 m yr^{-1} on the south and 1.4 m yr^{-1} on the north slope, respectively (Dubey et al. 2003).

Climate Change and Expansion of Alien Invasive Species

Climate change and invasive species are considered major contributors to global biodiversity loss. There are confirmed negative impacts of invasive alien species on native plant and animal species, which can cause several consequences, such as a decline in number or even sometimes extinction of native species because of their high fecundity, thus negatively affecting ecosystems. Alien invasive species can respond better to recent climate change by adjusting their phenological changes, this way, climate change helps the invasiveness of alien species (Willis et al. 2010).

Lantana camara, one of the most aggressive and destructive foreign weeds in India, is an efficient competitor against native species and is known to have a profound impact on biodiversity. It has the potential to spread rapidly in all vegetation types including natural forests, grassland, agricultural land, and become the dominant understorey species in the infested region, blocking natural succession, reducing the number of native species, and decreasing diversity, therefore act as a threat to endemics species and sometimes may lead to species extinction. The plant has already invaded larger forest patches in the subtropical region, and smaller disturbed forest patches in the western Himalayas (Mungi et al. 2018), and tropical to the subtropical forest in the eastern Himalayas. Climate change along with regional anthropogenic pressure could facilitate the more aggressive expansion of *Lantana* in near future in the Indian Himalayan Region (Mungi et al. 2018). Apart from the impact of climate changes on flora, we have discussed in brief about the impact on wildlife in Himalayan regions, this is to assess that climate change is affecting overall ecosystems in Indian Himalayan regions.

Climate Change Impact on Wildlife Animal Species

The large-sized mammals that occur in low densities in the high mountainous region of the Himalayas are globally threatened due to fragile climatic and ecological envelopes (Mukherjee et al. 2021). Himalayan brown bear (*Ursus arctos isabellinus*) is a top carnivore in the high-altitude Himalayan region whose distribution is restricted to high lands Himalayas mainly in northwestern and central Himalayas with relatively small and fragmented populations (Sharief et al. 2020). In a recent study on Himalayan brown bears, Mukherjee et al. (2021) have predicted a significant reduction in suitable habitat and biological corridors of the species due to climate change as it is getting warmer faster than elsewhere in the Himalayas. In their study, Mukherjee et al. (2021) predicted a massive decline of about 73% of the Himalayan brown bear habitat by the year 2050. In a recommendation to mitigate climate change, the authors suggested adopting spatial planning for the protection of natural habitats outside the protected areas.

Cold-water species like fishes are most vulnerable to climate change in the Himalayas due to their limited thermal range (Sharma et al. 2021). The snow trout

(*Schizothorax richardsonii*), the iconic cold-water fish species of Himalayan rivers, would lose their habitat by 16% in 2050 and nearly 27% by 2070 due to climate change as predicted recently by Sharma et al. (2021).

Conclusions

Globally, there is numerous evidence, including species distribution model studies, that support ongoing and future climate change impact on forest ecosystems. Fragile ecosystems like IH could be much affected by increasing temperature in response to global warming. As a result, future climate change could seriously impact the various forest types which are likely to lead to a change in biodiversity, community composition, population structure, and even productivity of the forests including the shift of numerous plants to higher latitudes and altitudes. To understand research trends, we need to identify knowledge gaps and suggest priority research areas in the climate change scenario and its impact on forest ecosystems of IHR. Moreover, the IHR has been a center point of the rich biodiversity of India, and also the limelight for new species discoveries in past decades. Future research is to focus on the conservation of forest and speciation which must potential for effective conservation interventions. There are a few major gaps in the Himalayan studies which are still required for better management of the forest ecosystem, and conserving the surrounding environment and habitat. Moreover, there is an urgent need to assess species distribution (importance of medicinal plants endemic to the Himalayas) and the potential for conservation corridors to support endemic fauna populations due to continuous discoveries of species in the Himalayan regions. Due to the impact of global warming, many species of plants and animals have been shifted to the threatened category in the IH and several species are facing the risk of extinction before someone has even recorded their existence (Tewari et al. 2017). Therefore, effective policies and management strategies along with coordinated efforts are required for the adaptation and mitigation of climate change to save the fragile landscape of the IH. For this purpose, all the stakeholders such as policymakers, decision management, biologists, ecologists, wildlife conservationists, and researchers must come together with directions for future biodiversity assessment, forest management for conservation of flora & fauna at the grass-root level to obviate the eminent impacts of climate change.

Acknowledgements The authors are grateful to Shiv Nadar University, Greater Noida, Uttar Pradesh, India for support.

References

Alamgir M, Pretzsch J, Turton SM (2014) Climate change effects on community forests: finding through user's lens and local knowledge. Small-Scale 13:445–460. https://doi.org/10.1007/s11842-014-9264-8

Anand A, Pandey PC, Petropoulos GP, Palvides A, Srivastava PK, Sharma JK, Malhi RKM (2020) Use of hyperion for mangrove forest carbon stock assessment in Bhitarkanika forest reserve: a contribution towards blue carbon initiative. Remote Sens MDPI 12(4):597. https://doi.org/10.3390/rs12040597

Bhutiyani MR, Kale VS, Pawar NJ (2007) Long-term trends in maximum, minimum and mean annual air temperatures across the northwestern Himalaya during the twentieth century. Clim Chang 85:159–177

Bhutiyani MR, Kale VS, Pawar NJ (2010) Climate change and the precipitation variations in the northwestern Himalaya: 1866–2006. Int J Climatol J Roy Meteorol Soc 30(4):535–548

Champion HG, Seth SK (1968) A revised survey of the forest types of India. Govt, of India, New Delhi

Chaturvedi RK, Gopalakrishnan R, Jayaraman M, Bala G, Joshi NV, Sukumar R, Ravindranath NH (2010) Impact of climate change on Indian forests: a dynamic vegetation modeling approach. Mitig Adapt Strateg Glob Change. https://doi.org/10.1007/s11027-010-9257-7

Chaukiyal SP (2011) Shifting in flowering behavior of *Myrica esculenta* Buch.-Ham. Ex D. Don-a Report. Indian For 137:1350–1352

Chen IC, Hill JK, Ohlemüller R, Roy DB, Thomas CD (2011) Rapid range shifts of species associated with high levels of climate warming. Science 333:1024–1026

Chitale VS, Behera MD (2012) Can the distribution of sal (*Shorea robusta* Gaertn. f.) shift in the northeastern direction in India due to changing climate? Curr Sci 102:1126–1135

Dash SK, Jenamani RK, Kalsi SR, Panda SK (2007) Some evidence of climate change in twentieth-century India. Clim Chang 85:299–321

Deshingkar P, Bradley PN, Chadwick MJ, Leach G (1997) Adapting to climate change in a forest based land use system–a case study of Himachal Pradesh, India. In: Atmospheric Environmental Issues in Developing Countries. Stockholm Environment Institute

Dimri AP, Dash SK (2012) Wintertime climatic trends in the western Himalayas. Clim Chang 111:775–800. https://doi.org/10.1007/s10584-011-0201-y

Dubey B, Yadav RR, Singh J, Chaturvedi R (2003) Upward shift of Himalayan pine in Western Himalaya, India. Curr Sci 85:1135–1136

Gaira KS, Rawal RS, Rawat B, Bhatt ID (2014) Impact of climate change on the flowering of *Rhododendron arboreum* in central Himalaya, India. Curr Sci 106:1735–1738

Guhathakurta P, Rajeevan M (2008) Trends in the rainfall pattern over India. Int J Climatol 28:1453–1469

IPCC (2014) Climate change 2014: impacts, adaptation, and vulnerability. Part A: Global and Sectoral aspects. In: Field CB et al (eds) Contribution of working group II to the fifth assessment report of the intergovernmental panel on climate change. Cambridge University Press, Cambridge

Kelly AE, Goulden ML (2008) Rapid shifts in plant distribution with recent climate change. Proc Natl Acad Sci USA 105:11823–11826

Kirschbaum MUF, Cannell MGR, Cruz RVO, Galinski W, Cramer WP (1996) Climate change impacts on forests. In: Watson RT, Zinyowera MC, Moss RH, Dokken DJ (eds) Climate change 1995. Impacts, adaptation and mitigation of climate change: scientific-technical analyses. Cambridge University Press, Cambridge

Kueppers LM, Snyder MA, Sloan LC, Zavaleta ES, Fulfrost B (2005) Modeled regional climate change and California endemic oak ranges. Proc Natl Acad Sci USA 102:16281–16286

Kumar P (2012) Assessment of impact of climate change on Rhododendrons in Sikkim Himalayas using Maxent modelling: limitations and challenges. Biodivers Conserv 21:1251–1266

Loarie SR, Carter BE, Hayhoe K, McMahon S, Moe R et al (2008) Climate change and the future of California's endemic flora. PLoS ONE 3:e2502. https://doi.org/10.1371/journal.pone.0002502

Majumdar K, Dutta BK, Shankar U (2014) Lower altitudinal variation in habitat associations, tree diversity and co-dominant population structures along moist deciduous forests dominated by sal (Shorea robusta) in northeast India. For Sci Technol 10:201–212

Malhi RKM, Anand A, Mudaliar AN, Pandey PC, Srivastava PK, Sandhya Kiran G (2020) Synergetic use of in situ and hyperspectral data for mapping species diversity in Shoolpaneshwar Wildlife Sanctuary, Gujarat. Trop Eco (Springer) Special issue (Ecosystem Function and Remote Sensing). 61:106–115. https://doi.org/10.1007/s42965-020-00068-8

Mao AA, Dash SS, Singh P (2017) Rhododendron of North East India: a pictorial handbook. Botanical Survey of India

Mittermeier RA, Gil PR, Hoffmann M, Pilgrim J, Brooks T, Mittermeier CG, Lamoreux J, Da Fonseca GAB (2004) Hotspots revisited: Earth's biologically richest and most endangered terrestrial ecoregions. CEMEX, Mexico City

Mukherjee T, Sharma LK, Kumar V et al (2021) Adaptive spatial planning of protected area network for conserving the Himalayan brown bear. Sci Total Environ. https://doi.org/10.1016/j.scitotenv.2020.142416

Mungi NA, Coops NC, Ramesh K, Rawat GS (2018) How global climate change and regional disturbance can expand the invasion risk? Case study of *Lantana camara* invasion in the Himalaya. Biol Invasions 20:1849–1863. https://doi.org/10.1007/s10530-018-1666-7

NMSHE (2015–16) National mission for sustaining Himalayan ecosystem. Department of Science and Technology, Ministry of Science and Technology, Government of India. http://www.knowledgeportal-nmshe.in/Home.aspx

NOAA (2009) The state of the climate. US National Oceanic and Atmospheric Administration, US Department of Commerce.

NOAA (2020) Climate change impacts. National Oceanic and Atmospheric Administration, US Department of Commerce. https://www.noaa.gov/education/resource-collections/climate/climate-change-impacts

Padma TV (2014) Himalayan plants seek cooler climes—race is on to record mountain biodiversity before it is lost. Nature 512:359

Pandey PC, Baltzer H, Tate N (2014) Mapping tree species in coastal portugal using statistically segmented principal component analysis and other methods. IEEE Sens J 14(12):4434–4441

Pandey PC, Akash A, Srivastava PK (2019a) Spatial distribution of mangrove forest species and biomass assessment using field inventory and earth observation hyperspectral data, biodiversity and conservation 28(8–9):2143–2162. ISSN: 1572-9710, https://doi.org/10.1007/s10531-019-01698-8

Pandey PC, Srivastava PK, Chetri T, Choudhary BK, Kumar P (2019b) Forest biomass estimation using remote sensing and field inventory: a case study of Tripura, India. Environ Monit Assess 191(593) https://doi.org/10.1007/s10661-019-7730-7

Ravindranath NH, Joshi NV, Sukumar R, Saxena A (2006) Impact of climate change on forests in India. Curr Sci 90:354–361

Schickhoff U, Bobrowski M, B J et al (2015) Do Himalayan treelines respond to recent climate change? An evaluation of sensitivity indicators. Earth Syst Dynam 6:245–265

Shankar U, Tripathi AK (2017) Rainforests north of the Tropic of Cancer: physiognomy, floristics, and diversity in 'lowland rainforests' of Meghalaya, India. Plant Divers 39:20–36

Sharief A, Joshi BD, Kumar V et al (2020) Identifying Himalayan brown bear (*Ursus arctos isabellinus*) conservation areas in Lahaul valley Himachal Pradesh. Global Ecol Conserv 21:e00900. https://doi.org/10.1016/j.gecco.2019.e00900

Sharma A, Dubey VK, Johnson JA, Rawal YK, Sivakumar K (2021) Is there always space at the top? Ensemble modeling reveals climate-driven high-altitude squeeze for the vulnerable snow trout *Schizothorax richardsonii* in Himalaya. Ecol Indic 120:106900

Shetty D (2018) In the Himalayas, living the crisis that the IPCC report warns of. https://www.indiaspend.com/in-the-himalayas-living-the-crisis-that-the-ipcc-report-warns-of/

Shrestha UB, Gautam S, Bawa KS (2012) Widespread climate change in the Himalayas and associated changes in local ecosystems. PLoS ONE 7:e36741. https://doi.org/10.1371/journal.pone.0036741

Telwala Y, Brook BW, Manish K, Pandit MK (2013) Climate-induced elevational range shifts and increase in plant species richness in a Himalayan biodiversity epicentre. PLoS ONE 8:e57103. https://doi.org/10.1371/journal.pone.0057103

Tewari VP, Verma RK, Gadow KV (2017) Climate change effects in the Western Himalayan ecosystems of India: evidence and strategies. Ecosyst 4:13. https://doi.org/10.1186/s40663-017-0100-4

Tripathi AK, Shankar U (2014) Patterns of species dominance, diversity and dispersion in 'Khasi hill sal' forest ecosystem in northeast India. Ecosyst 1:1–20

Trisos CH, Merow C, Pigot AL (2020) The projected timing of abrupt ecological disruption from climate change. Nature 580:496–501. https://doi.org/10.1038/s41586-020-2189-9

Willis CG, Ruhfel BR, Primack RB et al (2010) Favorable climate change response explains non-native species' success in Thoreau's woods. PLoS ONE 5:e8878

Yadav RR, Singh J, Dubey B, Chaturvedi R (2004) Varying strength of relationship between temperature and growth of high-level fir at marginal ecosystems in western Himalaya, India. Curr Sci 86:1152–1156

Zomer RJ, Trabucco A, Metzger MJ, Wang M, Oli KP, Xu J (2014) Projected climate change impacts on spatial distribution of bioclimatic zones and ecoregions within the Kailash Sacred Landscape of China, India, Nepal. Clim Chang 125:445–460

Chapter 11
Climate Change and Its Impact on Indian Himalayan Forests: Current Status and Research Needs

Hukum Singh and **Manoj Kumar**

Abstract Climate change is affecting global natural resources, including forests. It has also affected the Indian Himalayan forests by influencing ecosystem services derived from it. The majority of the Indian population depends directly or indirectly on these services, ultimately affecting Himalayan communities. The climate change impacts may alter the structure, function, and composition of the Himalayan forest and are expected to influence the region's biodiversity. The assessment suggests a more significant rise in temperature of the western parts compared to the eastern part of the Indian Himalayas. Climate change effects are manifested as species range shift, phenological changes, changes in growth patterns, host-parasite interactions, insect pest incidence, habitat adaptability, biogeochemical interactions, and plant-animal-resource interactions, and hydrological behavior, etc. This chapter focuses on the issues of climate change and its implications on the forest ecosystem of the Indian Himalayas. It also covers the key issues, research gaps, and future research needs for the region concerning climate change studies. The constructed state of knowledge about the climate change impacts may provide insight into the forest ecosystem of the region. It will help researchers and decision-makers to formulate and prioritize adaptation and mitigation-related research to reduce climate change effects in the present and future.

Keywords Climate change · Forest ecosystems · Phenology · Ecosystem services · Vulnerability · Indian Himalayan region

Introduction

Climate change has morphed into a worldwide crisis caused by multiple anthropogenic activities, especially the blazing of fossil fuels and land-use changes.

H. Singh (✉)
Plant Physiology Discipline, Genetics and Tree Improvement Division, Forest Research Institute, Dehradun, Uttarakhand, India

M. Kumar
GIS Centre, Forest Research Institute, Dehradun, Uttarakhand, India

S. Rani and R. Kumar (eds.), *Climate Change*, Springer Climate,
https://doi.org/10.1007/978-3-030-92782-0_11

Human-induced greenhouse gases (GHGs) emissions since the pre-industrial epoch have driven an immense accumulation of concentrations of carbon dioxide (CO_2—400 ppm), methane (CH_4—1800 ppb), and nitrous oxide (N_2O—330 ppb), and other GHGs in the atmosphere (IPCC 2014). It is predicted that if the increase in CO_2 emissions continues, the atmospheric CO_2 concentration will reach up to 720–1000 ppm (IPCC 2014). This may enhance the air temperature to around 2.6–5.4 °C by 2100 (IPCC 2014). Given the current rate of fossil fuel consumption, it is predicted that the Earth's surface temperature will rise by 0.4–0.6 °C on an average during the twenty-first century (Norby et al. 2003). For India, it is expected to warm by 0.5 °C by 2030 (around equivalent to the warming in the twentieth century) and 2–4 °C by the twenty-first century, with the most significant rise in the northern part of India (Shukla 2003; National Intelligence Council 2009). The accumulation rate of CO_2 in the atmosphere is 3.5 Pg per year (Pg = 1015 g or billion tons) (Albrecht and Kandji 2003). The immense contribution of anthropogenic CO_2 emissions into the environment comes from blazing non-renewable energy sources, particularly fossil fuels, and converting tropical forests to agricultural production (www.fao.org). The remaining contribution is mainly because of land-use changes, especially deforestation (Paustian et al. 2000; IPCC 2001).

Globally, climate change is perceived as a considerable danger and is at the focal point of scientific and political debate in recent years. India has a rationale to be worried about the issues of climate change. India's large population relies upon the agriculture and forestry sectors which are climate-sensitive sectors (Upgupta et al. 2015). The adverse effects on water resources such as the decline of glaciers, reduced rainfall, and increased flooding in certain areas also threaten food security (Smadja et al. 2015; Jain and Singh 2018). Climate change is having negative impacts on natural ecosystems, including the Himalayan ecosystems (Tewari et al. 2017). Compared with other hilly regions in the world, the Himalayas face a relatively high-temperature rise. According to reports, the Himalayas' rate of temperature rise (0.06 °C/year) is about three times the global average. Due to the rapid warming of the Himalayas, mountain ecosystems have shown changes in the diversity as well as distribution of flora and fauna (Shrestha et al. 2012; Chhetri et al. 2021).

Climate change is perceived as one of the critical threats to biodiversity. India's climate is highly assorted, varied from the sub-frozen Himalayan winter towards the tropical climate of the south (Krishnan et al. 2020). The Intergovernmental Panel on Climate Change (IPCC) has recognized the Himalayan ecosystem as fragile and highly vulnerable to changing climate (IPCC 2007). The Indian Himalayan Region (IHR) is a mega hotspot of biodiversity and a repository of valuable medicinal plants (Rana et al. 2017). The region offers various ecosystem services essential for the sustenance of humans. Climate change is influencing these services mainly attributed to increasing atmospheric CO_2 concentration, an anomaly in rainfall pattern, rising air/surface temperature, melting and shrinkage of glaciers, etc. Himalaya's immense water reservoir in ice (also referred to as the third pole) is also under threat of climate change (Smadja et al. 2015; Jain and Singh 2018). The impact of climate change on the Himalayan forest ecosystem has been manifested as species range shift, phenological shifts in plants' life cycle, forest growth patterns, changes in ecosystem boundaries, and other biological and non-biological responses/stresses (Upgupta et al. 2015; Singh et al. 2017; Zheng et al. 2021). The physiological response of forests under climate change influence ultimately changes region's biodiversity (Bellard et al. 2012; Saxena and Rao 2020; Kumar et al. 2021b, 2021f). All of these alterations affect endemic species' habitats and the region's biodiversity (Mehta et al. 2020).

The continuous and systematic observation of climate change influence on the structure, function, composition, phenology, etc., is not adequately addressed for the region. The systematic and continuous long-term observations are essential to developing a clear understanding of climate change's impact on the Himalayas forest ecosystems (Negi et al. 2019). In the absence of adequate information, the IPCC lists the Himalayas as a white spot in its documents and emphasizes the need for systematic observation of this area (IPCC 2007).

The response of forest ecosystems to disturbances ranges from a year to many years or even hundreds of years. It largely depends on the system's state, conditions, nature, intensity, and disturbance duration. It is to consider the current state of forest ecosystems as a consequence of both the present and past events (Tewari et al. 2017). Besides, the belowground processes are the key drivers that regulate the biodiversity of the ecosystem. The belowground system of vegetation to changing climatic conditions is very complex and has not been adequately dealt with. Vegetation affects the type and quantity of carbon inflowing the soil system and affects the plant's root area physical structure (Watham et al. 2020; Saxena and Rao 2020). This effect is considered to be an indirect effect on the microbial community and its composition. Factors such as water, temperature, nitrogen, and other nutrients directly impact the microbial community because organisms respond to temperature or drought stress (Kumar et al. 2020a; Singh et al. 2021; Verma et al. 2021). At the same time, the microbial community is also under the influence of changes in resource availability. Plants would be affected by microbial responses to climate change, whether directly or indirectly, due to input in the form of nutrient supply made available to plants by microbial action (Pugnaire et al. 2019). Therefore, the critical step in understanding the ecosystem response to climate change must consider understanding the microbial community's role (Heath et al. 2005; Leaky et al. 2009).

The vegetation responses to changing climatic conditions such as elevated CO_2 and temperature and other climatic variables show a wide range of patterns (Saxena and Rao 2020; Scheiter et al. 2020). These changes could be either structural modulation, a functional response such as rate of accumulation of primary and secondary metabolites, or the changes in health-promoting substances or the biomass production (Tewari et al. 2017; Apurva et al. 2017; Apurva and Singh 2017; Singh et al. 2018; Sharma et al. 2018; Sharma et al. 2019; Yadav et al. 2019a, 2019b; Kumar et al. 2021c; Singh et al. 2021). There is evidence that dominant factors, especially CO_2 and air temperature rise, cause noticeable impacts on life cycle, species distribution, physiological behavior, and plants' biochemical components (Gairola et al. 2010; Singh et al. 2014; Gupta et al. 2018b, Kumar et al. 2020c; Singh et al. 2021; Sharma and Singh 2021).

The IHR's higher altitude is more vulnerable to climate change's effects (Apu and Ghimire 2015). It has been stated that changes in alpine ecosystems, fragmentation of habitats, changes in the distribution range of species, changes in phenological patterns, changes in secondary metabolites (Kumar et al. 2012), and the invasion of alien species will impact the biodiversity of the region (Thapa et al. 2018). Climate change will also affect the Himalayas' medicinal plants (Kumari and Singh 2018; Kumar et al. 2019a, 2019b; Gaira and Dhar 2020; Yadav et al. 2021; Dhyani et al. 2021; Kumar et al. 2021d). Telwala et al. (2013) reported that Sikkim's endemic plant species in the Himalayas had moved upward by 27.53 m per decade during 1849–2010. Changes in the habitat range of trees (Kelly and Goulden 2008), birds (Freeman and Freeman 2014), moths (Chen et al. 2009), and butterflies have been observed globally (Konvicka et al. 2003). Several organisms may face extinction as a result of these changes, which may also affect population structure and function. Moreover, not only the floral species are impacted due to climate change, but similar effects are evident on faunal species in the IHR (Singh et al. 2020, 2020c).

The Himalayan biodiversity is also expected to change as a result of projected climate change scenarios (changes in temperature, rainfall, and so on) (Gilani et al. 2020). Species may adapt to new environmental conditions or adjust their distribution to follow suitable habitat conditions or face extinction if they cannot move or adapt (Singh et al. 2020a).

The scientific community has paid increasing attention to the effect of climate change on global biodiversity hotspots in recent decades; however, scientific information concerning the Himalayas is limited (Shrestha et al. 2012; Telwala et al. 2013; Kumar et al. 2018; Singh et al. 2020a). As a result of the Himalayan habitats' unique evolutionary history and complexity, it is necessary to conduct a more systematic and continuous observation of the biological response to changing climatic circumstances (Pandit et al. 2007; Kumar et al. 2018; Pandey et al. 2020; Rashid and Romshoo 2020). Research on these topics is crucial for understanding forest ecosystem processes and functioning to establish the knowledge base required to assess and predict climate change impacts (Rawat et al. 2020). Although little is known about the timing and extent of specific evolutionary processes linked to climate change, developing scientific literature is essential to understand such processes. It is of utmost requirement to undertake long-term and systematic scientific observation and monitoring to understand the potential effects of climate change on the terrestrial structure. This will improve climate change's knowledge and experience and its linkages with forest ecosystem function and processes (Pandit et al. 2007; Rawat et al. 2020).

Having discussed these facts in mind, the present study attempts to understand the current state of knowledge, and future research needs to highlight issues of the IHR related to (i) the biodiversity of the region, (ii) key drivers of changes, (iii) changes in the various attributes of Himalayan forests, (iv) phenological changes, (iv) ecosystem services, (v) species range shift and (vi) response of alpine tree line. The compilation of information is expected to help to address climate change-related issues. Simultaneously, the study also highlights significant research questions directly linked to climate change and biological diversity, including flora, fauna, and microbes in the IHR.

Biodiversity of the IHR

The IHR covers an area of about 750,000 km^2 (spread over 3000 km in length and 250–300 km in width), which lies between ~300 and 8000 m above mean sea level (AMSL) (https://nmhs.org.in/BCM.php). It has a variety of landscapes and various soil forms with varying climatic conditions. It is a dynamic landscape with wealthy biodiversity. The IHR has a high degree of endemism and is noted for its rare flora and fauna (https://nmhs.org.in/BCM.php). This region occupies nearly 10,000 species, 300 mammals species, 979 bird species, 176 reptiles species, 105 amphibian species, and 269 freshwater fish species (Kumar and Chopra 2009; https://nmhs.org.in/BCM.php). Their various levels of endemism are presented in Table 11.1. This region has about 1748 (23.4% of India) species of medicinal plant with more than 675 wild edible species (Kumar and Chopra 2009; https://nmhs.org.in/BCM.php). Besides, this landscape hosts diverse ethnic groups inhabiting the remote and challenging terrains. The communities of the IHR have traditionally been dependent on bio-resources to support their livelihood (Kumar and Chopra 2009; Rautela and Karki 2015; https://nmhs.org.in/BCM.php).

It can be seen that changing climatic circumstances alter the habitats of numerous species, including plants and animals. To find suitable conditions for their survival, they must adapt or migrate to areas with favorable conditions. Air temperature is one of the key climate drivers that strongly affect the ecosystems of this region. Even small fluctuations in average air temperature can significantly affect habitat structure, functioning, and biodiversity. The interconnected nature of ecosystems means that the loss of species may affect a series of ecosystem functions. Climate change would significantly impact the physiological changes of many species (Kumar et al. 2021f). There is evidence that certain species are physiologically susceptible to temperature spikes (Gupta et al. 2018a; Kumar et al. 2021a; Kumar et al. 2021f).

The species diversity improves the ecosystem's ability to sustain multiple functions such as soil binding, maintaining the soil's health, retaining soil fertility, supplying clean water to streams and rivers, nutrient cycling, etc. All such benefits are sometimes referred to as "ecosystem services" while the role is referred to

Table 11.1 Biodiversity of the Indian Himalayan Region

S. No	Group	Species	
		Total	Endemic
1	Mammals	300	12
2	Plants	10,000	3160
3	Birds	979	15
4	Amphibians	105	42
5	Reptile	176	48
6	Fishes (freshwater)	269	33

(Kumar and Chopra 2009; retrieved on 27/11/2020 from https://nmhs.org.in/BCM.php)

as "ecosystem function". Species loss may reduce these services and functioning significantly when environmental conditions change rapidly. Wherever a species disappears, it is evident that the functioning of ecosystems and their services change. These changes are linked to land degradation, changes in the forest, agricultural productivity, and a decline in the water supply's quantity and quality.

Climate change-induced variations such as tree phenology, tree growth dynamics, and species range shift including shifting tree line in the alpine, and changes in the species' habitat influence the landscape's biological diversity (Schickhoff et al. 2015; Lü et al. 2020; Bagaria et al. 2021; Zheng et al. 2021). Recent studies of climate modeling suggest a significant shift in the habitat and distribution of floral and faunal species leading to range expansion towards higher altitudes (Walther et al. 2005; Telwala et al. 2013; Subha et al. 2018; Singh et al. 2020; Adhikari and Kumar 2020; Kumar et al. 2021e; Mishra et al. 2021). Nevertheless, these observations highlight the uniqueness of species' feedback to climate change, which is expected to affect the ecosystem's floral and faunal diversity.

Key Drivers of Changes in the Biodiversity

The Himalayan mountain ecosystems have been recognized as one of the hotspots of biodiversity. The Himalayan mountains have a unique climate compared to other climatic zones (Korner 2002; Telwala et al. 2013). Various changes have been witnessed in the distribution pattern of flora and fauna of the region due to climate change (Singh et al. 2020a; Negi and Mukherjee 2020). The upward expansion and range shift of various species has been reported in the Himalayan ecosystem (Negi et al. 2017; Telwala et al. 2013). The population of different biological organisms has declined due to anthropogenic interventions besides climate change (Bellard et al. 2012). Climate change is thought to have contributed more to population loss in some cases than other factors; as a result, climate change has emerged as one of the prominent drivers of change in this region (Bellard et al. 2012). The multiple drivers well-known in the region are (i) climate change, (ii) encroachment of habitats, (iii) land-use/land cover (LULC) change, (iv) fragmentation of land, (v) forest fire, (vi) livestock grazing and fodder collection, (vii) deforestation, (viii) harvesting of biomass by local communities, (ix) expansion of agricultural land into forest lands, (x) overexploitation of medicinal and aromatic plants, (xi) use of chemical fertilizers as part of modern agriculture, (xii) introduction of invasive alien species, (xiii) unsustainable patterns of ecotourism, etc. Various researchers who have worked in the IHR have reported climate change as the main driver or threat to Himalayan biodiversity (Jetz et al. 2007; Salick et al. 2014; Subha et al. 2018; Singh et al. 2020a).

Forest Cover and Phenological Changes Owing to Climate Change

In recent decades, the world's total forest cover has faced degradation and deforestation because of the growing pressure of human population and its demands (Chakraborty et al. 2018; Kumari et al. 2019). Additional pressure from climate change on forest ecosystems is leading to forest degradation (Singh et al. 2020b). The climate change impacts on forest ecosystems are long-term and irreversible. The forest cover area had decreased by approximately 11 million km^2 by 1990 (Ramankutty and Foley 1999). Primarily, most of the deforestation happened in temperate regions until the mid-twentieth century. However, land abandonment in Western Europe and North America has increased in recent decades, while deforestation in tropical areas has increased exponentially. The rate of net loss of tropical forest cover in America slowed in the 1990s as compared to the 1980s, but it increased in Africa and Asia (IPCC 2007). In most areas of Asia, climate change is expected to exacerbate the threats to biodiversity posed by LULC changes and population pressures. The IPCC reported this concern with high confidence. Many species in Asia could face extinction due to the synergistic effects of climate change and habitat degradation (Holyoak and Heath 2016). There will be more threats to the ecological health of wetlands, mangroves, and coral reefs in Asia. The extent and scale of potential forest fires in North Asia are likely to increase in the future due to climate change and severe weather conditions that could restrict forest expansion (Backlund et al. 2008; Savita et al. 2017).

Climate change is widely acknowledged to have a significant effect on organisms' phenological behavior. Phenological behavior is one of the biological indicators of climate change. Phenology is looking at the recurring life cycle of plants and animals affected by environmental and climatic conditions, especially seasonal changes (Cleland et al. 2007). These variations consist of seasonal changes in temperature and rainfall forced by weather and climate, so phenological events' timing is an excellent indicator of climate change (Malik et al. 2020a, 2020b). When researchers integrate biological clocks into climate monitoring, another term, i.e., seasonality, is used (Aryal et al. 2020). Seasonality is a concept that refers to non-biological phenomena that are identical, such as the timing of when the fall forms on a freshwater lake in the autumn and when it breaks in the spring (Malsawmkima and Sahoo 2020). Some examples of flower phenology events include budding, leafing, plants' flowering in spring, leaf color changes in autumn, etc. (Gupta and Singh 2017; Kumar et al. 2019b; Thapliyal et al. 2020). As far as animal phenology events are concerned, a few important events are bird migration and nesting, insect hatching, and animal emergence from hibernation. Phenological monitoring provides independent measures of the climate change impact on organisms (Bagaria et al. 2020). At the ecosystem level, phenological monitoring at different stages of the food chain (plant growth, insect hatching, and bird feeding/nesting) can shed light on the "rippling effect" of climate change (Naylor et al. 2007).

It is believed that changing climate with increasing concentration of atmospheric CO_2 would enhance plant growth (Singh et al. 2018; Sharma et al. 2018; Yadav et al. 2019a). According to ecological research from a diverse variety of habitats, species adapt to climate change in extremely heterogeneous ways (Donner et al. 2005; Parmesan 2006; Khanduri et al. 2008; Singh et al. 2010; Willis et al. 2010).

Climate change will force species to either adjust their phenological characteristics (the timing of seasonal events including leaf sprouting, leaf shedding, planting, and fruiting, as well as seed germination) to adapt to new environments or migrate to more suitable ecosystem conditions to avoid extinction (Willis et al. 2010; Prajapati et al. 2020; Basnett and Devy 2021). Well-established ecological records indicate that plant phenological events have changed, such as early leaves, flowering, fruiting, and the increase in the growing season's length in recent decades (Khanduri et al. 2008; Singh et al. 2010; Kumar et al. 2019a). Although research on species' adaptation responses is growing, it is expected that the effect of climate change will be so gradual that most species' life history characteristics will not be able to keep up (Donner et al. 2005). Therefore, to survive, most species will be forced to change their range (Parmesan 2006).

The phenological changes in various plant species of the Himalayas in the Uttarakhand state have been reported. A study by Negi et al. (2017) reported early flowering, fruiting, and leaf emergence (about 20–25 days) of various Himalayan plants such as *Berginia ligulata*, *Allium stracheyi*, *Prunus cerasoides*, *Rhododendron arboreum*, *Bauhinia variegate*, and *Bombax cieba*. An important plant species, i.e., *Rhododendron arboreum* was observed to have early flowering in January, whereas normal flowering is from February to March (Negi et al. 2017). Most findings report that changing climatic variability, influences plant species' phenological behavior, including crops (Salick et al. 2014; Kumar et al. 2018, 2019a). The tree lines in the Himalayas have shifted as a result of phenological behavior and species adaptation (Schwab et al. 2018). Negi et al. (2017) and Singh et al. (2020) traced changes in distribution, population density, and regeneration patterns of tree species in Uttarakhand's Himalayan ecosystem. Moreover, faunal species such as birds including bats and insects have demonstrated changes in their phenological behavior, such as the prior beginning of the migration, egg-laying, and breeding (Thapa et al. 2021). Such changes influence the composition of ecosystems, thereby affecting ecosystem functioning and services (Upgupta et al. 2015; Negi and Rawal 2019).

Ecosystem 'Services' Response to Climate Change

As per the Millennium Ecosystem Assessment (MEA), climate change can be one of the key causes of biodiversity loss by the end of the twenty-first century (MEA 2005). Bellard et al. (2012) conducted that changing climate conditions force biodiversity to respond to changing habitat conditions, life cycles, or the evolution of novel physical characteristics (Negi and Rawal 2019). The global average temperature of the Earth has risen by 0.74 °C (Hannah et al. 2007). Besides, other factors such as rainfall patterns and the occurrence of extreme events have also increased. These changes have not been consistent at a spatial or temporal scale, and the range of variability in the climate has been witnessed in the Himalayan region (Singh et al. 2020a). Climate change has influenced the biological and physical system of the Himalayas, particularly commencement, length and end of the season, glacier melting and retreat, and habitat shift of the species (Upgupta et al. 2015). Such changes also have resulted in changes in the regime of water availability affecting biological survival (Hannah et al. 2007).

Sometimes such changes affect biological diversity and ecosystem services positively (Upgupta et al. 2015). The changes linked with increasing temperature and CO_2 concentration might have improved the net primary productivity (NPP) of the Himalayan region because NPP had an increasing trend for 2004–2014 (Kumar et al. 2018).

The model-based studies advocate that climate change may impact the Himalayan ecosystem's functioning (Hannah et al. 2007; Singh et al. 2020). Climate change may also alter nutrient cycling and biogeochemical cycle under changes in litterfall pattern and decomposition rate (Kumar et al. 2018). Therefore, it is required to investigate climate change impacts on other ecosystem services, including supply of food, fiber, timber, carbon sequestration, water regulation and supply, host-parasite interactions, etc. (Joshi and Singh 2020; Joshi et al. 2021). Conversely, there is uncertainty regarding the magnitudes and extent of such impact influencing ecosystem services flow. Furthermore, there is ambiguity concerning how potential climate change becomes permanent in terms of ecosystem conditions and resources, which needs to be studied.

Climate Change Affects Alpine Tree Line and Range Shift of Species

The observation of the alpine tree line response can be used as one of the biological indicators to trace the changes in the Himalayan ecosystem's functioning due to climate change (Schickhoff et al. 2015; Singh et al. 2020). Various terms are used for such studies among the scientific community. To have a better understanding, these terms have been explained in Table 11.2. Among all the terminology, the term "Alpine tree line" has widely been used for climate change-related studies impacting the alpine vegetation of the Himalayas.

People have high confidence that recent warming has had an intense impact on terrestrial biological environments such as early spring activities, the timing of leaf unfolding, bird migration and spawning, and the transfer of habitat range to higher elevations (Gaire et al. 2020). According to satellite-based observations since the early 1980s, it is believed that due to global warming, many areas have seen a trend of early vegetation greening in spring, which is related to the long hot season (IPCC 2007). Approximately 20–30% of the animal and plant species measured so far

Table 11.2 Terminologies used for studying ecological dynamics in the mountain ecosystem and their explanation

Term	Definition	Interpretation
Treeline	It expresses the maximum elevation range up to which tree exists	This term must not be inferred as the timberline, which is usually referred to as the commercially important timber species
Timberline	It denotes the ecological identity of the upper altitudinal limit of tree growth in the mountain and the Himalayan ecosystem (Negi 2012)	The term timberline refers to the highest limit for commercial timber species (Negi 2012)
Alpine treeline	It represents the maximum elevation range to which tree exists in a mountain ecosystem such as Himalayas (Negi 2012)	The term "Alpine treeline" is the best-fitted ecological term for forest trees at a higher altitude, reflecting the science behind ecological principles (Negi 2012)
Forest line	It is the uppermost elevations of the nearest forest stands in the mountain or Himalaya (Singh et al. 2009)	This term considers all the species including trees; hence sometimes, it creates confusion to define a treeline (Negi 2012)
Upper treeline	Existence of treeline at the limit of the highest altitude in the mountain ecosystem (Negi 2012)	It creates a lower line's utopian existence, which has no ecological relevance (Negi 2012). It's challenging to determine the upper treeline in high mountains or higher altitudes (Miehe et al. 2007)
Ecotonal zone	It refers to the highest mountain vegetation's transitional boundary, including herbs, shrubs, and trees (Negi 2012)	This zone may occur at more than one elevation at the upper and lower borderline. It could not express climatic circumstances existing similarly at the treeline, even within an ecotonal area (Negi 2012). Hence, this term diverges from ecology principles that do not excuse the strength of treeline in ecological studies (Negi 2012)

would pose a greater risk of extinction if the global average temperature increases more than 1.5–2.5 °C (IPCC 2007). Species range shift under the influence of climate change has also been supported by the paleo ecological records (e.g., fossils) (Coope and Wilkins 1994). Climate change affects community structure, species abundance, species interactions, and the habitat's shifting (Sala et al. 2020; Jetz et al. 2007; Bhandari et al. 2019, 2020).

Telwala et al. (2013) reported that about 90% of the alpine endemic plant species in the Sikkim Himalayas shifted their range of existence in the known history of period 1850–1909 by 23–998 m during the assessment of 2007–2010. The upward displacement rate per decade has been reported as 27.53 m in the IHR (Telwala et al. 2013), while in the Alps, this is reported as 34.26 m (Walther et al. 2005). It is predicted that the suitable habitat of birch (*Betula utilis*), the main tree species in the natural tree line of the Himalayas, will move upwards in the eastern Himalayas by the years 2050 and 2070 under the influence of climate change while there will be a net decrease in suitable habitat in the western Himalayas (Hamid et al. 2019; Roy and Rathore 2019). The endemic species of the Himalayan ecosystem are susceptible and more vulnerable to climate change (Bhattacharjee et al. 2017; Ahmad et al. 2021). It is predicted that Himalayan native angiosperms are expected to lose 16% and 18% of their viable habitat by 2050 and 2070, respectively (Manish et al. 2016). However, it is expected that by 2050 and 2070, the alpine meadow area will decrease by 1% and 3%, respectively. The expansion of shrub habitat has been witnessed towards the northern part of the Sikkim Himalaya (Manish et al. 2016).

Future Research Needs for Addressing Climate Change-Related Issues

As very little information is available concerning the time and extent of the ecological impact in the IHR. Thus, there is an urgency to improve scientific understanding. The dynamics of forest ecosystems operate on multiple time scales, so long-term observations must determine the key factors that control ecosystem structure and function (Sekar et al. 2017; Rawat et al. 2020). Retrospective experiments expand the period of long-term studies by producing reference data and allowing calibration (Singh and Thadani 2015).

The (United Nations Framework Convention on Climate Change) (UNFCCC 34) Subsidiary Science and Technology Advisory Agency (SBSTA) dialogue on research needs and priorities stressed the importance of maintaining ongoing, comprehensive measurements and expanding the reach of observations that might be employed for climate change impact studies (such as the Himalayas) (Rawat et al. 2020; Verma et al. 2020). Accordingly, various dimensions of the Himalayan ecosystem have been identified where systematic and continuous long-term observations are required to investigate the climate change effects (Singh and Thadani 2015; Sekar et al. 2017). Some of the thematic areas of research relevant in the light of climate change for the IHR are (i) biological diversity, (ii) species distribution and composition, (iii) species range shift, (iv) shifting of the alpine tree line, (v) phenological changes, (vi) biogeochemical interactions, (vii) insect pest interaction, (viii) biochemical constituents of medicinal plants, (ix) soil microbial dynamics, and other soil fauna under climate change, etc. These thematic areas need to address the following prominent questions.

- What impact will climate change have on biodiversity (including plants, animals, insects, fungi, microbes, etc.) species populations, distribution, and range?
- What will be the impact of climate change on floral and faunal phenology?
- What effect will climate change have on the abundance, distribution, and migration of insect pests?

- How will the host-parasite interaction change due to climate change?
- How will the pollinators respond to different climate change scenarios?
- What will be the impact of climate change on biogeochemical cycling?
- What will be its influence on different plant functional traits (such as regeneration, morphology, transpiration, photosynthesis, growth, etc.)?
- How will climate change impact bioactive ingredients of medicinal and aromatic plants?
- How will the climatic variables impact the hydrological services?
- What will be its impact on the snow and glacier reservoirs of the Himalayas?
- What will be its impact on forest fire incidences?
- What will be the spatial and temporal variation of climate change vulnerability, and how can this be minimized?
- How will the adaptation and mitigation potential of the Himalayan ecosystem be influenced?
- How will the Himalayan species respond to different gradients of CO_2, temperature, soil moisture, and humidity?
- What will be its influence on the resource use efficiency of Himalayan plants?
- What will be its impact on spatial and temporal variation of productivity?
- What will be its impact on the multiple ecosystem services derived from the Himalayas?
- What is the paleo-climatological evidence, and how could this be linked to climate change studies?
- What is the effective feedback between the atmosphere and Himalayan forests, and how are they governed under changing climate?
- How could multiple complex forces be integrated simply and scientifically to represent them in a computer-based model?

Conclusions

It is concluded that climate change would have multiple effects on the Himalayan forest ecosystem (endemic species, host-parasite interactions, ecosystem boundaries and range shift, changes in alpine tree line, habitat alterations, phenological modulations, carbon sequestration potential, genetic diversity). Its related disturbances such as floods, droughts, wildfires, etc. are probably to become more frequent soon. Hence, it is necessary to prioritize central research questions emerging due to climate change to ensure the Himalayan forest ecosystem's sustainability. Forest ecosystem response, understanding, and linkages with the climatic and other variables are essential for formulating any plan to mitigate climate change impacts. This will further assist in developing computer-based models to simulate the responses. The compiled facts under this chapter would help policymakers, planners, and researchers understand various climate change implications relevant to the Himalayan forest ecosystem. The thematic research areas' prioritization to address the emerging prominent research questions highlighted in this chapter is essential to have a sustained flow of the entire ecosystem of the IHR. There is also a need for international and national interdisciplinary collaborations to deal with the mentioned research questions, funding facilities, and better equipment.

References

Adhikari BS, Kumar R (2020) Effect of snowmelt regime on phenology of herbaceous species at and around treeline in Western Himalaya, India. Notulae Scientia Biologicae 12(4):901–919

Ahmad M, Sharma P, Rathee S, Singh HP, Batish DR, Lone GR, Kaur S, Jaryan V, Kohli RK (2021) Niche width analyses facilitate identification of high-risk endemic species at high altitudes in Western Himalayas. Ecol Ind (126). https://doi.org/10.1016/j.ecolind.2021.107653

Albrecht A, Kandji ST (2003) Carbon sequestration in tropical agroforestry systems. Agr Ecosyst Environ 99:15–27

Apu KC, Ghimire A (2015) High-altitude plants in era of climate change: a case of Nepal Himalayas. In: Öztürk M, Hakeem K, Faridah-Hanum I, Efe R (eds) Climate change impacts on high-altitude ecosystems. Springer, Cham. https://doi.org/10.1007/978-3-319-12859-7_6

Apurva K, Singh H (2017) Impact of heat and drought stress on the physiological response of *Terminalia arjuna* grown under elevated temperature conditions: an adaptive mechanism to climate change. J Pharmacognosy Phytochem 7(1):1374–1378

Apurva K, Singh H, Kumar H (2017) Elucidating morphological and growth changes in Arjun tree (*Terminalia arjuna*) grown under elevated temperature condition. Trends Biosci 10(1):332–336

Aryal S, Gaire NP, Pokhrel NR, Rana P, Sharma B, Kharal DK, Poudel BS, Dyola N, Fan Z-X, Grießinger J, Bräuning A (2020) Spring season in western Nepal Himalaya is not yet warming: a 400-year temperature reconstruction based on tree-ring Widths of Himalayan Hemlock (*Tsuga dumosa*). Atmosphere 11(2):132. https://doi.org/10.3390/atmos11020132

Backlund P, Janetos A, Schimel D (2008) The effects of climate change on agriculture, land resources, water resources, and biodiversity in the United States. In: Synthesis and assessment product 4.3. US Environmental Protection Agency, Climate Change Science Program, Washington, DC, p 240

Bagaria P, Sharma LK, Joshi BD, Kumar H, Mukherjee T, Thakur M, Chandra K (2020) West to east shift in range predicted for Himalayan Langur in climate change scenario. Glob Ecol Conserv. https://doi.org/10.1016/j.gecco.2020.e00926

Bagaria P, Thapa A, Sharma LK, Joshi BD, Singh H, Sharma CM, Sarma J, Thakur M, Chandra K (2021) Distribution modelling and climate change risk assessment strategy for rare Himalayan Galliformes species using archetypal data abundant cohorts for adaptation planning. Clim Risk Manag (31). https://doi.org/10.1016/j.crm.2020.100264

Basnett S, Devy SM (2021) Phenology determines leaf functional traits across Rhododendron species in the Sikkim Himalaya. Alp Bot 131:63–72. https://doi.org/10.1007/s00035-020-00244-5

Bellard C, Bertelsmeier C, Leadley P, Thuiller W, Courchamp F (2012) Impacts of climate change on the future of biodiversity. Ecol Lett 15(4):365–377

Bhandari MS, Meena RK, Shankhwar R, Shekhar C, Saxena J, Kant R, Pandey VV, Barthwal S, Pandey S, Chandra G, Ginwal HS (2019) Prediction mapping through Maxent modeling paves the way for the conservation of *Rhododendron arboreum* in Uttarakhand Himalayas. J Indian Soc Rem Sens 48:411–422. https://doi.org/10.1007/s12524-019-01089-0

Bhandari MS, Shankhwar R, Meena RK, Pandey S, Kant R, Barthwal S, Ginwal HS, Chauhan JS (2020) Past and future distribution pattern of *Myrica esculenta* in response to climate change scenario. Model. Model Earth Syst Environ. https://doi.org/10.1007/s40808-020-00902-x

Bhattacharjee A, Anadón JD, Lohman DJ, Doleck T, Lakhankar T, Shrestha BB, Thapa P, Devkota D, Tiwari S, Jha A, Siwakoti M, Devkota NR, Jha PK, Krakauer NY (2017) The impact of climate change on biodiversity in Nepal: current knowledge, lacunae, and opportunities. Climate 5(4):80. https://doi.org/10.3390/cli5040080a

Chakraborty A, Saha S, Sachdeva K, Joshi PK (2018) Vulnerability of forests in the Himalayan region to climate change impacts and anthropogenic disturbances: a systematic review. Reg Environ Change 18:1783–1799. https://doi.org/10.1007/s10113-018-1309-7

Chen IC, Shiu HJ, Benedick S, Holloway JD, Chey VK, Barlow HS, Hill JK, Thomas CD (2009) Elevation increases in moth assemblages over 42 years on a tropical mountain. Proc Natl Acad Sci 106:1479–1483

Chhetri B, Badola HK, Barat S (2021) Modelling climate change impacts on distribution of Himalayan pheasants. Ecol Ind 123.https://doi.org/10.1016/j.ecolind.2021.107368

Cleland EE, Chuine I, Menzel A, Mooney HA, Schwartz MD (2007) Shifting plant phenology in response to global change. Trends Ecol Evol 22(7):357–365

Coleman DC, Odum EP, Crossley DA Jr (1992) Soil biology, soil ecology and global change. Biol Fertil Soils 14:104–111

Coope GR, Wilkins AS (1994) The response of insect faunas to glacial-interglacial climatic fluctuations. Philos Trans Royal Soc London B Biol Sci 344:19–26

Dhyani A, Kadaverugu R, Nautiyal BP, Nautiyal MC (2021) Predicting the potential distribution of a critically endangered medicinal plant *Lilium polyphyllum* in Indian Western Himalayan Region. Reg Environ Change 21(30). https://doi.org/10.1007/s10113-021-01763-5

Dhyani S, Kadaverugu R, Pujari P (2020) Predicting impacts of climate variability on Banj oak (*Quercus leucotrichophora* A. Camus) forests: understanding future implications for Central Himalayas. Reg Environ Change 20(113). https://doi.org/10.1007/s10113-020-01696-5

Donner SD, Skirving WJ, Little CM, Oppenheimer M, Hoegh-Guldberg O (2005) Global assessment of coral bleaching and required rates of adaptation under climate change. Glob Change Biol 11:2251–2265

Freeman BG, Freeman AMC (2014) Rapid upslope shifts in New Guinean birds illustrate strong distributional responses of tropical montane species to global warming. Proc Natl Acad Sci 111:4490–4494

Gaira KS, Dhar U (2020) Phenological change modelling for selected Himalayan medicinal herbs using herbarium records: a case study. Ecol Inform (60). https://doi.org/10.1016/j.ecoinf.2020.101177

Gaire NP, Fan Z, Bräuning A, Panthi S, Rana P, Shrestha A, Bhuju DR (2020) *Abies spectabilis* shows stable growth relations to temperature, but changing response to moisture conditions along an elevation gradient in the central Himalaya. Dendrochronologia (60). https://doi.org/10.1016/j.dendro.2020.125675

Gairola S, Shariff NM, Bhatt A, Kala CP (2010) Influence of climate change on production of secondary chemicals in high altitude medicinal plants: Issues needs immediate attention. J Med Plants Res 4(18):1825–1829

Gilani H, Goheer MA, Ahmad H, Hussain K (2020) Under predicted climate change: Distribution and ecological niche modelling of six native tree species in Gilgit-Baltistan, Pakistan. Ecol Ind 111:106049. https://doi.org/10.1016/j.ecolind.2019.106049

Gupta SK, Ram J, Singh H (2018) Comparative study of transpiration in cooling effect of tree species in the atmosphere. J Geosci Environ Prot 6:151–166

Gupta SK, Ram J, Singh H, Kumar P (2018) Comparative study of cooling under shades of some forestry tree species with respect to ambient temperatures. J Geosci Environ Prot 6:51–66

Gupta SK, Singh H (2017) Observations on sporadic flowering in *Ficus benjamina* L. and *Peltophorum africanum* L. in new forest. Indian Forester 143:290–291

Hamid M, Khuroo AA, Charles B, Ahmad R, Singh CP, Aravind N (2019) Impact of climate change on the distribution range and niche dynamics of Himalayan birch, a typical treeline species in Himalayas. Biodivers Conserv 28:2345–2370

Hannah L, Midgley G, Andelman S, Araújo MI, Hughes G, Martinez-Meyer E, Pearson R, Williams P (2007) Protected area needs in a changing climate. Frontier Ecol Environ 5(3):131–138

Heath J, Ayres E, Possell M, Bardgett RD, Black HIJ, Grant H, Ineson P, Kerstiens G (2005) Rising atmospheric CO_2 reduces sequestration of root-derived soil carbon. Science 309:1711–1713
Holyoak H, Heath SK (2016) The integration of climate change, spatial dynamics, and habitat fragmentation: a conceptual overview. Integrative Zoology 11:40–49
IPCC (2001) Climate change 2001: impacts, adaptation and vulnerability. In: McCarthy JJ, Canziani OF, Leary NA, Dokken DJ, White KS (eds) Contribution of working group II to the third assessment report of the intergovernmental panel on climate change. Cambridge University Press, Cambridge, UK, and New York, USA, p 1032
IPCC (2007) Climate change 2007: the physical science basis. In: Solomon S, Qin D, Manning M, Chen Z, Marquis M, Averyt KB, Tignor M, Miller HL (eds) Contribution of working group I to the fourth assessment report of the intergovernmental panel on climate change. Cambridge University Press, Cambridge, United Kingdom and New York, NY, USA
IPCC (2014) Climate change 2014: synthesis report. In: Contribution of working groups I, II and III to the fifth assessment report of the intergovernmental panel on climate change
Jain CK, Singh S (2018) Impact of climate change on the hydrological dynamics of River Ganga, India. J Water Clim Change 11(1):274–290
Jetz W, Wilcove DS, Dobson AP (2007) Projected impacts of climate and land-use change on the global diversity of birds. PLoS Biol 5:1211–1219
Joshi R, Singh H (2020) Carbon sequestration potential of disturbed and non-disturbed forest ecosystem: a tool for mitigating climate change. Afr J Environ Sci Technol 14(11):385–393. https://doi.org/10.5897/AJEST2020.2920
Joshi R, Singh H, Chhetri R, Poudel SR, Rijal S (2021) Carbon sequestration potential of community forests: a comparative analysis of soil organic carbon stock in community managed forests of Far-Western Nepal. Eurasian J Soil Sci 10(2):https://doi.org/10.18393/ejss.825066
Kelly AE, Goulden ML (2008) Rapid shifts in plant distribution with recent climate change. Proc Natl Acad Sci 105:11823–11826
Khanduri VP, Sharma CM, Singh SP (2008) The effects of climate change on plant phenology. Environmentalist 28:143–147. https://doi.org/10.1007/s10669-007-9153-1
Konvicka M, Maradova M, Benes J, Fric Z, Kepka P (2003) Uphill shifts in distribution of butterflies in the Czech Republic: effects of changing climate detected on a regional scale. Glob Ecol Biogeogr 12:403–410
Korner C (2002) Mountain biodiversity, its causes and functions. Mountain biodiversity: a global assessment. Parthenon Publishing, London, pp 3–20
Krishnan R, Gnanaseelan C, Sanjay J, Swapna P, Dhara C, Sabin TP, Jadhav J, Sandeep N, Choudhury AD, Singh M, Mujumdar M, Parekh A, Tewari A, Mehajan R (2020) Introduction to climate change over the Indian Region. In: Krishnan R, Sanjay J, Gnanaseelan C, Mujumdar M, Kulkarni A, Chakraborty S (eds) Assessment of climate change over the Indian Region. Springer, Singapore. https://doi.org/10.1007/978-981-15-4327-2_1
Kumar A, Kumar P, Singh H, Kumar N (2021) Adaptation and mitigation potential of roadside trees with bio-extraction of heavy metals under vehicular emissions and their impact on physiological traits during seasonal regimes. Urban Forestry Urban Greening. https://doi.org/10.1016/j.ufug.2020.126900
Kumar A, Kumar P, Singh H, Kumar N (2021) Modulation of plant functional traits under essential plant nutrients during seasonal regime in natural forests of Garhwal Himalayas. Plant Soil. https://doi.org/10.1007/s11104-021-05003-x
Kumar A, Kumar P, Singh H, Bisht S, Kumar N (2021) Relationship of physiological plant functional traits with soil carbon stock in the temperate forest of Garhwal Himalaya. Curr Sci 120(8):1368–1373. https://doi.org/10.18520/cs/v120/i8/1368-1373
Kumar A, Tewari S, Singh H, Kumar P, Kumar N, Bisth S, Kushwah S, Nidhi KR (2021) Biomass accumulation and carbon stocks in different agro forestry system prevalent in Himalayan foothills, India. Curr Sci 120(6):1083–1188. https://doi.org/10.18520/cs/v120/i6/1083-1088

Kumar D, Rawat S, Joshi (2021d) Predicting the current and future suitable habitat distribution of the medicinal tree *Oroxylum indicum* (L.) Kurz in India. J Appl Res Med Aromatic Plants. https://doi.org/10.1016/j.jarmap.2021.100309

Kumar M, Kalra N, Khaiter P, Ravindranath NH, Singh V, Singh H, Sharma S, Rahnamayan (2019a) PhenoPine: a simulation model to trace the phenological changes in *Pinusroxhburghii* in response to ambient temperature rise. Ecol Model 404:12–20

Kumar M, Kalra N, Singh H, Sharma S, Rawat PS, Singh RK, Gupta AK, Kumar P, Ravindranath NH (2021e) Indicator-based vulnerability assessment of forest ecosystem in the Indian Western Himalayas: an analytical hierarchy process integrated approach. Ecol Ind. https://doi.org/10.1016/j.ecolind.2021.107568

Kumar M, Padalia H, Nandy S, Singh H, Khaiter P, Kalra N (2019b) Does spatial heterogeneity of landscape explain the process of plant invasion? A case study of *Hyptissuaveolens* from Indian Western Himalaya. Environ Monit Assess 191:794. https://doi.org/10.1007/s10661-019-7682

Kumar M, Rawat SPS, Singh H, Ravindranath NH, Kalra N (2018) Dynamic forest vegetation models for predicting impacts of climate change on forests: an Indian perspective. Indian J For 41(1):1–12

Kumar N, Kumar A, Jeena N, Singh R, Singh H (2020c) Factors influencing soil ecosystem and agricultural productivity at higher altitudes. In: Goel R, Soni R, Suyal D (eds) Microbiological advancements for higher altitude agro-ecosystems & sustainability. In: Rhizosphere biology. Springer, Singapore

Kumar N, Singh H, Sharma SK (2020b) Antioxidants: responses and importance in plant defense system. In: Roychowdhury R, Choudhury S, Hasanuzzaman M, Srivastava S (eds) Sustainable agriculture in the era of climate change. Springer, Cham. https://doi.org/10.1007/978-3-030-45669-6_11

Kumar P, Singh R, Singh H, Chand T, Bala N (2020) Assessment of soil carbon dioxide efflux and its controlling factors in moist temperate forest of West Himalayas. Curr Sci 19(4):661–669

Kumar RK, Sruajani MS, Arya JC, Joshi GC (2012) Impact of climate change on diversity of Himalayan medicinal plant: a threat to Ayurvedic system of medicine. Int J Res Ayurveda Pharmacy 3(3):327–331

Kumar V, Chopra AK (2009) Impact of climate change on biodiversity of India with special reference to Himalayan region—an overview. J Appl Natural Sci 1(1):117–122

Kumari A, Singh H (2018) Impact of heat and drought stress on physiological response of Terminalia Arjuna grown under elevated temperature condition: an adaptive mechanism to climate change. J Pharmacognosy Phytochem 7(1):1374–1378

Kumari R, Banerjee A, Kumar R, Kumar A, Saikia P, Khan ML (2019) Deforestation in India: consequences and sustainable solutions. IntechOpen. https://doi.org/10.5772/intechopen.85804

Leakey ADB, Ainsworth EA, Bernacchi CJ, Rogers A, Long SP, Ort DR (2009) Elevated CO_2 effects on plant carbon, nitrogen, and water relations: six important lessons from FACE. J Exp Bot 60:2859–2876

Lü X, Paudayal KN, Uhl D, Zhu L, Yao T, Mosbrugger V (2020) Phenology and climatic regime inferred from airborne pollen on the northern slope of the Qomolangma (Everest) region. J Geophys Res Atmospheres 125:e2020JD033405. https://doi.org/10.1029/2020JD033405

Malik R, Rossi S, Sukumar R (2020a) Variations in the timing of different phenological stages of cambial activity in *Abies pindrow* (Royle) along an elevation gradient in the north-western Himalaya. Dendrochronologia 59:125660. https://doi.org/10.1016/j.dendro.2019.125660

Malik R, Rossi S, Sukumar R (2020b) Cambial phenology in *Abies pindrow* (Pinaceae) along an altitudinal gradient in northwestern Himalaya. IAWA J 41(2):186–201. https://doi.org/10.1163/22941932-bja10007

Malsawmkima B, Sahoo UK (2020) Flowering phenology of *Rhododendron arboreum* Sm. at two elevations in Phawngpui National Park, Mizoram: climate change implications. Int J Ecol Environ Sci 46(3–4):259–267

Manish K, Telwala Y, Nautiyal DC, Pandit MK (2016) Modelling the impacts of future climate change on plant communities in the Himalaya: a case study from Eastern Himalaya, India. Model Earth Syst Environ 2:92. https://doi.org/10.1007/s40808-016-0163-1

MEA (Millennium Ecosystem Assessment) (2005) Ecosystems and human wellbeing: synthesis. Island Press, Washington DC

Mehta P, Sekar KC, Bhatt D, Tewari A, Bisht K, Upadhyay S, Negi VS, Soragi B (2020) Conservation and prioritization of threatened plants in Indian Himalayan Region. Biodivers Conserv 29:1723–1745. https://doi.org/10.1007/s10531-020-01959-x

Miehe G, Miehe S, Vogel J, Co S, Duo L (2007) Highest treeline in the northern hemisphere found in Southern Tibet. Mt Res Dev 27(2):169–173. https://doi.org/10.1659/mrd.0792

Mishra SN, Gupta HS, Kulkarni N (2021) Impact of climate change on the distribution of Sal species. Ecol Inform 61. https://doi.org/10.1016/j.ecoinf.2021.101244

National Intelligence Council (2009) India: the impact of climate change to 2030 (A Commissioned Research Report) prepared by Joint Global Change Research Institute and Battelle Memorial Institute, Pacific Northwest Division, Richland, Washingto, p 54

Naylor RL, Liska AJ, Burke MB, Falcon WP, Gaskell JC, Rozelle SD, Cassman KG (2007) The ripple effect: biofuels, food security, and the environment. Environ Sci Pol Sustain Dev 49(9):30–43

Negi GCS, Mukherjee S (2020) Climate change impacts in the Himalayan Mountain ecosystems. In: Goldstein MI, DellaSala DA (eds) Encyclopedia of the World's biomes. Elsevier, pp 349–354. https://doi.org/10.1016/B978-0-12-409548-9.12056-1

Negi GCS, Rawal RS (2019) Himalayan biodiversity in the face of climate change. In: Garkoti S, Van Bloem S, Fulé P, Semwal R (eds) Tropical ecosystems: structure, functions and challenges in the face of global change. Springer, Singapore. https://doi.org/10.1007/978-981-13-8249-9_14

Negi PS (2012) Climate change, alpine treeline dynamics and associated terminology: focus on Northwestern Indian Himalaya. Trop Ecol 53:371–374

Negi VS, Maikhuri RK, Pharswan D, Thakur S, Dhyani PP (2017) Climate change impact in the Western Himalayan: 'purple's perception and adaptive strategies. J Mt Sci 14(2):403–416

Negi VS, Pathak R, Rawal RS, Bhatt ID, Sharma S (2019) Long-term ecological monitoring on forest ecosystems in Indian Himalayan Region: criteria and indicator approach. Ecol Ind 102:374–381

Norby RJ, DeLucia EH, Gielen B, Calfapietra C, Giardina CP, King JS, Ledford J, McCarthy HR, Moore DJP, Ceulemans R, Angelis PD, Finzi AC, Karnosky DF, Kubiske ME, Lukac M, PregitzerKS S-M, Schlesinger WH, Oren R (2003) Forest response to elevated CO_2 is conserved across a broad range of productivity. Proc Natl Acad Sci 102:18052–18056

Pandey S, Cherubini P, Saurer M, Carrer M, Petit G (2020) Effects of climate change on treeline trees in Sagarmatha (Mt. Everest, Central Himalaya). J Veg Sci. https://doi.org/10.1111/jvs.12921

Pandit MK, Sodhi NS, Koh LP, Bhaskar A, Brook BW (2007) Unreported yet massive deforestation driving loss of endemic biodiversity in Indian Himalaya. Biodivers Conserv 16:153–163

Parmesan C (2006) Ecological and evolutionary responses to recent climate change. Ecol Evol 37:637–669

Paustian K, Six J, Elliott ET, Hunt HW (2000) Management options for reducing CO_2 emissions from agricultural soils. Biogeochemistry 48:147–163

Prajapati N, Sharma R, Singh H (2020) Unraveling the bio-herbicidal potential of eucalyptus species aqueous leaf extract and leaf oil on germination and initial growth performance of weed *Partheniumhysterophorus*. Int J Curr Microbiol Appl Sci 9(9). https://doi.org/10.20546/ijcmas.2020.909.xx

Pugnaire FI, Morillo JA, Peñuelas P, Reich PB, Bardgett RD, Gaxiola A, Wardle DA, van der Putten WH (2019) Climate change effects on plant-soil feedbacks and consequences for biodiversity and functioning ofterrestrial ecosystems. Sci Adv 5(11):eaaz1834

Ramankutty N, Foley JA (1999) Estimating historical changes in global land cover: croplands from 1700 to 1992. Glob Biogeochem Cycles 13(4):997–1027. https://doi.org/10.1029/1999GB900046

Rana SK, Rana HK, Ghimire SK, Shrestha KK, Rajnitkar S (2017) Predicting the impact of climate change on the distribution of two threatened Himalayan medicinal plants of Liliaceae in Nepal. J Mt Sci 14(3):558–570. https://doi.org/10.1007/s11629-015-3822-1

Rashid I, Romshoo SA (2020) Impact of climate change on vegetation distribution in the Kashmir Himalaya. In: Dar G, Khuroo A (eds) Biodiversity of the Himalaya: Jammu and Kashmir State. Topics in biodiversity and conservation (18). Springer, Singapore. https://doi.org/10.1007/978-981-32-9174-4_40

Rautela P, Karki B (2015) Impact of climate change on life and livelihood of indigenous people of higher Himalaya in Uttarakhand, India. Am J Environ Prot 3(4):112–124

Rawat AS, Kalra N, Singh H, Kumar M (2020) Application of vegetation models in india for understanding the forest ecosystem processes. Indian Forester 146(2):93–100

Roy A, Rathore P (2019) Western Himalayan forests in climate change scenario. In: Navalgund R, Kumar A, Nandy S (eds) Remote sensing of Northwest Himalayan ecosystems. Springer, Singapore. https://doi.org/10.1007/978-981-13-2128-3_12

Sala OE, Chapin FS, Armesto JJ, Berlow E, Bloomfield J, Dirzo R, Huber-Sanwald E, Huenneke LF, Jackson RB, Kinzig A, Leemans R, Lodge DM, Mooney HA, Oesterheld M, Poff NL, Sykes MT, Walker BH, Walker M, Wall DH (2020) Global biodiversity scenarios for the year 2100. Science 287:1770–1774

Salick J, Ghimire SK, Fang Z, Dema S, Koncha KM (2014) Himalayan alpine vegetation, climate change and mitigation. J Ethnobiol 34(3):276–293

Savita, Sharma SK, Singh H, Singh O (2017) Forest fire: a review. Indian Forester 143(8):719–728

Saxena KG, Rao KS (2020) Climate change and vegetation phenology. In: Tandon R, Shivanna K, Koul M (eds) Reproductive ecology of flowering plants: patterns and processes. Springer, Singapore. https://doi.org/10.1007/978-981-15-4210-7_2

Scheiter S, Kumar D, Corlett RT, Gaillard C, Langan L, Lapuz RS, Martens C, Pfeiffer M, Tomlinson KW (2020) Climate change promotes transitions to tall evergreen vegetation in tropical Asia. Glob Change Biol 26:5106–5124. https://doi.org/10.1111/gcb.15217

Schickhoff U, Bobrowski M, Böhner J, Bürzle B, Chaudhary RP, Gerlitz L, Heyken H, Lange J, Müller M, Scholten T, Schwab N, Wedegärtner R (2015) Do Himalayan treelines respond to recent climate change? An evaluation of sensitivity indicators. Earth Sys Dyn 6(245–265):2015. https://doi.org/10.5194/esd-6-245-2015

Schwab N, Kaczka RJ, Janecka K, Böhner J, Chaudhary RP, Scholten T, Schickhoff U (2018) Climate change-induced shift of tree growth sensitivity at a Central Himalayan Treeline Ecotone. Forests (9): 267. https://doi.org/10.3390/f9050267

Sekar KC, Rawal RS, Chaudhery A, Pandey A, Rawat G, bajpai O, Josh B, Bisht K, Mishra BM (2017) First GLORIA Site in Indian Himalayan Region: towards addressing issue of long-term data deficiency in the Himalaya. Natl Acad Sci Lett 40:355–357.https://doi.org/10.1007/s40009-017-0584-z

Sharma R, Singh H (2021) Alteration in biochemical constituents and nutrients partitioning of *Asparagus racemosus* in response to elevated atmospheric CO_2 concentration. Environ Sci Pollut Res. https://doi.org/10.1007/s11356-021-16050-3

Sharma R, Prajapati N, Singh H (2019) Elevated carbon dioxide impacts on bioactive compounds or nutraceuticals properties of medicinal plants. J Pharmacognosy Phytochem 8(1):1924–1926

Sharma R, Singh H, Kaushik M, Nautiyal R, Singh O (2018) Adaptive physiological response, carbon partitioning and biomass production of *Withania somnifera* (L.) Dunal grown under elevated CO_2 regimes. 3 Biotech. 8:267. https://doi.org/10.1007/s13205-018-1292-1

Shrestha UB, Gautam S, Bawa KS (2012) Widespread climate change in the Himalayas and associated changes in local ecosystems. PLoS One 7:e36741. https://doi.org/10.1371/journal.pone.0036741

Shukla PR (2003) Climate change and India: vulnerability assessment and adaptation. Orient Blackswan. ISBN 8173714711, 9788173714719

Singh H, Kumar N, Kumar M, Singh R (2020a) Modelling habitat suitability of western tragopan (*Tragopan melanocephalus*) a range-restricted vulnerable bird species of the Himalayan region, in response to climate change. Clim Risk Manag 29.https://doi.org/10.1016/j.crm.2020.100241

Singh H, Sharma R, Savita SMP, Kumar M, Verma A, Ansari MW, Sharma SK (2018) Adaptive physiological response of *Parthenium hysterophorus* to elevated atmospheric CO_2 concentration. Indian Forester 144:1–14

Singh H, Verma A, Ansari, MW, Shukla A (2014) Physiological response of rice (*Oryza sativa* L.) genotypes to elevated nitrogen applied under field conditions. Plant Sig Behav 9:e29015. https://doi.org/10.4161/psb.29015

Singh H, Verma A, Kumar M, Sharma R, Gupta R, Kaur M, Negi M, Sharma SK (2017) Phytoremediation: a green technology to clean up the sites with low and moderate level of heavy metals. Austin Biochem 2(2):1–8

Singh H, Yadav M, Kumar N, Kumar A, Kumar M (2020c) Assessing adaptation and mitigation potential of roadside trees under the influence of vehicular emissions: a case study of *Grevillea robusta* and *Mangifera indica* planted in an urban city of India. PLoS One 15(1):e0227380. https://doi.org/10.1371/journal.pone.0227380

Singh N, Tewari A, Shah S (2020) Treeline species: regeneration status and seedling dynamics in Western Himalayan region. Environ Ecol 38(3B):725–732

Singh PB, Mainali K, Jiang Z, Thapa A, Subedi N, Awan MN, Ilyas O, Luitel H, Zhou Z, Hu H (2020) Projected distribution and climate Refugia of endangered Kashmir musk deer *Moschuscupreus* in greater Himalaya, South Asia. Sci Rep 10(1511). https://doi.org/10.1038/s41598-020-58111-6

Singh RK, Sinha VS, Joshi PK, Kumar M (2020b) Modelling agriculture, forestry and other land use (AFOLU) in response to climate change scenarios for the SAARC nations. Environ Monit Assess 192(4):1–8

Singh S, Singh H. Sharma SK, Nautiyal R (2021) Seasonal variation in biochemical responses of bamboo clones in the sub-tropical climate of Indian Himalayan foothills. Heliyon 7(4):e06859.https://doi.org/10.1016/j.heliyon.2021.e06859

Singh SP, Thadani R (2015) Complexities and controversies in Himalayan research: a call for collaboration and rigor for better data. Mt Res Dev 35(4):401–409

Singh SP, Sharma S, Dhyani PP (2009) Himalayan arc and treeline: distribution, climate change responses and ecosystem properties. Biodivers Conserv 28:1997–2016

Singh SP, Singh V, Skutsch M (2010) Rapid warming in the Himalayas: ecosystem responses and development options. Clim Dev 2:221–232

Smadja J, Aubriot O, Puschiasis O, Duplan T, Grimaldi J, Hugonnet M, Buchheit P (2015) Climate change and water resources in the Himalayas. J Alpine. https://doi.org/10.4000/rga.2910

Subba B, Sena S, Ravikantha G, Nobis MP (2018) Direct modelling of limited migration improves projected distributions of Himalayan amphibians under climate change. Biol Cons 227:352–360

Telwala Y, Brook BW, Manish K, Pandit MK (2013) Climate-induced elevational range shifts and increase in plant species richness in a Himalayan biodiversity epicentre. PLoS One 8(2):e57103. https://doi.org/10.1371/journal.pone.0057103

Tewari VP, Verma RK, von Gadow K (2017) Climate change effects in the Western Himalayan ecosystems of India: evidence and strategies. Forest Ecosyst 4(13). https://doi.org/10.1186/s40663-017-0100-4

Thapa S, Baral S, Hu Y, Huang Z, Yue Y, Dhakal M, Jnawali SR, Chettri N, Racey PA, Yu W, Wu Y (2021) Will climate change impact distribution of bats in Nepal Himalayas? A case study of five species. Glob Ecol Conserv (26):e01483. https://doi.org/10.1016/j.gecco.2021.e01483

Thapa S, Chitale V, Rijal SJ, Bisht N, Shrestha BB (2018) Understanding the dynamics in distribution of invasive alien plant species under predicted climate change in Western Himalaya. PLoS One 13(4):e0195752. https://doi.org/10.1371/journal.pone.0195752

Thapliyal G, Vemanna RS, Pawar PM, Barthwal S, Meena RK, Pandey S, Bhandari MS (2020) First record of off-season flowering in *Populusdeltoides* from India: paradigm of climate change indicator. Int J Biometeorol 64:1629–1634.https://doi.org/10.1007/s00484-020-01915-y

Upgupta S, Sharma J, Jayaraman M, Kumar V, Ravindranath NH (2015) Climate change impact and vulnerability assessment of forests in the Indian Western Himalayan region: a case study of Himachal Pradesh, India. Clim Risk Manag 10:63–76

Verma A, Ansari MW, Singh H, Kumar N, Anwar MS, Mudila H, Tuteja N (2021) Proteomics for brassinosteroid signalling: understanding brassinosteroids mediated stress responses through advanced proteomics. Plant Gene 26. https://doi.org/10.1016/j.plgene.2021.100282

Verma S, Kumar N, Verma A, Singh H, Siddique KHM, Singh P (2020) Novel approaches to mitigate heat stress impacts on crop growth and development. Plant Physiol Rep. https://doi.org/10.1007/s40502-020-00550-4

Walther GR, Beißner S, Burga CA (2005) Trends in the upward shift of alpine plants. J Veg Sci 16:541–548

Watham T, Srinet R, Nandy S, Padalia H, Sinha SK, Patel NR, Chauhan P (2020) Environmental control on carbon exchange of natural and planted forests in Western Himalayan foothills of India. Biogeochemistry 151:291–311. https://doi.org/10.1007/s10533-020-00727-x

Willis CG, Ruhfel BR, Primack RB, Miller-Rushing AJ, Losos JB, Davis CC (2010) Favourable climate change response explains non-native species' success in Thoreau's woods. PLoSOne 5:e8878. https://doi.org/10.1371/journal.pone.0008878

Yadav S, Bhattacharya P, Areendran G, Sahana M, Raj K, Sajjad H (2021) Predicting impact of climate change on geographical distribution of major NTFP species in the Central India Region. Model Earth Syst Environ. https://doi.org/10.1007/s40808-020-01074-4

Yadav SK, Singh H, Ginwal HS, Barthwal S (2019b) Elevated CO_2 enhanced growth and physiological process of *Populus deltoides* Bartr. ex Marsh. Indian Forester 145(1):23–27

Yadav SK, Singh H, Nautiyal R, Ginwal H, Ansari SA, Barthwal S (2019a) Modulation of morpho-physiological responses in *Populus deltoides* by elevated carbon dioxide and temperature. Forest Science fxz048. https://doi.org/10.1093/forsci/fxz048

Zheng L, Gaire NP, Shi P (2021) High-altitude tree growth responses to climate change across the Hindu Kush Himalaya. J Plant Ecol rtab035. https://doi.org/10.1093/jpe/rtab035

Chapter 12
Assessment of Agroforestry Area and its Carbon Storage Potential Along Altitudes of Tehri Garhwal District, Uttarakhand, in North-Western Indian Himalaya

K. K. Vikrant, D. S. Chauhan, and R. H. Rizvi

Abstract Agroforestry is an integration of trees and crops on the same plot of land to escalate the sustainability and productivity of the farming system and its income. It is a form of land use management system of farming that is essential for the sustainable growth of the region. Hence, this present work has been carried out in the Tehri Garhwal district of Uttarakhand which is located in the North-Western Indian Himalaya (NWIH) region. The sub-pixel classifier method was adopted for forest area assessment by using satellite remote sensing data (RS-2/LISS IV, 2014). Stratified random sampling was conducted in the study area for estimating agroforestry area and carbon storage potential. Non destructive and destructive methods were followed for biomass estimation. Carbon percentages were determined by the conversion factor. Results show that the highest area under agroforestry was observed in the middle latitude/sub-temperate zone (3707.36 ha) followed by the lower altitude/subtropical zone (2231.26 ha). Maximum carbon stock was comparatively higher at lower altitude (2.07 Mg ha^{-1}) followed by middle altitude (1.63 Mg ha^{-1}). It had contributed more agroforestry area and carbon stock as compared to other species which are mostly adopted by the farmers in agroforestry. It can be concluded that agroforestry acts a crucial role in climate change amelioration, soil enrichment, and improvement of the economic status of people's livelihood of the state in the NWIH.

Keywords Agroforestry area · Carbon storage · Altitude · Climate change · Remote sensing

K. K. Vikrant (✉)
Forestry Department, Himalayan College, Roorkee, Haridwar, Uttarakhand, India

D. S. Chauhan
Department of Forestry, Hemvati Nandan Bahuguna Garhwal University, Srinagar, Pauri Garhwal, Uttarakhand, India

R. H. Rizvi
Central Agroforestry Research Institute, Jhansi, Uttar Pradesh, India

S. Rani and R. Kumar (eds.), *Climate Change*, Springer Climate,
https://doi.org/10.1007/978-3-030-92782-0_12

Introduction

Ecosystem services are the important functions of agroforestry for the growth of the Indian economy. Although agroforestry is a primitive land use practice in India, the modern farming system tool was initiated after the emphatic suggestion of the National Commission on Agriculture in 1976 (Dhyani 2009). Because of the great variation in the altitude, topography, climate, tree resource, availability of irrigation of water, socioeconomic, and culture, a variety of land use patterns exist in the region. The National Action Plan for Climate Change (2008) proposed that under agroforestry, 0.80 m ha of the area would involve improved agroforestry practices on the existing lands and 0.70 m ha would involve additional lands under agroforestry. Under its green Indian mission, the focus was put on the agroforestry system and practices. Agroforestry has been traditionally the art of living in India for a long time. Nowadays, sustainable management of trees and crops on farms is demanded.

Carbon sequestration is the above-ground biomass and underground biomass in the form of root biomass as carbon storage in a standing tree. Moreover, the agroforestry system creates viable options to fulfill the requirement of livelihood, carbon sequestration, and socioeconomic issues. The total agroforestry area is estimated at 25.3 mha and carbon sequestration potential is estimated between 0.25–19.14 and 0.01–0.60 Mg C/ha/year in India (Dhyani et al. 2016).

In the emerging scenario, there is a need for the plantation of fast-growing tree species for carbon sequestration which has also been mentioned in the National Agroforestry Policy (2014). The Himalayan region is vulnerable to climate change due to its altitude and complex climate. Being a source of the largest rivers in the world, it requires immediate attention for carbon management. In Garhwal Himalaya of India, agroforestry is a stable character of agriculture and forestry landscape. It forms an integral relationship with the farmer over the farms. But there is still a paucity of information on the area of agroforestry in the Himalayan region to understand its carbon storage potential. Therefore, this study has been examined the area and carbon storage potential of agroforestry in Tehri Garhwal district of Uttarakhand in North-Western Indian Himalaya (NWIH) using remote sensing data.

Study Area

The Tehri Garhwal district of Uttarakhand is situated between 30° 03′ and 30° 53′ North latitude and 77° 56′ and 79° 04′ East longitude having a geographical area of 3642 km^2 (FSI 2015) (Fig. 12.1).

There are three agroclimatic zones in the Tehri Garhwal district, viz., sub-tropical zone (300–1200 m), sub-temperate zone (1200–2000 m), and temperate zone (2000–2800 m) (Singh and Singh 1992). Six blocks were selected for the present study in the district representing three zones including Kritinagar, Pratap Nagar, Jakhnidhar,

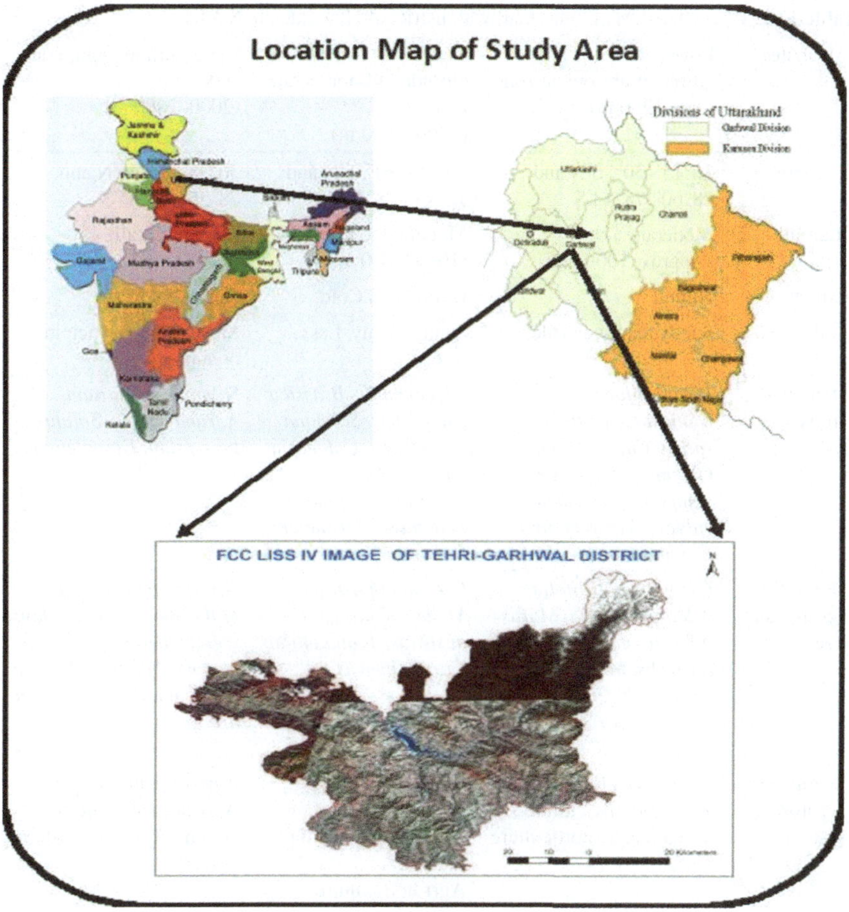

Fig. 12.1 Location map of the study area

Thauldhar, Chamba, and Devprayag. The characteristics feature of the study area are provided in Table 12.1.

Materials and Methods

Resourcesat-2 LISS IV remote sensing data of the study area have been collected from the National Remote Sensing Center (NRSC), Hyderabad, for the year 2013. Geo-referenced forest cover map (2013) of the district has been procured from the Forest Survey of India (FSI), Dehradun, India. Cartosat (version 3) digital elevation model (DEM) (2014) has been taken from BHUVAN, India (http://bhuvan.nrsc.gov.in/data/download/index.php).

Table 12.1 Characteristics of Tehri Garhwal district of Uttarakhand, NWIH

Attributes	Lower altitude/Subtropical zone (286–1200 m)	Middle altitude/Sub-temperate zone (1200–2000 m)	Upper altitude/temperate zone (2000–2800 m)
Location	30°15′–30°26′N and 78°60′–78°43′S	30°26′–30°35′N and 78°43′–78°40′S	30°38′–30°40′N and 78°36′–78°35′S
Rainfall	Moderate (Approx. 1000 mm)	Moderate to heavy (1000–2000 mm)	Heavy rainfall (<2000 mm)
Climate	Humid	Humid and Cold	Heavy cold
Soil type	Sandy loamy, Fertile	Sandy loamy, Less fertile	Mostly loamy, Rich in organic matter
Dominant crops	*Echinochloa frumentacea, Oryza sativa, Cajanus spp, Glycine max, Brassica campestris, Hordeum vulgare, Vigna mungo, Cajanus cajan,*	*Glycine max, Brassica campestris, Solanum melongena, Colocasia antiquorum, Amaranthus spinosus, Zea mays, Coriander sativum*	*Solanum tuberosum, A.frumentacea, Solanum tuberosum, F.esculentum*
Dominant agroforestry tree	*Grewia oppositifolia, Adina cordifolia, Morus alba, Toona ciliata, Celtis australis, Melia azedarach, Citrus limon, Bauhinia variegata, Pyrus pashia*	*Grewia oppositifolia, Melia azedarach, Celtis australis, Toona ciliata, Citrus sinensis*	*Grewia oppositifolia, Malus domestica, Juglans regia, Quercus leucotrichophora, Myrica esculenta, Rhododendron arboreum*
Dominant agroforestry system	Agri-silviculture system, Agri-hortisilviculture system, Agri-horticulture system	Agri-silviculture system, Agri-hortisilviculture system, Agri-horticulture system	Agri-silviculture system, Agri-hortisilviculture system, Agri-horticulture system

**Source*: Negi et al. 2009

Assessment of Agroforestry Area

For assessment of agroforestry area, six blocks were selected for field survey, viz., Kritinagar, Devprayag, Chamba, Jakhnidhar, Pratapnagar, and Thaulidhar (Fig. 12.1). The village's name, elevation, GPS point, tree species, and crop species were also recorded from the agroforestry area of study sites. Villages were surveyed from each selected block up to 10%. Plots with a size of 100 m^2 were randomly laid out in the agroforestry area in each village and also tracked using GPS. Stratified random sampling has been done in the study area for the estimating agroforestry area and its carbon storage potential.

The forest area has been masked from the district area. Resourcesat-2 LISS IV data were used for land use/land cover (LULC) analysis by the supervised method.

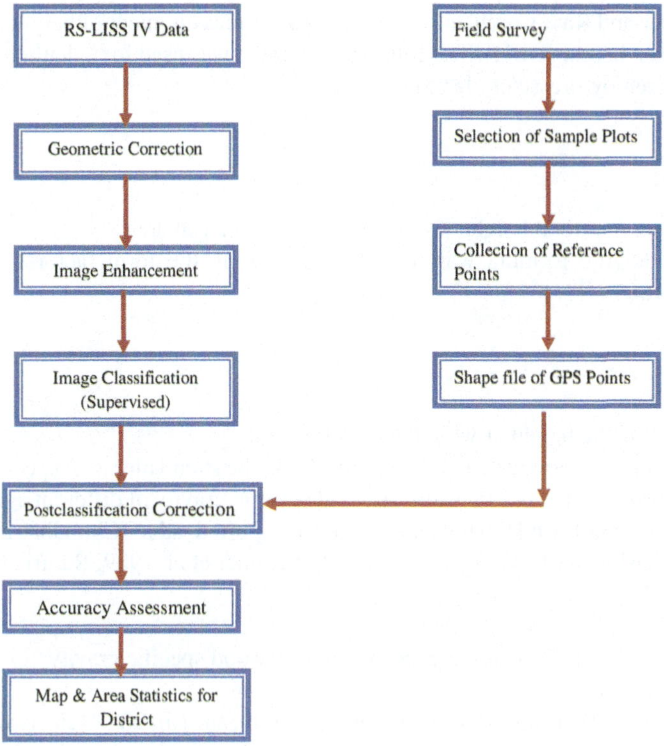

Fig. 12.2 Methodology of agroforestry mapping of the Tehri Garhwal district of Uttarakhand, India

Signatures were generated by GPS points collected during the field survey. The post-classification correction was done on the classified image using ERDAS Imagine software. A geo-referenced forest cover map of the FSI has been used for masking of the forest area of the study region (Fig. 12.1). Cartosat DEM of the district has been divided into four elevation zones, viz., 286–1200, 1200–2000, 2000–2800, and more than 2800 m (Fig. 12.2). LULC, DEM, and agroforestry area maps have been prepared GIS environment.

Assessment of Carbon Stock

Assessment of carbon stock, density in each study site was studied using the Quadrat method (Mishra 1968). All individuals of the tree, shrubs, and fruit species were recorded within the quadrate size of 100 m^2 and 1 m^2 size quadrate for the crop. Marked 10 sample plots of 100 m^2 size each were randomly laid out in each agroforestry system on each altitude of each block for tree (woody perennials) enumeration. 1 × 1 m size plot was used for annuals, i.e., agricultural crop, grass, and weeds.

Tree caliper and Ravi's multi-meter were used for measurement of height and diameter at breast height (DBH). The following formula was used for calculating the tree volume given by Presseler (1865) (Eq. 12.1).

$$v = f \times h \times g \quad (12.1)$$

v = volume; h = total height; f = form factor; g = basal area.

The following formula was used for calculating the form factor in Eq. 12.2 (Pressler 1865).

$$f = 2h_1/3h \quad (12.2)$$

where h_1 = is the height at which diameter is half.

Stem biomass was determined by multiplying the stem volume of specific gravity (IPCC 2006) (Eq. 12.3). The value of wood specific gravity of different agroforestry species in the Garhwal Himalayas was used as reported value (Choudhry and Ghosh 1958; Purkashyatha 1982; Rajput et al. 1985; Kumar et al. 1989; Raturi et al. 2002) (Table 12.2).

$$\text{Stem biomass} = \text{Stem volume} \times \text{wood specific gravity} \quad (12.3)$$

Based on DBH groups (i) 0–10 cm, (ii) 10–20 cm, and (iii) 20–30 cm, two branches were randomly selected from each group and weighed fresh weight. Each subsample was being dried to a constant weight at 65 °C. Branch and leaves biomass was being estimated by the formula of Chidumayo (1990) (Eqs. 12.4 and 12.5).

$$B_{\text{dwi}} = B_{\text{fwi}}/1 \times M_{\text{cbdi}} \quad (12.4)$$

where B_{dwi}—oven dry weight of the branch, B_{fwi}—fresh/green weight of the branch, M_{cbdi}—moisture content on dry weight basis.

$$L_{\text{dwi}} = L_{\text{fwi}}/1 \times M_{\text{cbdi}} \quad (12.5)$$

where L_{dwi}—oven dry weight of leaf, L_{fwi}—fresh/green weight of leaf, M_{cbdi}—moisture content on a dry weight basis.

A factor of 0.25 was multiplied by the above-ground biomass for determining below-ground biomass (IPCC 1996). Crops were harvested and collected samples were weighed and oven dried at 65 °C for estimation of biomass. Total biomass carbon stock of agroforestry was the sum of total biomass carbon of trees and total biomass carbon of crops. The biomass carbon was estimated from multiply with total biomass by a factor of 0.45 (Woomer 1999).

Table 12.2 Specific gravity values of different agroforestry tree species

S. No	Species	Specific gravity	Source
1	*Quercus leucotrichophora*	0.826	Raturi et al. (2002)
2	*Grewia oppositifolia*	0.606	Purkayastha (1982)
3	*Melia azedirach*	0.491	Raturi et al. (2002)
4	*Celtis australis*	0.444	Rajput et al. (1985)
5	*Toona ciliata*	0.424	Rauri et al. (2002)
6	*Adina cardifolia*	0.583	Raturi et al. (2002)
7	*Mangifera indica*	0.588	Chowdhury and Ghose (1958)
8	*Citrus limon*	0.91	Ting and Blair (1965)
10	*Pyrus communis*	0.676	Tumen (2014)
11	*Ficus roxburghii*	0.443	Sheikh et al. (2011)
12	*Prunus cerasoides*	0.69	Kumar (1989)
13	*Anogeissuslatifolia*	0.757	Purkayastha (1982)
14	*Psidiumguajava*	0.59	Sheikh et al. (2011)
15	*Morus alba*	0.603	Purkayastha (1982)
16	*Citrus sinensis*	0.916	Joseph and Abdullahi (2016)
17	*Juglanseregia*	0.59	Wani et al. (2014)
18	*Bahuniaverigata*	0.55	Kanawajia et al. (2013)
19	*Ficus palmate*	0.578	Sheikh et al. (2011)
20	*Malusdomestica*	0.67	Miles and Smith (2009)
21	*Prunusarmenica*	0.50	Miles and Smith (2009)
22	*Prunuspersica*	0.90	Babu et al. (2014)
23	*Myrica esculenta*	0.737	Sheikh et al. (2011)
24	*Pyrus pashia*	0.70	Kumar (1989)
25	*Ficus auriculata*	0.443	Sheikh et al. (2011)
26	*Punica granatum*	0.99	Felter and Lloyd (1898)
27	*Carica papaya*	0.918	Afolabi and Ofobrukweta (2011)
28	*Bombax ceiba*	0.33	Troup (1921)
29	*Rhododendron arboreum*	0.512	Rajput et al. (1985)
30	*Pinusroxburghii*	0.491	Rajput et al. (1985)
31	*Embilicaofficenalis*	0.614	Sheikh et al. (2011)
32	*Psidiumguajava*	0.59	Kanawjia et al. (2013)
33	*Albizialeeback*	0.69	Mani and Parthasarathy (2007)
34	*RhusParviflora*	0.620	Chowdhury and Ghose (1958)
35	*Wood fructicosa*	0.55	Chaturvedi et al. (2012)
36	*Musa Paradisica*	0.29	Omotosa and Ogunsile (2010)
37	*Acacia catechu*	0.825	Purkayastha (1982)

Results and Discussion

Area Assessment

According to Fig. 12.3, the agroforestry area was estimated to be 7029.06 ha (1.93%) in Tehri Garhwal district. The cropping area was estimated to be about 20.0% of the total area of the district. It has been examined that area obtained under agriculture and agroforestry was 72,876.42 ha (20.01%) and 7029.06 ha (1.93%) in the study area, respectively (Fig. 12.4).

Maximum area under agroforestry was found in 1200–2000 m elevations (3707.36 ha) followed by 288–1200 m elevation (2231.26 ha) (Fig. 12.5).

Mahto et al. (2016) calculated the maximum agroforestry area in 1200–1600 m and followed by 1600–2000 m altitudinal zone in the Tehri Garhwal district. The agroforestry area obtained in the Tehri Garhwal district was classified as dominant species based on existing agroforestry systems in the district. (Fig. 12.6). The maximum

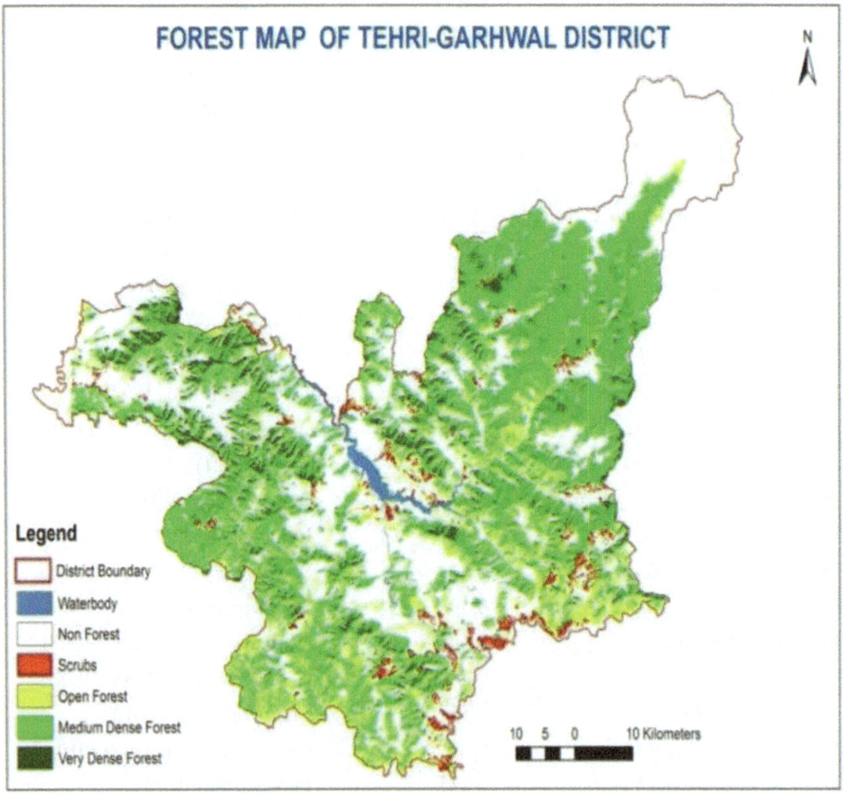

Fig. 12.3 Forest cover of Tehri Garhwal district of Uttrarakhand, India. *Source* FSI, Dehradun, India, 2013

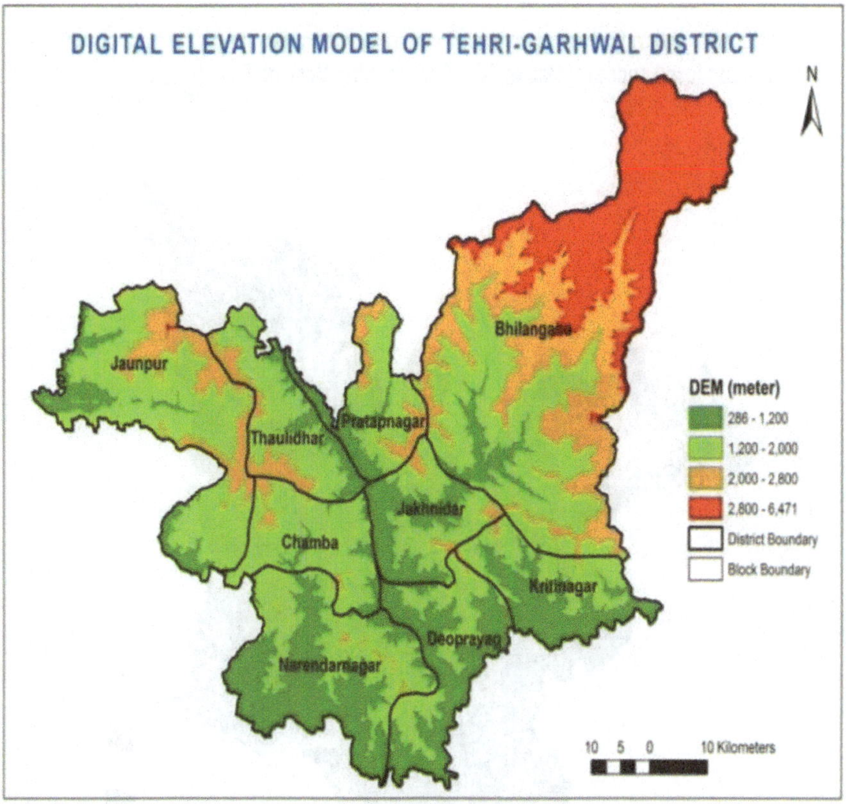

Fig. 12.4 Elevation (m) map of Tehri Garhwal district (Cartosat-DEM)

area was found under *Grewia oppositifolia* (2330.82 ha) followed by *Celtis australis* (1456.80 ha) and *Quercus leucotrichophora* (1129.10 ha). The agroforestry area of the remaining species systems in the district was estimated to be about 2112.36 ha (Fig. 12.7).

Carbon Stock Assessment

Considering the effect of elevations on tree density, the variation reveals that mean tree density differed significantly ($p \leq 0.01$). At the lower elevation, maximum mean tree density (220.0 tree ha^{-1}) was recorded followed by middle elevation (195.0 tree ha^{-1}) (Table 12.3). Due to altitudinal variation, climate variation stimulated the diversity of plant species (Brown 2001). The above-ground biomass was significantly ($p \leq 0.01$) different with the elevations. It has been found maximum (3.25 ton ha^{-1}) at lower elevation followed by middle elevation (2.54 ton ha^{-1}) (Table 12.3). The

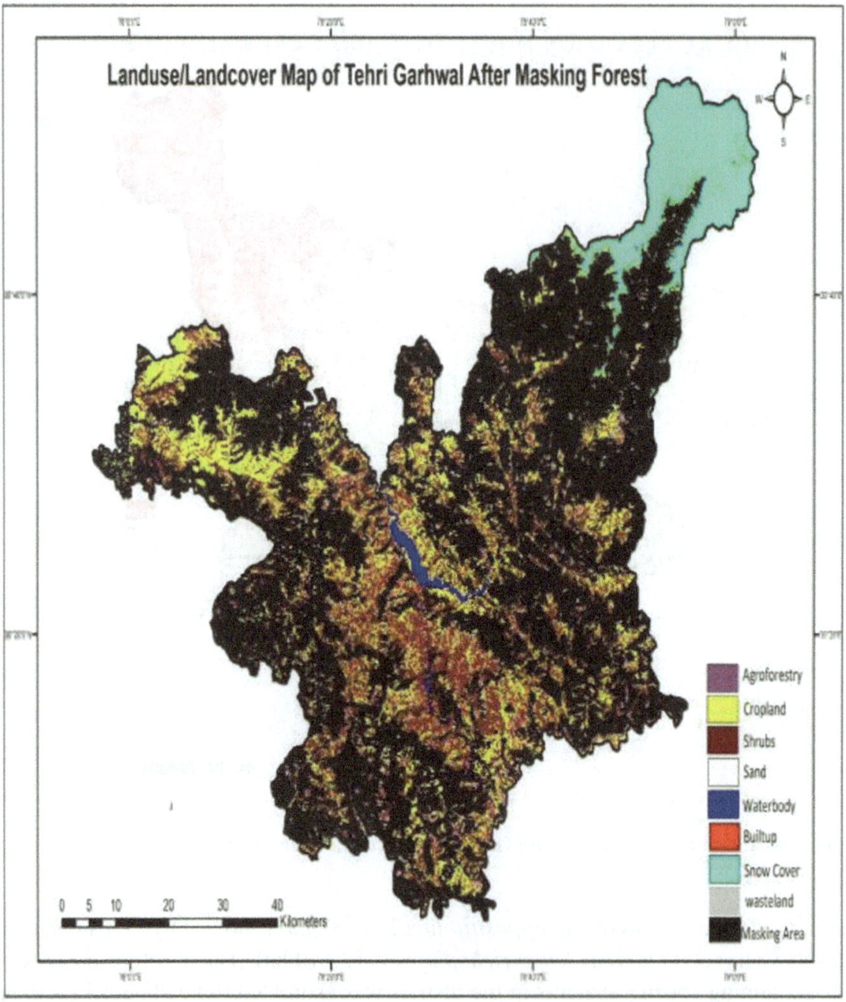

Fig. 12.5 Distribution of LULC of Tehri Garhwal district of Uttarakhand, India

present values are supported by the previous studies (Kumar et al. 2012; Bijalwan 2013). Table 12.3 represents that the below-ground biomass which has been found maximum at lower elevation (0.81 ton ha^{-1}) followed by middle elevation (0.63 ton ha^{-1}).

Total tree biomass is significantly ($p \leq 0.01$) different with elevations. It was recorded maximum at the lower elevation (4.09 ton ha^{-1}) followed by the middle elevation (3.13 ton ha^{-1}) (Table 6). It was observed that the total tree biomass reduces approximately by 50% at the upper elevation. The present finding is similar to the findings reported by the previous studies (Kumar et al. 2012; Bijalwan 2013). Effect of variation of crop biomass was significantly ($p \leq 0.01$) different on elevations

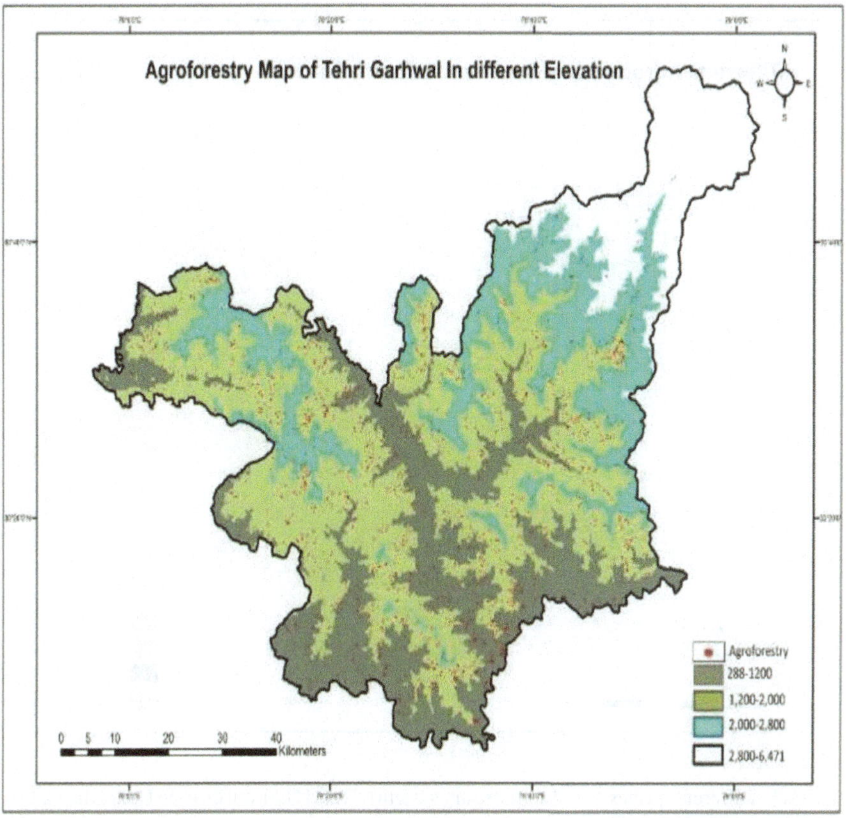

Fig. 12.6 Agroforestry under different elevation zones of Tehri Garhwal district (Cartosat-DEM)

because values were more or less similar in all the elevations but a little bit higher at lower elevation (0.51 ton ha^{-1}) (Table 12.3). The variation in annual crop biomass can be explained due to the tree-crop interaction effects. The reduced biomass may be due to more competition for resources like nutrients, moisture, and light compared to the tree species (Kumar et al. 2012). Total biomass differs significantly ($p \leq 0.01$) with elevations. Table 12.3 represents that the total biomass was higher at the lower elevation (4.61 ton ha^{-1}) followed by middle elevation. However, total biomass in the area has been decreasing with the increasing elevation. Tree density of *Grewia oppositifolia, Celtis australis, Ficus roxbughii, Morus alba, Citrus spp, Malus domestica,* and *Psidium guajava* was also showed higher at this elevation. The data showed that the carbon stock was significantly ($p \leq$ 0.01) different across the elevations which were recorded maximum at lower elevation (2.07 ton C ha^{-1}) while minimum at upper elevation (1.29 ton C ha^{-1}) (Table 6). Carbon stock has been found maximum at lower elevation because of the presence of the tree species, namely *Grewia oppositifolia, Celtis australis, Ficus roxbughii, Morus alba, Citrus spp, and Psidium guajava* which are planted at high density.

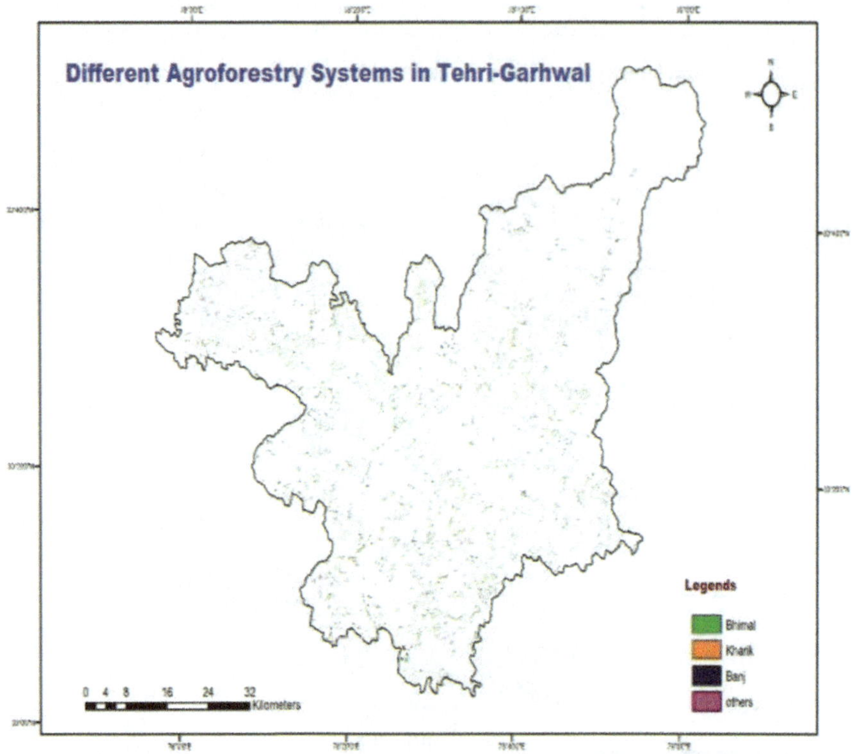

Fig. 12.7 Different species-based agroforestry system of Tehri Garhwal district, Uttarakhand, India

Table 12.3 Density, AGB, BGB, CB, TB, and TC by agroforestry along elevations

Elevation (m)	tree density	AGB	BGB	TTB	CB	TB (Tree + crop)	TC
286–1200	220.0	3.25	0.81	4.06	0.55	4.61	2.07
1200–2000	195.0	2.53	0.60	3.13	0.49	3.62	1.63
2000–2800	177.2	1.92	0.45	2.37	0.50	2.87	1.29
CD	11.05	0.449	0.115	0.564	0.089	0.581	0.261

Note: AGB = Above-ground biomass; BGB = Below-ground biomass; TTB = Total tree biomass; CB = Crop biomass; TB = Total biomass; TC = Total carbon; CD = Critical difference

Density Contribution by Tree Species in Agroforestry

In the district, across elevations, a total of thirty-seven agroforestry tree species were observed. Figure 12.8 and Table 12.4 show that among the dominant tree species, *Grewia oppositifolia* has contributed maximum density (9.43%) followed by *Quercus leucotrichophora* (5.88%) and the rest of the species have contributed 39.48% density. In the present study, *Grewia oppositifolia* contributed maximum

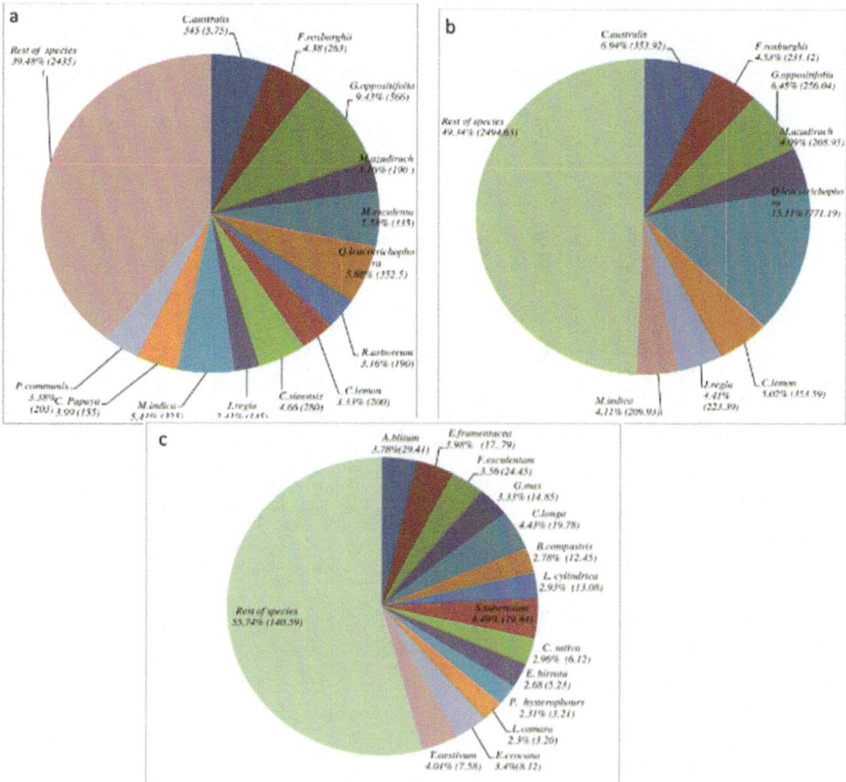

Fig. 12.8 **a** Density (No. of trees ha^{-1}), **b** carbon stock (kg ha^{-1}), and **c** carbon stock (kg ha^{-1}) contributed by crops of Tehri Garhwal district (Vikrant et al. 2019)

density as compared to other species across the agroforestry area of the district. It indicated that *Grewia oppositifolia*-based agroforestry occupied most of the agroforestry area in Tehri Garhwal district. The present data are supported by Vikrant et al. (2018) who reported that *Grewia oppositifolia* occupied 0.64% area followed by *Celtis australis* (0.40%) in the Tehri Garhwal district. Dhadhwal et al. (1989) also reported that it is the most common multipurpose tree which is best preferred for fodder, fuel, and fiber in the Garhwal Himalaya. Due to the multipurpose nature of these species, these are most adopted and preferred by local farmers.

Total Carbon Stock Contribution by Trees in Agroforestry

In terms of carbon stock, a total of thirty-seven agroforestry trees species have been contributing biomass carbon in the district. Figure 12.8 and Table 12.4 represent that among the dominant tree species, *Quercus leucotrichophora* was contributing

Table 12.4 Contribution (%) of density (individual's ha^{-1}), biomass, and total carbon stock (kg ha^{-1}) of agroforestry tree species and crop species across the Tehri Garhwal district, Uttarakhand

S. No	Tree Species	Density	Contribution	AGB	BGB	TB	TC	Contribution
1	*Adina cardifolia (Roxb.) Hook.f*	6	1.00	218.6	54.65	273.25	122.96	2.41
2	*Albizia leeback (Linn.) Benth*	8	1.33	56.7	14.17	70.87	31.89	0.62
3	*Anogeissus latifolia (Roxb.ex DC.) Wall.ex Bedd.)*	6	1.00	33.2	8.3	41.5	18.675	0.37
4	*Acacia catchu (L.F.) Wild*	8	1.33	179.5	44.87	224.37	100.96	1.98
5	*Bahunia verigata (L.)Benth*	5	0.83	132.5	33.12	165.62	74.53	1.46
6	*Bombax ceiba (L.)*	8	1.33	151.1	37.77	188.87	84.99	1.67
7	*Celtis australis (L.)*	35	5.75	629.2	157.3	786.5	353.92	6.94
8	*Ficus palamata Forssk*	11	1.83	189.8	47.45	237.25	106.76	2.09
9	*Ficus roxburghii Wall*	27	4.38	410.89	102.72	513.62	231.12	4.53
10	*Ficus semicordata Buch-Ham.ex Sm*	6	1.00	72.77	18.19	90.96	40.93	0.80
11	*Grewia oppositifolia Roxb*	57	9.43	455.18	113.79	568.98	256.04	5.02
12	*Holptelia integrifolia (L.)*	8	1.25	161.58	40.39	201.97	90.88	1.78
13	*Melia azedirach (L.)*	19	3.16	371.48	92.87	464.35	208.95	4.09

(continued)

Table 12.4 (continued)

S. No	Tree Species	Density	Contribution	AGB	BGB	TB	TC	Contribution
14	*Myrica esculenta Buch.-Ham*	34	5.58	35.38	8.845	44.22	19.9	0.39
15	*Morus alba (L.)*	8	1.25	130.23	32.55	162.79	73.25	1.44
16	*Pinus roxburghii Sargent*	7	1.08	130.1	32.52	162.62	73.18	1.43
17	*Prunus cerasoides D.Don*	6	1.00	150.2	37.55	187.75	84.48	1.66
18	*Pyrus pashia Buch.-Ham.ex D.Don*	9	1.42	118.74	29.68	148.42	66.79	1.31
19	*Qurecus leucotrichophora A.Camus*	36	5.87	1371.01	342.75	1713.76	771.19	15.11
20	*Wood fructicosa (L.) Kurz*	4	0.58	58.4	14.6	73	32.85	0.64
21	*Toona cilata M.Roem*	6	0.92	176.005	44	220	99	1.94
22	*Rhododendrone arboreum Sm*	19	3.16	37.88	9.47	47.35	21.3	0.42
23	*Rhus parviflora Roxb*	5	0.75	115.1	28.77	143.87	64.74	1.27
Fruit tree species								
1	*Carica papaya L*	16	2.58	98.165	24.54	122.7	55.21	1.08
2	*Citrus aurentium L*	12	1.92	95.37	23.84	119.21	53.64	1.05
3	*Citrus limon Brum*	20	3.33	593.06	148.26	741.32	333.5963	6.54
4	*Citrus sinensis Osbeck*	28	4.66	204.03975	51	255.04	114.77	2.25

(continued)

Table 12.4 (continued)

S. No	Tree Species	Density	Contribution	AGB	BGB	TB	TC	Contribution
5	*Embilica officinalis Gaertn*	13	2.10	217.45	54.36	271.82	122.319	2.40
6	*Juglanse regia L*	15	2.41	397.15	99.28	496.43	223.39	4.38
7	*Malus domestica Borkh*	30	5.00	314	78.5	392.5	176.62	3.46
8	*Mangifera indica.L*	33	5.41	373.21	93.3	466.51	209.93	4.11
9	*Musa paradisica L*	24	4.00	218.72	54.68	273.4	123.03	2.41
10	*Prunus armenica L*	11	1.83	154.3	38.57	192.87	86.79	1.70
11	*Pyrus communis L*	21	3.38	242.39	60.59	302.99	136.34	2.67
12	*Punica granatum L*	11	1.78	151.38	37.84	189.22	85.15	1.67
13	*Psidium guajava L*	13	2.17	96.51	24.12	120.63	54.28	1.06
14	*Prunus armenica L*	11	1.83	227.66	56.91	284.57	128.05	2.51
15	*Prunus persica Batsch*	15	2.36	302.91	75.72	378.63	170.38	3.34
	Agriculture crop species							
1	*Amarnathus blitum L*	48,066	5.58	NA	NA	65.35	29.41	8.77
2	*Ehinochloa frumentacea Link*	38,233	4.44	NA	NA	39.53	17.79	5.31
3	*Eleusine coracana (L.) Gatern*	43,000	5.00	NA	NA	18.05	8.12	2.42
4	*Fagopyrum esculentum Mill*	31,633	3.68	NA	NA	54.45	24.5	7.31

(continued)

Table 12.4 (continued)

S. No	Tree Species	Density	Contribution	AGB	BGB	TB	TC	Contribution
5	*Hordeum vulagare L*	34,500	4.01	NA	NA	11.47	5.16	1.54
6	*Oryza sativa L*	21,100	2.45	NA	NA	10.42	4.69	1.40
7	*Panicum miliaceum L*	25,500	2.96	NA	NA	11.28	5.07	1.51
8	*Setaria italic L. P.Beauv*	19,750	2.29	NA	NA	16.06	7.22	2.15
9	*Sorghum vulgare L*	38,300	4.45	NA	NA	8.16	3.67	1.09
10	*Triticum aestivum L*	40,500	4.71	NA	NA	16.86	7.58	2.26
11	*Zea mays L*	13,200	1.53	NA	NA	13.26	5.96	1.78
	Leguminous crop species							
1	*Cajanus spp. (L.) Millsp*	20,233	2.35	NA	NA	16.53	7.44	2.22
2	*Cicer arientinum L*	15,966	1.85	NA	NA	17.04	7.67	2.29
3	*Glycine max (L.) Merr*	13,850	1.61	NA	NA	33	14.85	4.43
4	*Lens esculenta L*	12,300	1.43	NA	NA	0.35	0.15	0.04
5	*Pisum sativum L*	26,600	3.09	NA	NA	12.33	5.55	1.66
6	*Vigna mungo (L.) Hepper*	18,900	2.20	NA	NA	12.34	5.55	1.66
Other crop species (Oil, vegetables and rhizomatous etc.)								
1	*Cyclanthera pedata (L.) Schrad*	14,850	1.73	NA	NA	13.13	5.9	1.76
2	*Allium cepa L*	19,250	2.24	NA	NA	18.22	8.2	2.45
3	*Allium sativum L*	20,400	2.37	NA	NA	21.13	9.5	2.83

(continued)

Table 12.4 (continued)

S. No	Tree Species	Density	Contribution	AGB	BGB	TB	TC	Contribution
4	*Brassica compestris L*	22,875	2.66	NA	NA	27.62	12.43	3.71
5	*Brassica juncea L*	17,400	2.02	NA	NA	19.68	8.85	2.64
6	*Capsicum annuum L*	37,150	4.32	NA	NA	16.33	7.35	2.19
7	*Colocasia antiquorum Schott*	19,900	2.31	NA	NA	15.58	7.011	2.09
8	*Cucumis sativus L. Vari. Momodica*	21,733	2.52	NA	NA	13.06	5.88	1.75
9	*Curcuma longa L*	22,675	2.63	NA	NA	43.96	19.78	5.90
10	*Lagenaria siceraria Standl*	27,500	3.19	NA	NA	19.79	8.9	2.66
11	*Luffa acutangula Roxb*	28,050	3.26	NA	NA	18.64	8.39	2.50
12	*Luffa cylindrica Roem*	30,850	3.58	NA	NA	29.08	13.08	3.90
13	*Raphanus sativus L*	18,425	2.14	NA	NA	17.29	7.78	2.32
14	*Sesamum indicum L*	23,125	2.69	NA	NA	18.54	8.34	2.49
15	*Solanum melongena L*	15,580	1.81	NA	NA	16.26	7.31	2.18
16	*Solanum tuberosum L*	21,457	2.49	NA	NA	44.09	19.84	5.92
17	*Zingiber officinale Roxb*	17,600	2.04	NA	NA	11.42	5.13	1.53
18	*Spinacia oleracea L*	20,300	2.36	NA	NA	24.78	11.15	3.33

maximum (15.11%) biomass carbon stock followed by *Ceitis australis* (6.94%), by *Grewia oppositifolia* (6.45%), and the rest of the species were contributing about 49.34% in the district. In the present study, *Quercus leucotrichophora* contributed maximum biomass than other tree species. It is confirmed that *Q.leucotrichopora* had provided better tree attributes (bole, crown spread, crown length, leaf area) and a well-developed root system that may be contributing more biomass than other agroforestry species. Biomass in *Quercus leucotrichophora* was recorded higher as reported by Devi et al. (2013) and Sharma et al. (2010) for lower Western Himalaya. *Grewia opposoitifoila* contributed the maximum number of trees but biomass contribution was lower than *Quercus leucotrichophora.* It may be due to the continuous lopping off its branches for fuel and fodder during the lean period by local people. Therefore, stunting and bushy growth were noticed in the agroforestry field. Kumar et al. (2012) reported that overexploitation of resources from traditional agroforestry trees may reduce input biomass. This implies that the agroforestry trees have a significant carbon storage potential which may be helpful for climate change mitigation in the Tehri Garhwal district.

Carbon Stock Contribution by Crop in Agroforestry

As far as carbon stock is concerned, a total of forty agroforestry crops species were observed in the district. Out of forty, maximum biomass carbon-containing crop species are *Solanum tuberosum* (4.49%), *Curcuma longa* (4.43%), *Triticum aestivum* (4.01%), *Ehinochloa frumentacea* (3.98%), *Amarnathus blitum* (3.78%), *Fagopyrum esculenta* (3.56%), *Eleusine coracana* (3.4%), and *Glycine max* (3.33%), and rest of the species have contributed about 55.74% of carbon stock (Fig. 12.7 and Table 12.4). In the present study, *Solanum tuberosum* contributed maximum biomass among all the species. It may be occurring that *Solanum tuberosum* had maximum leaf area and dry weight as compared to other crop species. Due to the large leaf area, it would be capable of absorption of maximum sunlight and have a maximum amount of CO_2 fixation (Lakitan 2008).

Conclusions

Agroforestry participates in the climate change amelioration and improvement of the economic status of people's livelihood. This study is an attempt to estimate the area of agroforestry and its carbon storage potential by using remote sensing and field survey data. The findings show that the middle elevation/sub-temperate region (1200–2000 m) occupies maximum area under agroforestry which is a good indicator for the better potential to combat climate change. However, a lower elevation (286–1200 m)/subtropical zone was considered suitable for agroforestry areas in the district concerning density and biomass carbon stock because it provides favorable conditions for the growth of agroforestry species. In flat valley (near and along the river), *Grewia oppositifoila, Quercus leucotrichophora,* and *Celtis australis* were dominant agroforestry tree species in the district. It has occupied more agroforestry area, carbon stock, and density. These are mostly adopted by the local farmers of the district. Therefore, the present study suggests increasing area under these three species in the district. Hence, the agroforestry system needs to be promoted and encouraged for environmental security and biodiversity conservation. It would help to combat climate change in a more efficient manner.

Acknowledgements Authors are very grateful to the Director of Central Agroforestry Research Institute, Jhansi, Uttar Pradesh, India, for providing the laboratory for agroforestry area assessment and the Head of the Department of Forestry, Hemvati Nandan Bahuguna Garhwal University, Srinagar, Pauri Garhwal, Uttarakhand, India, for facilitating the research work. The first author is also thankful to the University Grant Commission, New Delhi, India, for financial assistance (Grant No. RGNF-2012-13-SC-BIH-30641).

References

Aflobi IS, Ofobrukweta K (2011) Physiochemical and nutritional qualities of *Caricapapaya*. J Med Pl Res 5(14):3113–3117

Babu KD, Patel R, Deka BC, Bujarburuah MK (2014) Maturity indices for harvesting of low chilling peach cultivars under mid-hill condition of Meghalaya. Acta Horticul 890(40):449–455

Bijalwan A (2013) Vegetation status of Agroforestry systems in Tehri district of Garhwal Himalaya, India. Asi J Sci Tech 4:11–14

Brown J (2001) Mammals on mountain sides: elevation patterns of diversity. Glo Eco Biogeo 10:101–109

Chaturvedi RK, Raghubansi AS, Singh JS (2012) Biomass estimation of dry tropical wood species at juveline stage. Scient Wood J 1–5

Chidumayo EN (1990) Above ground woody biomass structure and productivity in Zambezian woodland. For Eco Mngt 36:33–46

Chowdhury KA, Ghosh SS (1958) Indian woods: their identification, properties and uses. Manager of Publications, Delhi, India

Devi B, Bhardwaj DR, Panwar P, Pal S, Gupta NK, Thakur CL (2013) Carbon allocation, sequestration and CO_2 mitigation under plantation forest of North-Western Himalaya. India Annals Res 56(1):123–135

Dhyani SK (2009) Challenges and scope of agroforestry research in India. In: Chaturvedi OP, Venkutesh A, Yadav RS (eds) Natural Resource sustainability, livelihood and climate moderation. Dwivedi RP, Ramesh S, Dhyani SK, Satish serial publishing House Delhi, Alam Badre, pp 13–25

Dhyani SK, Ram A, Dev I (2016) Potential of agroforestry systems in carbon sequestration in India. I J Ag Sci 86(9):1103–1112

Felter HW, Lloyd JU (1898) Granatum (U. S. P.)-Pomegranate. King's American Dispensatory

FSI (2015) State of forest report. Forest survey of India, Ministry of Environment, Forest and Climate change, Government of India, Dehradun, Uttarakhand

IPCC (1996) Revised IPCC guidelines for national green house gas inventories. Cambridge University Press, Cambridge

Joseph G, Abdullahi P (2016) Physicochemical and proximate analysis of extracts from *Citrus sinenesis* of Dustinma state, Nigeria. Open Access Library J 3:1–6

Kanawjia A, Kumar M, Mehraj A (2013) Specific gravity of some woody species in the Srinagar valley of the Garhwal Himalaya. For Sci Pract 15(1):85–88
Kumar M, Anemesh K, Sheikh M, Raj AJ (2012) Structure and carbon stock potential in traditional agroforestry system of Garhwal Himalaya. J Agri Tech 8(7):2187–2200
Kumar P (1989) Evaluation of Biofuels of Solan district, Himanchal Pradesh, Ph.d thesis, pp 82–83
Lakitan B (2008) Dasar Fisiologi Tumbuhan. Rajawali Press, Jakarata
Mahto S, Dasgupta SS, Todaria NP, Sigh VP (2016) Agroforestry mapping and Characterization in four distirct of Garhwal Himalaya. Energy Eco Environ 1(2):86–97
Mani S, Parthasarathy N (2007) Above-ground biomass estimation in ten tropical dry evergreen forest sites of peninsular India. Bioma Bioene 31:284–290
Miles PD, Smith BW (2009) Specific gravity and other properties of wood and bark for 156 tree species found in N. America. USDA, Res. Note NRS-38. Newtown Square, PA: U.S. Department of Agriculture, Forest Service, Northern Research Station, pp 27–30
Mishra R (1968) Ecology work book. Oxford/IBH Publishing, New Delhi, p 243
NAFP (2014) National Agroforestry Policy, Government of India. Department of agriculture & cooperation. Ministry of agriculture, New Delhi, pp 1–13
Negi AK, Chauhan DS, Todaria NP, Rawat MSS (2009) Subsistence farming systems in Tehri district of Garhwal Himalaya in Uttarakhnad, India. Indian J Agrofores 11(2):45–52
Omostosa MA, Ogunsile BO (2010) Fibre and chemical properties of some Nigerian Musa species of Pulp production. Am J Mater Sci 23:160–167
Pressler M (1865) Das Gestz der Stammformbildung. Verlag Arnold, Leipzig
Purkayastha SK (1982) Indian woods: their identification, properties and uses, vol 4. The Controller of Publications, Delhi, India, p 172
Rajput SS, Shukla NK, Gupta VK (1985) Specific gravity of Indian timber. J Timber Dev Assoc India 31(3):12–41
Raturi RD, Chauhan L, Gupta S, Vijendra RR (2002) Indian woods: their identification, properties and uses, vol 6. ICFRE Publication, Dehradun, India
Sharma CM, Baduni NP, Gairola S, Ghildiyal SK, Suyal S (2010) Tree diversity and carbon stocks of some major forest type of Garhwal Himalaya, India. For Ecol Manage 260:2170–2179
Sheik M, Kumar M, Bussman W Rainer, Todaria NP (2011) Forest carbon stock and fluxes in physiographic zones of India. carbon Balance Manage 6:15
Singh JS, Singh SP (1992) Forests of Himalaya: structure, functioning and impact of man. Gyanodaya Prakashan, Nainital, Uttarakhand, India, p 294
Ting SV, Blair JG (1965) The reaction of specific gravity of whole fruit to the internal quality of orange. In: Proc. Florida state of Horticulture society pp 251–260
Troup RS (1921) The silviculture of Indian trees. Oxford Clarandon Press 3:11–95
Tumen I (2014) Anatomical physical and chemical properties of wild pear, *Pyrus communis L.* Zonguldak Kara Elamus University, Institute of natural and applied science, M.Sc thesis, p 84
Vikrant KK (2019) Carbon stock along altitude in traditional agroforestry system in Garhwal Himalaya. Department of forestry, HNB Garhwal University, Srinagar, Uttarakhand, Ph.d thesis, pp 50–55
Vikrant KK, Chauhan DS, Rizvi RH, Maurya A (2018) Mapping the extent of agroforestry area in different altitudes of Tehri district, North Western Himalaya, India through GIS and remote sensing data. J Indian Soc Remote Sens 46:1–12
Wani BA, Bodha RH, Khan A (2014) Wood specific gravity variation among five important hardwood species of Kashmir Himalaya. Pak J Biol Sci 17(3):395–401
Woomer PL (1999) Impact of cultivation of carbon fluxes in woody savannas of Southern Africa. Wate Air Soi Poll 70:403–412

Chapter 13
Climate Change: Concerns and Influences on Biodiversity of the Indian Himalayas

Damanjeet Kaur, Anupreet Singh Tiwana, Supinder Kaur, and Saurabh Gupta

Abstract The Himalayas are a diverse and imperative center of biodiversity due to their immense climatic, topographic, and geographic ascents. The Indian Himalayan Region (IHR) started from Siwaliks (foothills of the south) and extended up to Trans-Himalaya (Tibetan plateau in the north), covering 12 Indian states. The IHR harbors many rivulets, lakes, rivers, vegetation, and animal species thus serves as a rich repository of biodiversity due to its unique biogeography. Climate change impacts such as increasing temperature, melting glaciers, and extreme weather events are severely deteriorating the fragile ecosystem and natural resources of the Himalayas, thus inducing the loss of biodiversity at an alarming rate. Consequently, these are also altering the development, behavior, and interactions between different biological species. It leads to the adaptation of species by developing new traits or migration and even extinction of several species. The present chapter investigates the causes and consequences of continually varying climate conditions on the behavior and survival of the species in the IHR, along with various approaches to mitigate global climate change.

Keywords Climate change · Indian Himalayan Region · Biodiversity · Mitigation · Ecosystem

D. Kaur · S. Gupta (✉)
Department of Microbiology, Mata Gujri College (Autonomous), Fatehgarh Sahib, Punjab, India

D. Kaur
Department of Biotechnology, Punjabi University, Patiala, Punjab, India

A. S. Tiwana
Department of Geography, Mata Gujri College (Autonomous), Fatehgarh Sahib, Punjab, India

S. Kaur
University School of Law, Sri Guru Granth Sahib World University, Fatehgarh Sahib, Punjab, India

S. Rani and R. Kumar (eds.), *Climate Change*, Springer Climate,
https://doi.org/10.1007/978-3-030-92782-0_13

Introduction

India is a mega-diverse nation that covers merely 2.4% of the world's entire land area but harbors a relatively prosperous floral and faunal diversity, around 7–8% of all the recorded species from all over the globe (http://www.cbd.int/). The country's immense climatic conditions and geographic ascents have contributed towards diverse ecosystems that help to sustain affluent biodiversity (Samant 2021). Among thirty-four universally known biodiversity hotspots, the four hotspots, namely the Himalayas, the Western Ghats, the Indo-Burma and the Sunderland, are found in the Indian province, harboring numerous endemic species (www.iucn.org). The Himalayas are immense, and a significant portion of the Himalayan Mountains come under the Indian Territory. The Himalaya is a highly complex mountain system having an elevated level of geographical vulnerability, affluent biodiversity, variability of landscapes, dense coniferous forests, enduring rivers, lakes, glaciers and wetlands, holy places, diversified human races, religions and cultures. The Himalayas constitutes the immense, diversified mountains that separate the northern part of the Asian continent. Mountains are the absolute treasure of biological diversity, occupy around 24% of land across the globe and harbor 12% population of the world (Sharma et al. 2010). The extreme altitudinal variations, diverse climatic conditions, and unique physiography of mountains have significantly contributed towards richness and diversification in biodiversity components at all levels (Spehn et al. 2002). About 1/3rd of Earth's biodiversity entirely depends upon mountainous habitation, bearing a great level of floral and faunal endemism (Korner 2004; Chape et al. 2008). The supercilious and snow-roofed mountain crests, alpine pastures, scenic landscape of mid-altitude, forests, and rivers canyon make it geologically distinct from the world (Sati 2016). The Himalayan Mountains surround the Hindu Kush Mountain region and Tibetan Autonomous Region of China. The mountains are extended over Afghanistan, Pakistan, China, India, Nepal, Bangladesh, Myanmar, and Bhutan and cover 34.4 Lac km^2. The planet's highest peaks fall in the Himalayan realm, and the Kanchenjunga, K2, and Mount Everest are among 30 mountains having a height above 7500 m. The Himalayan region can be divided into three analogous precincts, i.e., greater Himalaya, middle or lesser Himalaya, and sub-Himalayan foothills, and contiguous Terai and Duar plain (Pramanik and Bhaduri 2016). While geographically, Himalayas are alienated into three different regions (i) Western Himalayas consist of Poonch and Jammu (sub-Himalayan Kashmir), Pir Panjal, Valley of Kashmir, Baltistan and Ladakh, Kohistan, and Gilgit regions (ii) Central Himalayan region comprises of Punjab and Himachal Himalayas, Kumaon and Garhwal Himalayas, and Nepal Himalayas (iii) Eastern Himalayan region is extended between Darjeeling and Sikkim Himalayas and Bhutan and Assam Himalayas (Pramanik and Bhaduri 2016). This chapter appraisal the causes and consequences of continually varying climate conditions on the behavior and survival of the species in the Indian Himalayan Region (IHR).

Indian Himalayan Region

The IHR holds a unique place in the world's mountainous ecosystem (Bhushan et al. 2016). These young mountains are most significant from a climate perspective and harbor an affluent diversity of flora, fauna, and human communities. The IHR extended over an area of around 5.37 Lac km^2, covering more than 16% of the total geographical area of India (Bhushan et al. 2016). The region stretched between the Indus and the Brahmaputra River, 21°57'–37°5' N latitudes and 72°40'–97°25' E longitudes (about 2500 km in length, 250–300 km in width and rise from low lying plains to above 8000 m AMSL). The region is fully spanning through the mountains in the North Indian states, particularly Jammu and Kashmir, Himachal Pradesh, Uttarakhand to northeast states, i.e., Arunachal Pradesh, Sikkim, Manipur, Mizoram, Tripura, Nagaland, Meghalaya, and partially through hilly regions of Assam (Dima Hasao and Karbi Anglong) and West Bengal (Darjeeling and Kalimpong). Geographically, the IHR started from Shiwalik foothills and stretched up to the Tibetan plateau, covering 95 districts (Sidhu 2016). The IHR shares boundaries with six neighboring nations with upstream and downstream geological connect (NITI Aayog 2018). The IHR encompasses two biogeographic zones, i.e., Trans-Himalaya and Himalaya. Further, seven different biotic provinces, i.e., Ladakh Mountains, Tibetan Plateau, Trans-Himalaya-Sikkim, Northwest Himalaya, West Himalaya, Central Himalaya, and East Himalaya, fall under these biogeographic zones (Rodgers et al. 2002). The Trans-Himalaya have cold climatic conditions, high elevation, and arid mountains in the regions of Jammu and Kashmir (Ladakh and Kargil), Himachal Pradesh (Lahaul and Spiti valleys and Pooh tehsil), Uttarakhand (Nanda Devi range) and Sikkim (Kangchenjunga range); this region is known as "high altitude cold desert" (Mehta and Julka 2001). The zone also comprises a composite network of desolate mountains, including the Zanskar, Ladakh, and Karakoram ranges, with an average altitude of about 4000 m. The Himalayan zone comprises a cluster of ranges such as the Siwalik and Lesser Himalayan Ranges, located in the south of the Greater Himalaya. The central mountain ranges found in the Indian region are Pir Panjal Range falling in Jammu and Kashmir and Himachal Pradesh, Dhauladhar Range in Himachal Pradesh, Nag Tibba, and Mussoorie Range in Uttarakhand (Chandra et al. 2018). The major peaks fall in the region are Kanchenjunga (8586 m), Dhaulagiri (8172 m), Annapurna (8078 m), Nanda Devi (7817 m), Kamet (7756 m), Trishul (7140 m), and Badrinath (7138 m), and distinguished valleys are Kashmir Valley in Jammu and Kashmir, Chamba and Kullu Valleys in Himachal Pradesh (Chandra et al. 2018). Around 17% region of the Indian Himalayas are comprised of glaciers and permanently covered with snow. In comparison, nearly 40% area is seasonally covered with snow and serves as an inimitable water reservoir for perennial rivers to fulfill the drinking, household, agriculture, hydropower, and industrial needs. Further, about 1,200,000 million m^3 of water flows from the Himalayan Rivers every year (Singh 2006).

The IHR has dynamic scenery with prosperous and remarkable biodiversity. Variations amongst geological features along with three-dimensional frameworks such as longitudinal: east–west, latitudinal: north–south, and altitudinal: low–high favor the diversification in climatic and habitual conditions in the area (Singh 2004). Further, diverse geographical orogeny also resulted in temporal and spatial variation in the ecological conditions and physiography of the region. Thus, the region with the loftiest mountain ranges with distinctive elevation, varied landscapes, steep gradient of slopes, diverse climatic conditions, and soil formation is distinguished for having unique flora, fauna, and an elevated level of endemism. The Indian Himalayan ecosystem harbors abundant forests consisting of over a thousand trees, herbs, shrubs, and climbers. The IHR hosts around 50% of India's total flowering plants; this 30% is endemic. The region harbors approximately 8000 angiosperms (40% endemic), 44 gymnosperms (16% endemic), 1737 bryophyte (33% endemic), 600 pteridophyte (25% endemic), 1159 lichen (11% endemic), and 6900 fungal (27% endemic) species (Singh and Hajra 1996). More than 1740 medicinal plant species having conventional and modern therapeutic values, 118 essential oil-yielding medicinal plant species, 675 wild edible plant species, 279 fodder, and 155 sacred plant species have been recorded in the region (www.envfor.nic.in). The region also serves as a sink for carbon dioxide as over 41% of its total geographical area is covered with forest, representing 1/3rd of India's total forest cover. Tropical wet evergreen forest (47,192 km^2), subtropical broad-leaved hill forest (20,623 km^2), subtropical dry evergreen forest (354 km^2), montane wet temperate forest (7457 km^2), tropical moist deciduous forest (207,649 km^2), subtropical pine forest (22,880 km^2), Himalayan moist temperate forest (27,510 km^2), and Himalayan dry temperate forest (7634 km^2) are the primary forests found in the Himalayan zone (Chaturvedi et al. 2011; Reddy et al. 2015). The IHR constitutes about half of outstanding quality forest cover and provides a surplus of products and services (Singh and Singh 1992, 2002).

Approximately, 30,377 species and subspecies of both animals and protozoan have been reported from the IHR, constituting around 30.16% of total Indian faunal diversity (Chandra et al. 2018). The IHR has been recorded with 372 protozoan species, including 206 free-living, 49 symbiotic, and 117 parasites, representing about 10.6% of the total protozoan diversity of India. The region comprises numerous species of mammals (280), birds (938), reptiles (200), amphibians (80), and fishes (316), thus constituting more than 27% of entire vertebrates of the country (Chandra et al. 2018). The major mammalian species found in the Trans-Himalayan region are *Pseudois nayaur* (Blue sheep), *Lepus oiostolus* (Tibetan woolly hare), *Procapra picticaudata* (Tibetan gazelle), *Ovis ammon hodgsoni* (Tibetan argali), *Pantholops hodgsoni* (Tibetan antelope), *Marmota himalayana* (Himalayan marmot), *Capra sibirica* (Himalayan ibex), *Panthera uncia* (snow leopard), *Bos grunniens* (wild yak), *Ovis vignei* (Ladakh urial), and *Equus kiang polyodon* (Tibetan wild ass) (Namgail 2009; Habib et al. 2015; Kumar et al. 2017). The region harbors 26,392 species and subspecies of Arthropoda, representing approximately 89.6% of the total diversity of the IHR. Likewise, the IHR also represents 24,784 species/subspecies of insect, accounts for 38.1% of total known Indian diversity. Amongst the biotic provinces, the highest faunal diversity has been reported in the Central Himalayan region (14,183), followed by the Western Himalayas (12,022), Northern-Western Himalayas (8731),

Eastern Himalayas (5542), Ladakh Mountains (1561), Tibetan Plateau (1320), and Trans-Himalaya-Sikkim (1112) (Chandra et al. 2018).

Climate Change

Climate is described as the general weather patterns such as temperature, wind, rainfall, humidity, etc., of a region over a particular time scale, usually years (Adedeji et al. 2014). Due to anthropogenic activities, the climate of the Earth is changing continuously. At present, the Earth's climate is immensely diverse. It was 100 million years ago when dinosaurs exist on the planet, and tropical plants flourish near the poles (Mac et al. 1998). Climate change is referred to as the significant alterations in average climatic conditions of a province over some decades. It is an extended process that distinguishes climate change from natural weather variability (Hegerl et al. 2007). Intergovernmental Panel on Climate Change (IPCC) has well-defined climate change as *"change in the state of the climate that can be identified by changes in the mean and/or the variability of its properties, and that persists for an extended period, typically decades or longer. It can manifest as extreme climatic events or as gradual systemic change leading to a slow deterioration in environmental conditions until customary practices or habitation become nonviable"* (IPCC 2007b). Global climate change is the average long-term change in the climate conditions over the entire Earth. These consist of temperature changes and alteration in how frequently the area experience extreme events like heat waves, droughts, floods, storms, etc. (Adedeji et al. 2014). Climate change is the potential menace to having a noticeable impact on both the natural ecosystem and biodiversity. Climate change has become an intense matter of hue and cries over the globe as it can alter and boost up the local anthropogenic disturbances.

The mean surface temperatures of the universe have increased by 0.7 °C over the last century and are expected to increase continuously. According to IPCC, the mean exterior temperature of the globe is likely to increase up to 1.1–6.4 °C by the last decade of the twenty-first century comparative to 1980–1999 (Sintayehu 2018). The era commencing 1983–2012 was the hottest phase of the previous 1400 years in the Northern Meridian. The collective global average land and ocean surface temperature as calculated by a linear trend showing an upsurge of 0.85 °C (0.65–1.06) between the periods of 1880 and 2012 (IPCC 2014a; Rao et al. 2016). From 1901 to 2010, the mean global sea level was raised by 19 cm (from 0.17 to 0.21) and has been further projected to increase by 0.09–0.88 m by 2100. Moreover, global average precipitations have increased 2% during the last century and are expected to rise shortly (IPCC 2013). These projections show that the increase in temperature varies by region and depends on the increase and decrease in the precipitations. However, a 0.7 °C increase in global average temperature seems a diminutive change but had already a provable effect on the natural resources. The increasing temperature of the Earth causes the melting of glaciers, rise in sea level, coral bleaching, tree lining shifting, inconsistent precipitation patterns, and other ecological challenges.

The variations in temperature and precipitation occur at different rates around the globe (Girvetz et al. 2009). As the climate is predicted to change continually in the near future, it is probably a foremost driver for biodiversity loss in the mountainous ecosystem (Bellard et al. 2012). Loss of biological diversity due to changing climate has altered the motif and flux of energy flow and material circulation either directly or indirectly, which further considerably affect the array and distribution of ecosystem and ecosystem services (Zhong and Wang 2017; Sintayehu 2018).

Causes of Climate Change

Global climate change is the major niggling hindrance that the planet is facing currently for its existence. The major causes of climate change are global warming, greenhouse gases, ozone depletion, and deforestation that affect the biological resources and life-supporting system both directly and indirectly. Further, climate change may arise as a result of both internal and external processes. Internal factors leading to climate change is being occurred on all time scales. Atmospheric processes that create internal variations operate instantaneously, such as condensation of water vapors in clouds, or may take up to years such as troposphere-stratosphere or inter-hemispheric exchange (Hegerl et al. 2007). At the same time, other components like the ocean and the vast ice sheets have a propensity to operate on a longer time scale. These climate system components generate internal variability according to their own and incorporate variability from the briskly changing atmosphere (Hasselmann 1976). Furthermore, internal variability has also been created due to interactions between components. Besides these external processes such as solar rays, volcanic eruptions, hydrological extremes, etc., are also contributing to variation in the climatic system. Other external influences include the alteration in the atmospheric composition that commences with the industrial rebellion, which is the outcome of human interferences (Hegerl et al. 2007).

Global Warming and Greenhouse Effect

The continuously escalating concentration of greenhouse gases in the atmosphere perturbs the environment to cause global warming, interrupting the weather and wind pattern and their exchange within the atmosphere (Pandey et al. 2011; Lone et al. 2017). The terms global warming and climate change are primarily used conversely (Cohen and Waddell 2009). Global warming is cited as an uprise in the global average temperature near the planet's surface (Singh and Singh 2012). Greenhouse gases are a crucial contributor to climate change and play a significant role in Earth's climate. Human-induced activities are changing the composition of the surroundings. Escalating anthropogenic activities such as agriculture, industrialization, burning of fossils, deforestation, and land use have risen the concentrations of greenhouse gases (GHGs), i.e., carbon dioxide (CO_2), nitrous oxide (N_2O), methane (CH_4), and F-gases and aerosols in the atmosphere (Berrou et al. 2009; Seinfeld 2011; Borduas and Donahue 2018). Further, random release of these GHGs intensified global warming, ocean acidification, and desertification and thus resulted in changes in the climatic conditions (Yoro and Daramola 2020). Different anthropogenic activities have contributed to different extents of GHGs in the atmosphere. The energy sector has been emerged as the principal contributor, emitting about 26% of total GHGs followed by industrial release (19%), use of land and deforestation (17%), farming practices (14%), transportation (13%), several commercial and domestic usages (8%), and effluent and garbage disposal (3%) (Chakraborty et al. 2014; Filho 2015). The upsurge of GHGs in the surroundings accelerates the greenhouse effect by trapping heat and escalating global warming. The atmospheric concentrations of these major greenhouse gases had attained the maximum recorded levels in the 1990s, resulting in the greenhouse effect (IPCC 2001). The Sun's radiations released from Earth's exterior are absorbed and retained in the subordinate atmosphere as heat (Darkwah et al. 2018). Water vapors, CO_2, CH_4, N_2O, and O_3, retain this heat and help to maintain the Earth's ambient temperature. Without these naturally occurring GHGs, the global mean temperature on Earth's surface will be about −19 °C, besides the current mean temperature, i.e., 14 °C (Cassia et al. 2018).

CO_2 is the principal GHG released by natural activities, including volcanic eruptions, animal respiration, plants and organic matter decay, and human-induced activities like cement manufacturing, deforestation, energy and power generation, and fossil fuels combustion. Due to brisk expansion in the combustion of fossil resources over the past century, the atmospheric concentrations of $CO_{2(atm)}$ have risen to 385 ppm (parts per million), about a 30% increase from the pre-industrial value (Meure et al. 2006) and is continuing to rise at a rate of 1.9 ppm yearly. The emission of global CO_2 has been reported to increase by 70% from 1970 to 2002. The global temperature and $CO_{2(atm)}$ are directly linked to each other; an increase in the concentration of atmospheric CO_2, CH_4, and other gases at a level of about 3 W m^{-2} directly lead to increased temperature in the lower atmosphere (IPCC 2007a). Over the period 1750–2011, the collective emission of anthropogenic CO_2 in the surroundings was 2040 ± 310 $GtCO_2$. Approximately, 880 ± 35 $GtCO_2$ of these discharges have been retained in the surroundings, and the rest get accumulated in the sea and on the land, i.e., in plants and soil—the uptake of anthropogenic CO_2 by the ocean results in acidification. The pH of the oceanic surface water has a decrease of 0.1, thus contributed to a 26% enhancement in the acidity (IPCC 2014a). Emissions of CO_2 as a result of blazing of fossil fuels and industrial activities have triggered around 70% increase in overall GHG's release from 1970 to 2002 (Alhorr et al. 2014). Despite a mounting number of mitigation policies on climate change, the total GHGs release from anthropogenic sources has a continual increase over the period 1970–2010, with maximum increases between 2000 and 2010. Anthropogenic GHGs emissions were 49 ± 4.5 GtCO2-eq/year in 2010 (IPCC 2014a).

Water exit in both vapors and clouds form in the troposphere and is reported as one of the most significant absorbers of alterations in infrared radiation. The atmospheric life span of water vapors is relatively short, usually in days compared to CO_2, i.e., mainly in years (IPCC 2014b). The concentrations of water vapors vary regionally, does not directly influence by anthropogenic activities. However, human interference indirectly increases the global temperature and vapor formation, resulting in warming by an approach called water vapor feedback (Soden et al. 2005).

N_2O and nitric oxide (NO) are greenhouse gases that are produced from various natural as well as anthropogenic activities, for instance, burning of fossil fuels, agriculture practices using commercial and organic fertilizers, biomass incineration, and nitric acid manufacturing (Medinets et al. 2015). The global emissions of N_2O have risen mainly due to human intervention during the last century (IPCC 2014b). The efficiency of N_2O towards contribution to global warming is about 300 times more than CO_2. N_2O acts as a strong GHG, while NO indirectly contributes to ozone production. In the stratosphere, N_2O carry out the removal of O_3 (IPCC 2014b). Methane (CH_4) is primarily the natural gas synthesized by the decay of wastes in landfills, agriculture, and domestic livestock. Being least in the atmosphere, CH_4 is extremely active GHG than CO_2. Irrespective of its small amount, CH_4 has 21 times as much global warming potential compared to CO_2.

Other GHGs are the F-gases which include synthetic halocarbons such as chlorofluorocarbons (CFCs), hydrofluorocarbons (HFCs), perfluorocarbons (PFCs), halons, and sulfur hexafluoride (SF_6) that are generated in the atmosphere from various industrial sectors (Chakraborty et al. 2014). HFCs have been produced as an alternate for CFCs and are widely used in refrigerators, coolers, semi-conductors, etc. Depending upon the category and life span of HFCs in the environment, the global warming potential of HFC's is 140–11,700 times superior to CO_2. If the indiscriminate release of these significant GHGs is not abridged, the planet will encounter destructive outcomes. The atmospheric temperature will rise 2 °C near 2050 and 4 °C by 2100 in coastal zones. Moreover, the inner temperature of the globe will also have a surge of 4 °C by 2050 and 7 °C by 2100 (Orimoloye et al. 2019; Hossain et al. 2019; Donat et al. 2020; Murdoch 2020).

Greenhouse Gases and Ozone Layer Depletion

Ozone (O_3) mainly originates in the stratosphere, but a small proportion is produced in the troposphere. Stratospheric ozone is produced by chemical catalysis between ultraviolet (UV) radiations and O_2. UV radiations break down one mole of O_2 into two oxygen atoms. Each of these immensely reactive oxygen atoms combines with O_2 and produces an O_3 molecule. The ozone layer absorbs about 99% of the Sun's moderate frequency UV rays, having wavelengths between 200 and 315 nm, thus protect living creatures, including plant and animal life, from destruction by harmful UV and infrared radiations (Whitefield et al. 1998; Cassia et al. 2018). Nevertheless, the rapid increase in GHGs, particularly CO_2 and F-gases, has significantly reduced the ozone layer's rapid depletion and created an ozone hole over Antarctica. The global average thickness of the ozone layer necessary for the survival of human beings is about 300 Dobson units, i.e., equal to 3 mm. Contrastingly, the ozone level fell to 90 Dibson units. Due to ozone depletion, the UV-B rays having a range between 280 and 320 nm are now reaching the Earth's surface and causing harm to the life forms (Chakraborty et al. 2014).

Deforestation and Climate Change

Forests are the natural reserves that cover around 31% of land on Earth. Forests play several crucial ecological functions; for instance, these provide shelter to numerous plant and animal species, food, medicine, livelihoods, and resources to human beings around the globe. Apart from these, forests also play a significant role in Earth's climate system, thus making these crucial ecological powerhouses inimitable. Forests also serve as natural carbon sinks as these can accumulate more than 3/4th of the carbon in terrestrial plants and nearly 40% of soil carbon. Plants absorb CO_2 and fix it by the process of photosynthesis while releasing the oxygen, thus helping to maintain the ecological balance. Despite these indispensable roles, forests are disappearing at an alarming rate. Between the years 1990 and 2016, about 1.3 million km^2 of forest cover was vanished worldwide due to human-induced and natural deforestation, thus significantly affecting the wildlife, ecosystem, weather pattern, and climate (www.nationalgeographic.com). Deforestation is the alteration of the forested area into non-forested for land use. It can be due to the intended exclusion of forest cover for agricultural purposes, urbanization, and industrialization, mining, or an outcome of overgrazing (Kumari et al. 2019). Further, populace pressures, profits, and social and political forces can also drive up the rate of forest loss. A considerable source of CO_2 emissions is deforestation, as it is predicted to cause about 1/4th of total anthropogenic carbon releases, deprivation of biological diversity and other ecological resources. The cutting and burning of the forests result in the massive release of stored atmospheric carbon in the form of CO_2. Between the years 2015 and 2017, the global loss of tropical forests has resulted in about 4.8 billion tons of CO_2 per year. Reducing deforestation seems to be one of the least expensive methods to lower CO_2 emissions and mitigate climate change (Kindermann et al. 2008). To a larger extent, deforestation results in drier and warmer weather conditions due to the synergetic impact of abridged evapotranspiration, elevated CO_2 concentration and reflective power triggering desertification, biodiversity loss and melting of glaciers, ultimately leading to food insecurity (Kumari et al. 2019).

Forests are vibrant reserves of biodiversity, which store an immense genetic pool. Exclusion or devastation of considerable areas of forest cover has resulted in a dissipated environment having diminished biological diversity. Deforestation without adequate reforestation usually leads to significant loss of biodiversity. The biodiversity loss is of global interest regardless of local and regional significance. If forest loss persists to the same extent, the dense forest cover (>40% canopy cover) in Indian Himalayas will be limited to only 10% of land area by 2100, and this may result in considerable loss of 366 endemic plants and 35 endemic vertebrates (Pandit et al. 2007). The need of the current time is to conserve the forest cover and the paramount carbon sink to preserve the crucial ecosystems and combat the consequences of global climate change to sustain the ecological balance.

Threats to Biodiversity

Biodiversity is crucial for the structure and functioning of the ecosystem and favors different products and services that human beings acquire from natural resources (Walther et al. 2002; Naeem et al. 2009). The chase between the nations for modernization has resulted in an excessive expansion in industrialization, transportation, and urbanization, leading to the devastation of ecological stability. The speedy growth of the human population causes excessive land use and anthropogenic activities, which further rigorously affect the flimsy landscape of the Himalayan ecosystem. Further prevalence of floods, forest fires, landslides, and earthquakes causing massive loss to biodiversity, assets, and natural resources has become a usual event. The overexploitation of natural resources continuously changes the weather pattern, thus creating a global problem (Kumar and Chopra 2009). Being one of the world's major biodiversity hotspots, the sensitive and fragile ecosystem of the Indian Himalayas is currently facing the menaces of instinctive expansion and changing weather conditions (Cruz et al. 2007). If the situation persists, the landscape of the Indian Himalayas will have only 10% dense forest cover and about 1/4th of endemic species from major taxonomic groups, including angiosperms, gymnosperms, and vertebrates, might have wiped out by 2100 (Pandit et al. 2007). More recently, the increased entrance to the worldwide market has enhanced the requirement for natural resources, thus favoring the bi-directional movement in the region. Regardless of evident remoteness and inaccessibility, the Himalayas are highly vulnerable to human-induced biodiversity loss. The loss of biological diversity and associated alterations in the surroundings are much faster nowadays than ever before in human history. Ecosystem disturbances resulting from climate change directly influence human beings, including abridged supply of water, loss of species and landscapes. Further, the resulting extreme events can devastate the regulating services of the ecosystem.

Climate Change and Biodiversity

The Himalayan region is of considerable significance in biological richness, biodiversity, socio-cultural diversity, and wealth. Climate change is most likely to have numerous impacts on biodiversity from ecological system to species level. The most apparent impact is temperature and precipitation on species, ranges, and ecosystem boundaries. Some species have been near the periphery of their ranges, and these may need to shift due to climate change (Lemoine and Bohning-Gaese 2003), while others may become extinct. Habitat and species loss and extinctions of fishes and aquatic animals have been probable as of collective effects of increased water removal and

Table 13.1 Threatened vertebrate's species reported from the Indian Himalayan Region (IUCN 2017)

Class/ group	Critically endangered	Endangered	Vulnerable	Threatened
Pisces	1	5	13	19
Amphibians	–	–	4	4
Reptiles	2	4	9	15
Birds	8	9	35	52
Mammals	3	17	23	43
Total	14	35	84	133

climate change (Spooner et al. 2011). The numbers of threatened vertebrates (critically endangered, endangered, and vulnerable) in the International Union for Conservation of Nature (IUCN) Red List comprises 596 species, including mammals (92), birds (87), reptiles (54), amphibians (75), and fishes (225) susceptible in different categories from India. Out of these, 133 species are present in the IHR, representing about 22.3% of the total Indian threatened vertebrate fauna (Table 13.1). About 43 species of mammals, 52 species of birds, 15 species of reptiles, 4 species of amphibians, and 19 species of fishes known from the IHR are under different threatened categories.

Catreus wallichi (Cheer pheasants), having a range primarily restricted to the Jammu and Kashmir, has been declining due to habitat loss (Wikramanayake et al. 1998). The genus *Schizothorax* represents six endemic species found in elevated mountain lakes and streams, while two other genera of these snow trout, *Ptychobarbus* and *Gymnocypris biswasi* unique to Himalaya Hotspot, have been thought to be extinct (IUCN 2004). A total of 713 threatened taxa belonging to 383 genera falling in 126 families comprising 663 angiosperms, 24 gymnosperms, 22 pteridophytes, and 4 bryophytes were reported as threatened under various categories in IHR. Among these, maximum threatened plants were found in the Fabaceae family (84 species) followed by Cyperaceae (65 species) and Poaceae (36 species) families (Nayar and Shastry 1987, 1988, 1990; Ved et al. 2003a, 2003b; IUCN 2016). Threatened *Angelica glauca, Coptis teeta, Lilium polyphyllum, Nardostachys jatamansi, Aconitum heterophyllum, and Gentiana kurroo* are of great concern for conservation (Mehta et al. 2020). Mountainous forests such as alpine, sub-alpine, dry temperate, and moist temperate forests are most susceptible to climate change impacts, attributing towards higher rates of climate change at higher elevations. Northwestern Indian Forests are highly susceptible to changing climate (Charturvedi et al. 2011).

Policies for Sustaining Himalayan Biodiversity

The Himalayan ecosystem is extremely vulnerable due to exploitation of natural resources, geographical reasons, and changing climate conditions. Climate change has adversely affected the Himalayan ecosystem via increased temperature, diverse precipitation patterns, drought, and biotic influences (Palni and Rawal 2010). There is a need to sustain the Himalayan ecosystem for the conservation of biodiversity. India has launched the National Environment Policy (NEP), National Action Plan on Climate Change (NAPCC), and National Biodiversity Action Plan (NBAP) for the conservation of mountains (www.envfor.nic.in).

NEP: Measures for the Conservation of Mountains (NEP 2006)

NEP specifies definite plans for progress and regulation of the province. Those plans must be devised and executed at the national and state levels.

"Appropriate land-use planning and watershed management practices for sustainable development of mountain ecosystem.

> "Best practice" norms for infrastructure construction in mountain regions avoid or minimize damage to sensitive ecosystems and despoiling of landscapes.
>
> Encourage cultivation of traditional crops and horticulture by promoting organic farming, enabling farmers to realize a price premium.
>
> Promote sustainable tourism by adopting "best practice" norms of eco-friendly and responsible tourism, creating appropriate facilities and access to ecological resources, and multi-stakeholder partnerships to enable local communities to gain livelihoods while leveraging the financial, technical, and technical managerial capacities of investors.
>
> Take measures to regulate tourist inflows into mountain regions to ensure that these remain within the carrying capacity of the mountain ecology."

NAPCC (NAPCC 2008)

The NAPCC has been issued by the MoEF in the year 2008, comprised of an inclusive set of mitigation and adaptation plans with intend to encourage the development of strategies while providing benefits for tackling climate change efficiently. NAPCC includes eight missions that demonstrate multi-pronged, comprehensive and coordinated approaches to accomplish key goals about climate change:

> "National Solar Mission.
>
> National Mission for Enhanced Energy Efficiency.
>
> National Mission on Sustainable Habitats.
>
> National Water Mission.
>
> National Mission for Sustaining the Himalayan Ecosystem.
>
> National Mission for a Green India.
>
> National Mission for Sustainable Agriculture.
>
> National Mission on Strategic Knowledge of Climate Change".

The National Mission for Sustaining the Himalayan Ecosystem (NMSHE)

The mission has been enacted to devise management practices for nurturing and conserving Himalayan glaciers and the mountainous ecosphere. This mission focus on four types of national capacities:

> "a) Human and knowledge capacities, b) Institutional capacities, c) Capacities for evidence-based policy building and governance and d) Continuous self-learning for balancing between forces of Nature and actions of mankind."

The mission aims to:

> "Understand whether and the extent to which the Himalayan glaciers are in recession and how the problem could be addressed.
>
> Establish an observational and monitoring network for the Himalayan environment.
>
> Promote community-based management of the ecosystems through incentives to community organizations and panchayats for protection of forested lands."

NBAP (NBAP 2008)

NBAP, formulated through a comprehensive inter-ministerial process, was approved by the Government of India in 2008. The NBAP identifies the threats and constraints in biodiversity convention, considering the existing legislation, implementation, mechanisms, strategies, plans, and programs, based on which action points have been designed.

The objectives of NBAP include:

> "Strengthening and integration of in situ, on-farm and ex-situ conservation.
>
> Augmentation of the natural resource base and its sustainable utilization: Ensuring inter and intra-generational equity.
>
> Regulation of introduction of invasive alien species and their management.
>
> Assessment of vulnerability and adaptation to climate change and desertification.
>
> Integration of biodiversity concerns in economic and social development.
>
> To prevent, minimize and abate the impacts of pollution.
>
> Development and integration of biodiversity databases.
>
> Strengthening implementation of policy, legislative and administrative measures for biodiversity conservation and management.
>
> Building of national capacities for biodiversity conservation and appropriate use of new technologies.
>
> Valuation of goods and services provided by biodiversity and use of economic instruments in decision-making processes.
>
> International cooperation to consolidate and strengthen bilateral, regional and multilateral cooperation on issues related to biodiversity."

Conclusions

The Himalayas are a highly diverse mountain system that serves as the richest source of biodiversity. The climate of the planet is changing continuously. Climate change is one of the most considerable aspects for the degradation of biological diversity in the Indian Himalayas these days, affecting the mountain abode communities in physical, social, and economic ways. Fragile landscapes of Himalayan ecosystems are highly vulnerable to climate conditions. Changing climate patterns adversely affects the floral and faunal diversity, resulting in species extinction in the Himalayan region. The chapter discussed various concerns and influences of climate change on biological species. India has framed different missions such as NEP, NAPCC, and NBAP and is working on these for sustainability and conservation of Himalayan Mountains and biodiversity.

References

Adedeji O, Okocha R, Olatoye O (2014) Global climate change. J Geosci Environ Protect 2:114–122

Alhorr Y, Eliskandarani E, Elsarrag E (2014) Approaches to reducing carbon dioxide emissions in the built environment: low carbon cities. Int J Sustain Built Environ 3:167–178

Bellard C, Bertelsmeier C, Leadley P et al (2012) Impacts of climate change on the future of biodiversity. Ecol Lett 15:365–377

Berrou A, Raybaut M, Godard A et al (2009) High-resolution photoacoustic and direct absorption spectroscopy of main greenhouse gases by use of a pulsed entangled cavity doubly resonant OPO. Appl Phys B Lasers Opt 98:217

Bhushan I, Kumar A, Patel JS et al (2016) The Indian Himalayan ecosystem as source for survival. In: Bisht J, Meena V, Mishra P et al (eds) Conservation Agriculture. Springer, Singapore, pp 367–380

Borduas N, Donahue NM (2018) The natural atmosphere. In: Dransfield T, Torok B (ed) Green chemistry. Elsevier, pp 131–150

Cassia R, Nocioni M, Correa-Aragunde N et al (2018) Climate change and the impact of greenhouse gasses: CO_2 and NO, friends and foes of plant oxidative stress. Front Plant Sci 9:273

Chakraborty S, Pattanayak A, Mandal S et al (2014) An overview of climate change: causes, trends and implications. In: Roychowdhury R (ed) Crop improvement in the era of climate change. International Publishing House, New Delhi, I.K, pp 1–29

Chandra K, Gupta D, Gopi KC et al (2018) Faunal Diversity of Indian Himalaya: an overview. In: Chandra K, Gupta D, Gopi KC et al (ed) Faunal Diversity of Indian Himalaya, pp 1–44 (Published by the Director, Zool. Surv. India, Kolkata)

Chape S, Spalding MD, Jenkins MD (2008) The world's protected areas: status, values and prospects in the 21st century. University of California Press, London in association with UNEP-World Conservation Monitoring Centre, Cambridge, pp 1–365

Chaturvedi RK, Gopalakrishnan R, Jayaraman M et al (2011) Impact of climate change on Indian forests: a dynamic vegetation modeling approach. Mitig Adapt Strateg Glob Change 16:119–142

Cohen SJ, Waddell MW (2009) Climate change in the 21st century. McGill-Queen's University Press

Cruz RV, Harasawa H, Lal M et al (2007) Asia climate change 2007: impacts, adaptation and vulnerability. In: Parry ML, Canziani OF, Palutikof JP et al (eds) Contribution of working group II to the fourth assessment report of the intergovernmental panel on climate change. Cambridge University Press, Cambridge, UK, pp 469–506

Darkwah WK, Odum B, Addae M et al (2018) Greenhouse effect: greenhouse gases and their impact on global warming. J Sci Res Reports 17:1–9

Donat MG, Sillmann J, Fischer EM (2020) Changes in climate extremes in observations and climate model simulations, from the past to the future. In: Sillmann J, Sippel S, Russo S (eds) Climate extremes and their implications for impact and risk assessment. Elsevier, Cambridge, MA, pp 31–57

Filho WL (2015) Handbook of climate change adaptation. Springer, Berlin, Heidelberg

Girvetz EH, Zganjar C, Raber GT et al (2009) Applied climate-change analysis: the climate wizard tool. PLoSONE 4:e8320

Habib B, Shrotriya S, Mahar N et al (2015) Field Sampling Protocol-Mammalian fauna in Trans-Himalayan Landscape, Uttarakhand, India. Wildlife Institute of India and Uttarakhand Forest Department, pp 1–30

Hegerl GC, Zwiers FW, Braconnot P et al (2007) Understanding and attributing climate change. In: Solomon S, Qin D, Manning M et al (eds) Climate change 2007: the physical science basis. Contribution of working group I to the fourth assessment report of the intergovernmental panel on climate change, Cambridge University Press, Cambridge, United Kingdom and New York, NY, USA

Hossain MS, Arshad M, Qian L et al (2019) Economic impact of climate change on crop farming in Bangladesh: an application of Ricardian method. Ecol Econ 164:106354

IPCC (2001) Climate change 2001: synthesis report. In: Watson RT and the Core Writing Team (ed) A contribution of working groups I, II, and III to the third assessment report of the Intergovernmental panel on climate change. Cambridge University Press, Cambridge, United Kingdom, and New York, NY, USA

IPCC (2007a) Climate change 2007: the physical science basis. In: Solomon S, Qin D, Manning M et al (ed) Contribution of working group I to the fourth assessment report of the intergovernmental panel on climate change. Cambridge University Press, Cambridge, United Kingdom and New York, NY, USA

IPCC (2007b) Climate change 2007: Synthesis Report. In: Bernstein L, Bosch P, Canziani O, Chen Z, Christ R, Davidson O et al Summary for policymakers, an assessment of the Intergovernmental panel on climate change

IPCC (2013) Climate change 2013: The physical science basis. Contribution of working group I to the fifth assessment report of the intergovernmental panel on climate change. Cambridge University Press, Cambridge, United Kingdom and New York, NY, USA

IPCC (2014a) Climate Change 2014: Synthesis Report. In: Pachauri RK, Meyer LA (ed) Contribution of working groups I, II and III to the fifth assessment report of the intergovernmental panel on climate change [Core Writing Team]. IPCC, Geneva, Switzerland

IPCC (2014b) Climate Change 2014: Mitigation of climate change. In: Edenhofer OR, Pichs-Madruga Y, Sokona E, Farahani S, Kadner K, Seyboth A et al (ed) Contribution of working group III to the fifth assessment report of the Intergovernmental panel on climate change. Cambridge University Press, Cambridge

IUCN (2004) The IUCN red list of threatened species. The World Conservation Union, Gland, Switzerland

IUCN (2015–2016). The IUCN red list of threatened species. www.iucnredlist.org

IUCN (2017) The IUCN red list of threatened species. Version 2017-3

Kindermann G, Obersteiner M, Sohngen B et al (2008) Global cost estimates of reducing carbon emissions through avoided deforestation. Proc Natl Acad Sci 105:10302–10307

Korner C (2004) Mountain biodiversity, its causes and function. Ambio Special Report 13:11–17

Kumar A, Adhikari BS, Rawat GS (2017) Biogeographic delineation of the Indian Trans-Himalaya: need for revision. Curr Sci 113:1032–1033

Kumar V, Chopra AK (2009) Impact of climate change on biodiversity of India with special reference to Himalayan region-an overview. J Appl Nat Sci 1:117–122

Kumari R, Banerjee A, Kumar R et al (2019) Deforestation in India: consequences and sustainable solutions. In: Suratman MN, Latif ZA, Oliveira GD et al (ed) Forest degradation around the world, IntechOpen

Lemoine N, Bohning-Gaese K (2003) Potential impact of global climate change on species richness of long-distance migrants. Conservation Biol 17:577–586

Lone B, Qayoom S, Singh P et al (2017) Climate change and its impact on crop productivity. British J Appl Sci Tech 21:1–15

Mac MJ, Opler PA, Puckett Haecker CE et al (1998) Status and trends of the Nation's biological resources. U.S. Department of the Interior, U.S. Geological Survey, pp 89–93

Medinets S, Skiba U, Rennenberg H et al (2015) A review of soil NO transformation: associated processes and possible physiological significance on organisms. Soil Biol Biochem 80:92–117

Mehta HS, Julka JM (2001) Mountains: Trans-Himalaya. In: Das AK, Sanyal AK (eds) Alfred JRB. Ecosystems of India, ENVIS Centre Zoological Survey of India, pp 73–92

Mehta P, Sekar KC, Bhatt D et al (2020) Conservation and prioritization of threatened plants in Indian Himalayan Region. Biodivers Conserv 29:1723–1745

Meure CM, Etheridge D, Trudinger C et al (2006) Law Dome CO_2, CH_4 and N_2O ice core records extended to 2000 years BP. Geophys Res Lett 33:L14810

Murdoch A, Mantyka-Pringle C, Sharma S (2020) The interactive effects of climate change and land use on boreal stream fish communities. Sci Total Environ 700:34518

NBAP (2008) National Biodiversity Action Plan, Ministry of Environment and Forest, GoI. Avilable at http://nbaindia.org/uploaded/Biodiversityindia/NBAP.pdf

Naeem S, Bunker DE, Hector A, Loreau M, Perrings C (2009) Biodiversity, ecosystem functioning and human wellbeing: an ecological and economic perspective. Oxford University Press, UK

Namgail T (2009) Mountain ungulates of the trans-Himalayan region of Ladakh, India. Int J Wild 15:35–40

National Action Plan on Climate Change, GOI (2008) http://pmindia.nic.in/Pg01-52.pdf

Nayar MP, Shastry ARK (1987) Red data book of Indian plants. Botanical Survey of India, Howrah

Nayar MP, Shastry ARK (1988) Red data book of Indian plants. Botanical Survey of India, Howra

Nayar MP, Shastry ARK (1990) Red data book of Indian plants. Botanical Survey of India, Howrah

NEP (2006) National Environment Policy, Ministry of Environment and Forest, GOI. Avilable at http://www.envfor.nic.in/nep/nep2006e.pdf

NITI Aayog (2018) Contributing to sustainable development in the Indian Himalayan region: a summary report

Orimoloye IR, Mazinyo SP, Kalumba AM et al (2019) Implications of climate variability and change on urban and human health: a review. Cities 91:213–223

Palni LMS, Rawal RS (2010) Conservation of Himalayan bioresources: an ecological, economical and evolutionary perspective. In: Sharma VP (ed) Nature at work: ongoing saga of evolution. Springer, New Delhi

Pandey D, Agrawal M, Pandey JS (2011) Carbon footprint: current methods of estimation. Environ Monit Assess 178:135–160

Pandit MK, Sodhi NS, Koh LP et al (2007) Unreported yet massive deforestation driving loss of endemic biodiversity in Indian Himalaya. Biod Cons 16:153–163

Pramanik P, Bhaduri D (2016) Impact of climate change on water resources in Indian Himalaya. In: Bisht J, Meena V, Mishra P et al (eds) Conservation agriculture. Springer, Singapore, pp 487–507

Rao CS, Gopinath KA, Prasad JVNS et al (2016) Climate resilient villages for sustainable food security in Tropical India: concept, process, technologies, institutions, and impacts. In: Sparks DL (ed) Advances in Agronomy, vol 140. Academic Press, Elsevier BV, pp 101–214

Reddy C, Jha CS, Diwakar PG, Dadhwal VK (2015) Nation-wide classification of forest types of India using remote sensing and GIS. Envir Mon Assess 187:1–42

Rodgers WA, Panwar HS, Mathur VB (2002) Wildlife protected area network in India: a review, executive summary. Wildlife Institute of India, Dehradun, pp 1–51

Samant SS (2021) Floristic diversity of the Himalaya in relation to climate change: status, values and conservation. J Graph Era Univ 9:31–54

Sati VP (2016) Geography of Himalaya. In: Singh RB (ed) Progress in Indian Geography, A country report, 2012–16. In: The 33rd international geographical congress, Beijing, China (August 21–25, 2016). Indian National Academy, New Delhi, pp 183–191

Seinfeld JH (2011) Insights on global warming. AIChE J 57:3259–3284

Sharma E, Chettri N, Oli KP (2010) Mountain biodiversity conservation and management: a paradigm shift in policies and practices in the Hindu Kush-Himalayas. Ecol Res 25:909–923

Sidhu GS (2016) Soil conservation of Northwestern Himalayas (NWH): their constraints and potentials for sustainable hill agriculture. In: Bisht J, Meena V, Mishra P et al (eds) Conservation agriculture. Springer, Singapore, pp 315–338

Singh BR, Singh O (2012) Study of impacts of global warming on climate change: rise in sea level and disaster frequency. In: Singh BR (ed) Global warming—impacts and future perspective, IntechOpen, pp 93–118

Singh DK, Hajra PK (1996) Floristic diversity. In: Gujral GS, Sharma V (ed) Changing perspectives of Biodiversity Status in the Himalaya. British Council, New Delhi

Singh JS (2004) Sustainable development of the Indian Himalayan region: linking ecological and economic concerns. Xth Pt. Govind Ballabh Pant Memorial Lecture

Singh JS (2006) Sustainable development of the Indian Himalayan region: linking ecological and economic concerns. Curr Sci 90:784–788

Singh JS, Singh SP (1992) Forests of Himalaya: structure, function and impact of man. Gyanodaya Prakashan

Singh SP (2002) Balancing the approaches of environmental conservation by considering ecosystem services as well as biodiversity. Curr Sci 82:1331–1335

Sintayehu DW (2018) Impact of climate change on biodiversity and associated key ecosystem services in Africa: a systematic review. Ecosystem Health Sust 4:225–239

Soden BJ, Jackson DL, Ramaswamy V et al (2005) The radiative signature of upper tropospheric moistening. Science 310:841–844

Spehn EM, Messerli B, Korner C (2002) A global assessment of mountain biodiversity: synthesis. In: Korner C, Spehn EM (ed) Mountain biodiversity. A global assessment, Parthenon, Boca Raton, pp 1–336

Spooner DE, Xenopoulos MA, Schneider C et al (2011) Coextirpation of host-affiliate relationships in rivers: The role of climate change, water withdrawal, and host-specificity. Global Change Biol 17:1720–1732

Ved DK, Kinhal GA, Haridasan K et al (2003a) Report of the conservation assessment and management prioritisation for the medicinal plants of Arunachal Pradesh, Assam, Meghalaya and Sikkim. 27 February—1 March 2003, Guwahati. Foundation for Revitalisation of Local Health Traditions and Himalayan Forest Research Institute, pp 1–153

Ved DK, Kinhal GA, Ravikumar K et al (2003b) Conservation assessment and management prioritisation for the medicinal plants of Himachal Pradesh, Jammu & Kashmir and Uttaranchal. Proceedings of the workshop held at Shimla during 19–24 May, 2003. Foundation for Revitalisation of Local Health Traditions, pp 1–24

Walther GR, Post E, Menzel A et al (2002) Ecological responses to recent climate change. Nature 416:389–395

Whitefield CP, Davidson AW, Ashenden TW (1998) The effects of nutrient limitation on the response of Plantago major to ozone. New Phytol 140:219–230

Wikramanayake E, Dinerstein E, Allnut T et al (1998) A biodiversity assessment and gap analysis of the Himalayas. World Wildlife Fund-U.S., Conservation Science Program/UNDP Report

Yoro KO, Daramola MO (2020) CO_2 emission sources, greenhouse gases, and the global warming effect. In: Farsi M, Makarem MA (eds) Rahimpour MR. Advances in carbon capture, Woodhead Publishing, pp 3–28

Zhong L, Wang J (2017) Evaluation on effect of land consolidation on habitat quality based on InVEST Model. Trans Chin Soc Agric Eng 33:250–255

Chapter 14
Lessons Learned from the COVID-19 Lockdown for Sustainable Northwestern Himalayan Region

Shashi Shekhar

Abstract This study tried to understand the changes in three air pollution parameters, namely absorbing aerosol index (AAI)—broadly referred to as dust and smoke, nitrogen dioxide (NO_2) and carbon monoxide (CO) column number density in the Western Indian Himalayas during the COVID-19 lockdown. The study used Sentinel 5P data over a 50 km radius of 11 non-attainment cities (NACs) of Northwestern Indian Himalayan states (7 of Himachal Pradesh, 2 of Uttarakhand and 2 of Union Territory of Jammu and Kashmir) during the lockdown phase-1 (March 24–April 14, 2020) with respect to the pre-lockdown period (March 24–April 14, 2019). Ground-measured data on particulate matter (PM_{10}) and nitrogen oxides (NO_x) has also been used for only 7 NACs of Himachal Pradesh for the lockdown phase-1 and pre-lockdown period. The average values of AAI, NO_2 and CO around 11 towns have been reduced by 55%, 19% and 7%, respectively, during lockdown phase-1, compared to the pre-lockdown period. The satellite observation is further complemented through ground-monitored data on air pollution. In the seven NACs of Himachal Pradesh, NO_x and PM_{10} mass concentrations have been substantially reduced during the March and April months of 2020 compared to the same months of 2019. This study would give an idea to environmentalists and policymakers to plan a sustainable emission policy to reduce the adverse impacts of air pollution on the physical aspects (e.g. snow and glaciers) of the Northwestern Himalayan region under the climate change conditions.

Keywords Sentinel 5P · Air quality · COVID-19 · Himalayas · Sustainability

Introduction

Coronavirus disease (COVID-19) caused enormous human death around the world. It is a respiratory disease that transmits from human to human. After analysing a large number of human deaths, an alarming rate of infections and critical illness, the World Health Organisation (WHO) declared COVID-19 as a global pandemic on March 11,

S. Shekhar (✉)
Himachal Pradesh State Pollution Control Board, Shimla, Himachal Pradesh, India

S. Rani and R. Kumar (eds.), *Climate Change*, Springer Climate,
https://doi.org/10.1007/978-3-030-92782-0_14

2020. Many countries including India imposed a lockdown to refrain the transmission rate of new infections by restricting the various types of anthropogenic activities (e.g. factories, non-essential goods market, commercial hubs, transport and personal vehicle movements) at varying scales. Several studies contemplated the effectiveness of COVID-19 lockdown on air pollution across cities around the globe using remote sensing air pollutants data and ground-measured (Kumari and Toshniwal 2020a; Mostafa, Gamal, and Wafiq 2021; Albayati et al 2021). These studies shared findings that reflected the universal decline in most of the air pollutants though the rate of its decline is different at varying spatial scales. Kumari and Toshniwal (2020) quantified air quality data of 162 monitoring stations from 12 cities across the globe and found positive impacts of COVID-19 on the environment. They found a significant reduction in particulate matter ($PM_{2.5}$, PM_{10}) and NO_2 in the lockdown phase as compared to the pre-lockdown phase. Mostafa et al. (2021) shown a reduction in NO_2 (~15–33%), Absorbing Aerosol Index (AAI) (30%), GHG emissions (4%), ozone level (2%) and CO (5%) over Egypt.

Furthermore, *Janta* curfew (people's curfew) was declared on March 22, 2020, by the Government of India. In India, there were 4 phases of lockdown in 2020, i.e. phase-1 (March 25–April 14), phase-2 (April 15–May 03), phase-3 (May 04–17) and finally phase-4 (May 18–31) (Pathakoti et al. 2020). During the lockdown phase-1 period, public transport came to a complete halt, which is considered to be the main source of air pollution in most of the Indian cities. All factories, markets, shops and construction works remained suspended across the world and Indian sub-continent. Yadav et al. (2020) assessed the major pollutants in 4 Megacities of India during the COVID-19 lockdown with the past several years (2013–2019). They reported a sharp decline in NO_2 levels (60–65%) and $PM_{2.5}$ and PM_{10} levels (30–50%). Similar results have also been shown by other studies over the different parts of India (Mahato et al. 2020; Sharma et al. 2020; Kumari and Toshniwal 2020b; Thomas et al. 2020; Nigam et al. 2021; Rani et al. 2021). Such findings are likely to have serious implications for policy makers in making strategies.

Pre-COVID studies in the Hindu Kush-Himalayan region have advocated the presence of anthropogenic black carbon (BC), fly ash, dust particles, nitrogen dioxide (NO_2 and surface ozone (O_3) (Ramanatham and Carmichael 2008; Bonasoni et al. 2021). The small amount of BC along with other air pollutants can modify the reflectance from the snow that can further enhance the rate of snowmelt and areal coverage of snow (Bonasoni et al. 2021). Against the backdrop of mentioned studies, the present study eventually establishes this direction to look upon the air quality scenario during the lockdown phase-1 (March 24–April 14, 2020) and pre-lockdown period (March 24–April 14, 2019) for 11 hill towns of the Northwestern Himalayas (NWH) (Himachal Pradesh, Uttarakhand and Jammu and Kashmir) falling under the non-attainment category of cities/towns of India. The ground-measured data on PM_{10}, NO_2 and Sentinel 5p measured data on AAI, NO_2 and carbon monoxide (CO) have been used to quantify the magnitude of change in these parameters of air quality during lockdown phase-1 compared to the pre-lockdown period.

Study Area

The Central Pollution Control Board of India had, in the first instance, identified 102 NACs across 23 Indian states and Union Territories based on their ground monitored ambient air quality data between 2011 and 2015. In India, cities were considered as NAC if they were consistently showing poorer air quality than the Indian National Ambient Air Quality Standards (NAAQS). NAC falling in the NWH comprise of seven towns of Himachal Pradesh (Baddi, Nalagarh, Parwanoo, Paonta Sahib, Kala Amb, Sundernagar and Damtal), two towns of Uttarakhand (Kashipur and Rishikesh) and two towns of Jammu and Kashmir UT (Jammu and Srinagar). The study area comprises geographical areas falling under a 50 km radius of 11 NACs falling in western Himalayan states of Himachal Pradesh, Uttarakhand and Jammu and Kashmir UT (Table 14.1 and Fig. 14.1).

Table 14.1 Geographical locations and elevation in above mean sea level (AMSL)

S. No	Name of the NAC	Latitude and longitude	Elevation (AMSL)
1	Baddi	30°56′30.89"N, 76°48′12.02"E	445
2	Nalagarh	31° 2′45.39"N, 76°42′9.72"E	346
3	Parwanoo	30°50′26.11"N, 76°57′30.15"E	779
4	Paonta Sahib	30°26′44.03"N, 77°36′15.47"E	397
5	Kala Amb	30°29′57.19"N, 77°12′25.59"E	337
6	Sundernagar	31°31′47.61"N, 76°53′20.20"E	878
7	Damtal	32°13′32.71"N, 75°39′36.30"E	313
8	Kashipur	29°12′37.56"N, 78°57′42.75"E	236
9	Rishikesh	30° 5′12.64"N, 78°16′3.64"E	371
10	Jammu	32°43′35.40"N, 74°51′25.77"E	310
11	Srinagar	34° 5′1.01"N, 74°47′50.23"E	1588

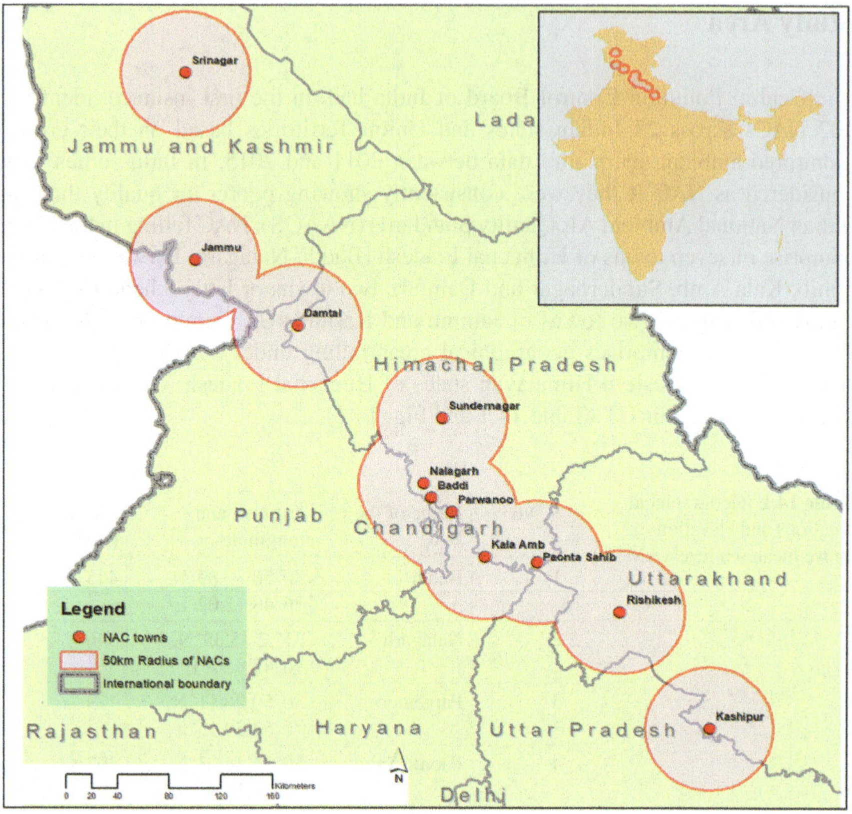

Fig. 14.1 Location map of 11 non-attainment towns/cities (NACs) of Northwestern India Himalaya considered in the present study

Materials and Methods

Multi-time-series Sentinel 5P Satellite data (March, 24 to April 14, 2020 and 2019) were retrieved from https://s5phub.copernicus.eu (via the Copernicus Data and Information Access). Further, these data were pre-processed and analysed/visualised in PanoplyWin-4.10.8 and Sentinel Application Platform (SNAP) 7.0 desktop applications. The spatial resolution of satellite data is 3.5 km × 7.5 km. The mean value of the aerosol index, NO_2 and CO of all sentinel 5p imageries for the period March 24 to April 14, 2019, was calculated for the pre-lockdown period and compared with same the period from March 24 to April 14, 2020, for the lockdown phase-1 period. Further, these mean images were clipped with the region of interest, i.e. around 50 km of point features of 11 towns for the bands of AAI, NO_2 column number density and CO column number density for the pre-lockdown period and post lockdown period. The list of Google Earth engine datasets and bands are as below:

- COPERNICUS/S5P/OFFL/L3 AER AI (absorbing aerosol index),
- COPERNICUS/S5P/OFFL/L3 NO_2 (NO_2 column number density),
- COPERNICUS/S5P/OFFL/L3 CO (CO column number density).

Subsequently, zonal statistics for mean value was derived for each region of NAC using GIS software. PM_{10} and NOx data were obtained from the Himachal Pradesh State Pollution Control Board for the lockdown phase-1 and pre-lockdown period. Changes in AAI, NO_2 and CO have been computed as follows:

$$\text{Absolute changes in air pollutants} = P_2 - P_1 \tag{14.1}$$

$$\%\ \text{change in air pollutants} = \frac{P_2 - P_1}{P_1} \times 100 \tag{14.2}$$

where P_1 is representing air pollutant concentrations during the reference and pre-lockdown period while P_2 is denoting the lockdown phase-1.

Results and Discussion

Reduction in AAI, NO_2 and CO

The main sources of NO_2 emissions are fertilisers and the combustion of fossils fuels. Besides this, CO is a good indicator of air pollution. The main sources of CO emissions are vehicles, industries and burning of forests and grasses. The average values of AAI, NO_2 and CO around 11 towns have been reduced by 55%, 19% and 7%, respectively, during lockdown phase-1 concerning the pre-lockdown period (2019) (Table 14.2). The range of reduction in AAI values varies from 32% (Sundernagar) to 92% (Kashipur). Substantial reduction in AAI values has been observed around all 11 cities. It is pertinent to note that positive values of the aerosol index generally represent absorbing aerosols (dust and smoke) while small or negative values represent non-absorbing aerosols. The range of reduction in tropospheric NO_2 density varies from 14% (Jammu) to 25% (Baddi). A substantial reduction in NO_2 has also been observed around all 11 towns. The range of reduction in CO density varies from 0 (Srinagar) to 10% (Kashipur) (Table 14.2). Similar findings of improvement in air quality have also been reported by other studies over the Himalayan region (Biswal et al 2020; Bahukhandi, Agarwal, and Singhal 2020; Moore and Semple 2021).

Table 14.2 Changes in AAI, NO_2 and CO in cities of NW Himalayas during the lockdown period phase-1

Town/city	Pre-lockdown (2019)	During lockdown (2020)	% change	Pre-lockdown (2019)	During lockdown (2020)	% change	Pre-lockdown (2019)	During lockdown (2020)	% change
		AAI			NO_2			CO	
Damtal	−0.98	−1.43	−46	0.00008	0.00006	−17	0.036	0.033	−8
Nalagarh	−0.86	−1.3	−51	0.00008	0.00006	−23	0.0356	0.033	−7
Baddi	−0.8	−1.25	−56	0.00009	0.00007	−25	0.0356	0.033	−7
Sundernagar	−1.17	−1.55	−32	0.00007	0.00006	−16	0.031	0.03	−3
Parwanoo	−0.84	−1.32	−57	0.00008	0.00007	-22	0.034	0.032	-6
Kala Amb	−0.72	−1.3	−81	0.00009	0.00007	−22	0.037	0.034	−8
Paonta Sahib	−0.77	−1.31	−70	0.00008	0.00006	−19	0.034	0.032	−6
Kashipur	−0.6	−1.15	−92	0.00008	0.00007	−16	0.04	0.036	−10
Rishikesh	−0.77	−1.36	−77	0.00008	0.00007	−17	0.036	0.033	−8
Srinagar	−1.05	−1.46	−39	0.00007	0.00006	−14	0.028	0.028	0
Jammu	−0.93	−1.3	−40	0.00008	0.00007	−14	0.035	0.033	−6
Average	−0.9	−1.3	−55.2	0.00008	0.00006	−19	0.035	0.032	−7

Reduction in Ground-Measured PM_{10} and NO_x

In the seven towns/cities of Himachal Pradesh, PM_{10} mass concentration has been substantially reduced during the March and April months of 2020 compared to the same months of 2019 (Fig. 14.2). PM_{10} concentration is almost nil in Kala Amb and Nalagarh towns during the March and April months of 2020 (Fig. 14.2). Reduction in nitrogen oxides (NO_x) is also showing a similar trend as PM_{10} during the March and April months of 2020 compared to the same months of 2019 (Fig. 14.3).

The study area receives a part of its air pollution from outside including the Indo-Gangetic Plain (IGP) due to trans-boundary air pollution movement. Further, high concentrations of pollutants in the IGP seem to have partial trans-boundary natural movement over the lower mountains of the Himalayas. These NACs are located along the transitional zone from mountain to plains, hence witness partial concentrations of air pollutants. Sentinel 5P data comprehensively establish that there has been a decline in air pollution concentration over the NACs of Northwestern Indian Himalaya (NWIH) coinciding with nationwide lockdown to stop the spread of the coronavirus (COVID-19) started during the last week of March 2020.

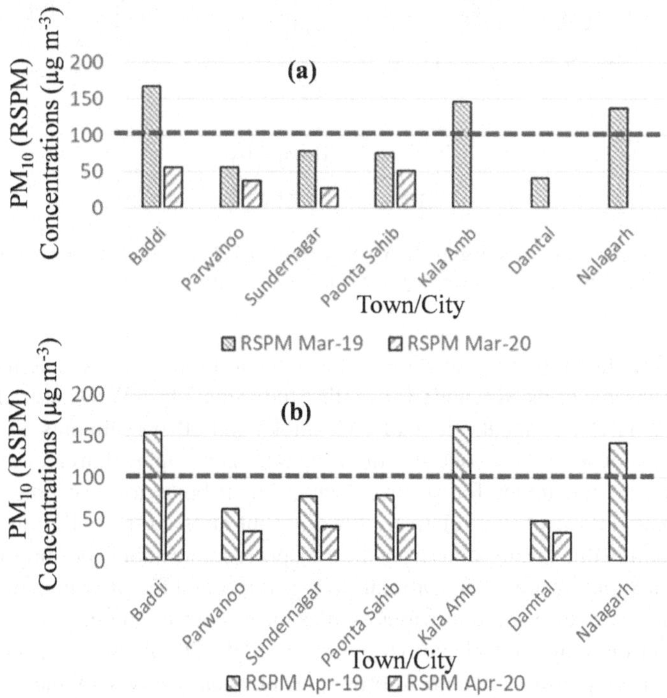

Fig. 14.2 PM_{10} (RSPM) concentrations ($\mu g\ m^{-3}$) in 7 NACs of Himachal Pradesh during **a** March 2019–2020 and **b** April 2019–2020 (24 h average standard = 100 $\mu g\ m^{-3}$)

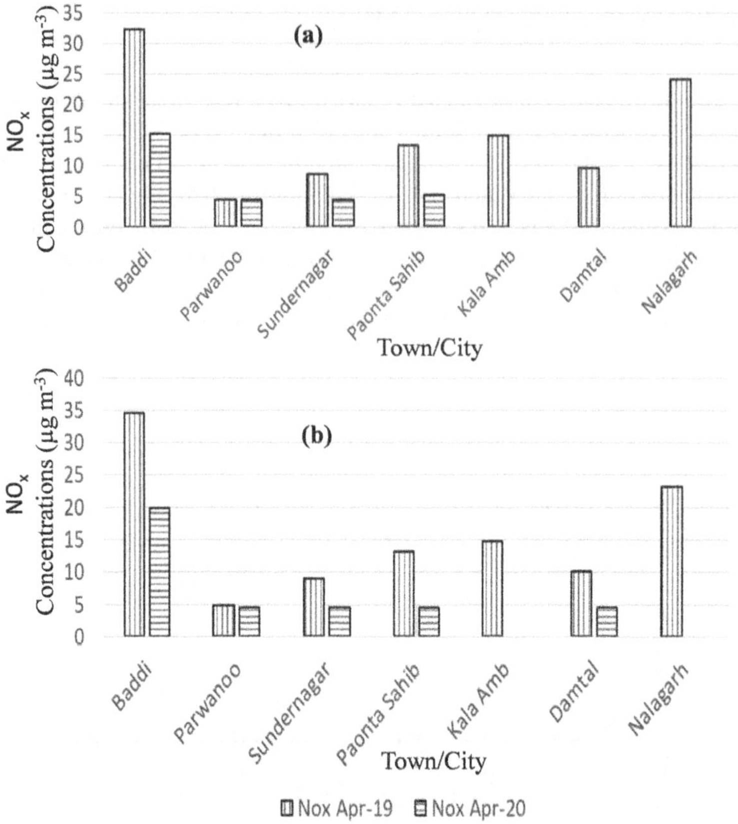

Fig. 14.3 NO_x concentrations (μg m^{-3}) in 7 NACs of Himachal Pradesh during **a** March 2019–2020 and **b** April 2019–2020 (24 h average standard = 80 μg m^{-3})

This led to the dramatic reduction in the concentration of particulate matter and other trace gases such as NO_2 and CO over the study area. The NWIH has traditionally been recording low concentrations of PM, smoke and other pollutants owing to its mountainous location. Over one-fourth of the total geographical area of the NWIH states seems to fall under forest cover, which contributes to absorbing localised air pollutants. However, the relatively low concentration of air pollution seems to originate mainly due to emissions from the transport sector, stubble burning and forest fire from surrounding areas. The present study is important for environmentalists and policy makers because some previous studies have reported the negative impact of air pollution on snow and glacier (Kang et al. 2019, 2020). In this context, such reduction in air pollution would improve the physical aspects of the Himalayan region and can be adopted as a strategy to minimise the impact of climate change as well. Guttikunda and Nishadh (2020) while reporting the positive influence of COVID-19 lockdown on air quality stated to achieve and maintain national ambient norms, we must employ an air shed strategy in which reductions in the city are supplemented by equal reductions throughout the area.

Such findings are likely to have many implications for policy makers in making strategies under the increasing climate change impacts. It would be useful to improve the sustainability of the fragile Himalayan ecosystem.

Conclusions

The air quality concerning aerosol, tropospheric NO_2, CO, PM_{10} and NO_x of the NACs of NWIH is observed to be low before and during lockdown phase-1 though with a declining trend during the lockdown. The anthropogenic activities and natural trans-boundary movement of pollutants both seem to contribute to whatever pollution level prevails in these towns. However, closure of all factories, markets, shops and construction works and suspension of vehicular movement seem to contribute towards further declining concentration of aerosol, tropospheric NO_2 and CO during lockdown phase-1. The improvement in the air quality of the study area has been established across all NACs in terms of AAI, NO_2 and CO parameters, which is well supplemented by ground-monitored values of the Pollution Control Board. Though, the size of cities and their geographical settings and other economic activities such as industrial activities and vehicular movement were strongly correlated with pollution concentrations. A substantial share of reduction of emissions seemed to come from the transport sector. But further research is needed in this respect and also on positive/negative feedbacks of air pollution reduction on snowmelt during the lockdown phase-1. This study would be beneficial for environmentalists and policymakers to plan a sustainable emission policy to reduce the adverse impacts of air pollution on the physical aspects (such as snow/glaciers) of the Western Himalayan region.

Acknowledgements The author is thankful to the Himachal Pradesh State Pollution Control Board and Central Pollution Control Board (CPCB), Ministry of Environment, Forest and Climate Change, Govt. of India for providing air quality data of Indian cities. The author is also grateful to the European Space Agency (ESA) for giving free and open access to Sentinel 5P data. Opinion expressed in the paper is independent of institutional affiliation of the author.

References

Albayati N, Waisi B, Al-Furaiji M, Kadhom M, Alalwan H (2021) Effect of COVID-19 on air quality and pollution in different countries. J Transp Health 101061

Bahukhandi K, Agarwal S, Singhal S (2020) Impact of lockdown Covid-19 pandemic on Himalayan environment. Int J Environ Anal Chem 1–5.https://doi.org/10.1080/03067319.2020.1857751

Biswal A, Singh V, Singh S, Kesarkar AP, Ravindra K, Sokhi RS, Chipperfield MP, Dhomse SS, Pope RJ, Singh T, Mor S (2020) COVID-19 lockdown induced changes in NO_2 levels across

India observed by multi-satellite and surface observations. Atmos Chem Phys 21:5235–5251. https://doi.org/10.5194/acp-21-5235-2021
Bonasoni P, Cristofanelli P, Marinoni A, Vuillermoz E, Adhikary B (2021) Atmospheric pollution in the Hindu Kush-Himalaya region: evidence and implications for the regional climate. Mt Res Dev 32:468–479
Guttikunda SK, Nishadh KA (2020) Air quality trends and lessons learned in India during the COVID lockdowns. Collaborative Clean Air Policy Centre, Delhi, India. Available via https://ccapc.org.in/policy-briefs/2020/aq-changes-lockdown. Accessed 30 July 2021
Kang S, Zhang Q, Qian Y, Ji Z, Li C, Cong Z, Zhang Y, Guo J, Du W, Huang J, You Q (2019) Linking atmospheric pollution to cryospheric change in the Third Pole region: current progress and future prospects. Natl Sci Rev 6(4):796–809
Kang S, Zhang Y, Qian Y, Wang H (2020) A review of black carbon in snow and ice and its impact on the cryosphere. Earth-Sci Rev 210:103346
Kumari P, Toshniwal D (2020a) Impact of lockdown on air quality over major cities across the globe during COVID-19 pandemic. Urban Clim 34:100719
Kumari P, Toshniwal D.(2020b) Impact of lockdown measures during COVID-19 on air quality—a case study of India. Int J Environ Health Res 1–8.https://doi.org/10.1080/09603123.2020.1778646
Mahato S, Pal S, Ghosh KG (2020) Effect of lockdown amid COVID-19 pandemic on air quality of the megacity Delhi, India. Sci Total Environ 730:139086
Moore GW, Semple JL (2021) Himalaya air quality impacts from the COVID-19 lockdown across the Indo-Gangetic Plain. GeoHealth 5(6):e2020GH000351
Mostafa MK, Gamal G, Wafiq A (2021). The impact of COVID 19 on air pollution levels and other environmental indicators—a case study of Egypt. J Environ Manag 277:111496
Nigam R, Pandya K, Luis AJ, Sengupta R, Kotha M (2021) Positive effects of COVID-19 lockdown on air quality of industrial cities (Ankleshwar and Vapi) of Western India. Sci Rep 11(1):1–2
Ramanathan V, Carmichael G (2008) Global and regional climate changes due to black carbon. Nature 1:221–227
Rani S, Kumar R, Acharya P, Maharana P, Singh R (2021) Assessing the spatial distribution of aerosols and air quality over the Ganga River Basin during COVID-19 Lockdown Phase-1. Rem Sens Appl Soc Environ 100546. https://doi.org/10.1016/j.rsase.2021.100546
Sharma S, Zhang M, Gao J, Zhang H, Kota SH (2020) Effect of restricted emissions during COVID-19 on air quality in India. Sci Total Environ 728:138878
Thomas J, Jainet PJ, Sudheer KP (2020) Ambient air quality of a less industrialized region of India (Kerala) during the COVID-19 lockdown. Anthropocene 32:100270
Yadav R, Korhale N, Anand V, Rathod A, Bano S, Shinde R, Latha R, Sahu SK, Murthy BS, Beig G (2020) COVID-19 lockdown and air quality of SAFAR-India metro cities. Urban Clim 34:100729

Chapter 15
Exploring Methodological Approaches for Climate Perception Studies in Himalaya

Prachita Arora, S. Nawaz Ali, and P. Morthekai

Abstract There is a whopping disconnect between the stakeholders, policymakers, and the scientific community towards climate change, anthropogenic contribution, risks and mitigation, and adaptation strategies. Therefore, to understand public perception of climate change and look into possible solutions, it is important to conduct climate change perception surveys that directly involve the public/stakeholders. The primary aim of these surveys is to understand the small-scale variabilities in weather and climate. This also allows us to understand the people's perceptions of critical changes and their willingness for a change when making choices. The perception studies are aimed at helping the policymakers to formulate better climate policies, by making them aware of the problems of locals. Since Himalayan dwellers are more susceptible to the adversities of climate change, documentation of their knowledge and opinions on climatic variability and its challenges becomes crucial. Therefore, it becomes important to conduct the surveys systematically. The present study provides a robust strategy to survey along with a brief description of the methodologies involved.

Keywords Climate perception · Indian Himalaya · Climate change · Sampling methods · Policymaking

Introduction

Climate change is a global phenomenon that is regarded as perhaps the most pressing issue of the modern era (Shi et al. 2015). However, the vulnerability it poses depends on who you are, where you are, and what you do. The repercussions of it are cataclysmic and differential in their reach or magnitude (Gentle et al. 2014). As a result, understanding public perception of climate change becomes critical for policymaking and its effectiveness in implementation, as it potentially boosts the public's confidence in the governance, which increases individuals' intentions to change their behavior and accept climate-conscious policy measures (Gadenne et al. 2011; Von

P. Arora (✉) · S. N. Ali · P. Morthekai
Birbal Sahni Institute of Palaeosciences, 53 University Road, Lucknow, Uttar Pradesh, India

S. Rani and R. Kumar (eds.), *Climate Change*, Springer Climate,
https://doi.org/10.1007/978-3-030-92782-0_15

Borgstede et al. 2013). It has been observed that the underprivileged, socioeconomically disadvantaged sections of the society, or those that are directly dependent on climate-sensitive resources are the most affected ones. Such populations include small farm holders, rural, coastal, mountainous populations, and indigenous communities. The poverty-stricken population seems to be more frequently exposed to climate hazards than their wealthier counterparts and has a propensity to lose relatively more when affected. This is because of the limited access to modern facilities and services of health and agriculture (Aryal et al. 2015). These marginalized communities however are not just simple sufferers of climate change, rather are excellent observers and interpreters of change due to their high dependencies on the ecosystem services (Raygorodetsky 2011; Macchi et al. 2014). The knowledge these communities possess is empirical, intuitive, usually asymmetrically distributed according to gender and age (Nawrotzki and Kadatska 2010). Thus, acquiring their opinions and perceptions (through climate change perception surveys) on various aspects related to climate change and its effects is imperative, as it offers valuable insights in discerning the climatic variability, risks, and challenges it accompanies.

'Climate change perception survey'—is a way of collecting information on the public perception of climate change and to identify key factors/barriers that hinder climate-conscious behavior. It seeks to understand how individuals and communities perceive climate change phenomena and their consequences and to serve as a foundation for motivating the public to make effective adjustments/adaptation as well as mitigation initiatives (Leiserowitz 2006; Lorezoni and Pidgeon 2006; Sullivan and White 2019). This becomes more important in ecologically fragile and transitional climatic zones where slight changes in the meteorological parameters have unprecedented impacts on the local ecosystem (Ali et al. 2018). Hence, perception studies in such areas help in measuring the intangible, facilitate citizen-state communication, rebalance information asymmetries, test policy assumptions, compare citizens and expert opinions as well as monitor state-society-environment relation over time.

Further, to determine long-term changes in the climate, it becomes essential to understand and estimate the variability of local weather and climate at a daily/monthly/yearly scale. In topographically heterogeneous terrains, like the Himalayas, extensive variations in the meteorological parameters occur across a fairly smaller spatial gradient (Bhutiyani et al. 2007). The Himalayan ecosystem is suggested to be the worst hit by climate change, despite the reason there is a scarcity of meteorological data from this region (Maurer et al. 2019). Hence, to fill the gaps in the meteorological data and understand local, small-scale variations in the climatic parameters, people's perceptions can be used as an effective tool. People's perceptions of climate change can be used to validate climate model simulations and identify climate change scenarios at a much larger temporal and spatial scales scale, supplementing scientific evidence with chronological and landscape-specific precision (Raygorodetsky 2011).

Understanding the climate change perception of any community is fundamental as it lays the foundation for environmentalists and governments to frame effective climate policies by including the individuals as stakeholders. It is important to have societal participation to curb climate change, because until the public recognizes its responsibility, developing effective communication strategies, common policies, and socially resilient technologies becomes quite critical (Hansen et al. 2012; Whitmarsh and Capstick 2018). Such studies enable the policymakers to easily assess public perceptions which help in shaping effective policies and adequately address the key concerns. Another reason for the need for perception studies is that they rely on a growing understanding of people who are good natural observers of their local environment which brings about an appreciation for their knowledge and experience (Salick and Byg 2007; Turner et al. 2009; Crona et al. 2013). This appreciation will positively motivate them to change their behaviors and to accept climate-friendly policy measures (Gadenne et al. 2011; Von Borgstede et al. 2013; Shi et al. 2015). Documenting and explaining the aspects of changing climate through perception studies is therefore critical, as it aids in understanding climate change and establishes local and global socio-political frameworks. The findings lay the groundwork for policymakers and scientists to operate within and devise appropriate policies with full support from stakeholders (O'Connor et al. 1999; Choi et al. 2005; Leiserowitz 2005; Brody et al. 2008; Agho et al. 2010; Crona et al. 2013).

Although several climate perception studies from the fragile Himalayan region have been conducted (Table 15.1), systematic surveys and methodologies need to be adopted to make the studies more fruitful. Hence, to understand more about the various sampling techniques, this present work compiles various sampling strategies used for understanding public perceptions. It also assesses various methods that can be used for examining them and discusses the strengths and weaknesses of each method.

Stages of Conducting a Survey

Survey research is the most common method for obtaining public knowledge as well as opinion (Morgan 1997). According to Kraemer (1991), any survey research has three distinct characteristics viz. quality, subjectivity, and generalisability. The quality and reliability of survey research are dependent on the way it is designed, conducted, analyzed, and reported (Lau 2017). As a preliminary to a brief description of various sampling methodologies, it is appropriate to briefly describe the steps involved in executing a survey.

Design parameters through clear survey goals and objectives are the important first steps in any survey process. The research objectives need to be specific, achievable, relevant, and measurable and should provide a framework to formulate the right questionnaire. With clear objectives, it is easy to design a detailed survey. Based on the objective of the survey, the target population needs to be identified. The target audience must correspond to the purpose of the survey. It is important to pre-decide if the requirement is to target only those with direct experience (information) with the survey topic or not. In the case of perception studies, such targeted surveys are not suitable because the unaffected (uniformed) section of people is not included. Subsequently, the method of data collection is chosen. Surveys can

Table 15.1 Sampling Strategies of Climate Perception Studies in the Indian Himalaya

Location	Site selection	Sampling strategy	Pilot study	Reference
South Sikkim	Purposive sampling method	Focus group discussion (4), Key informant interview, Census household survey (130 households)	Done (for both survey and FGD)	Barua et al. (2014)
Arunachal Pradesh	Random selection	Household survey (136), focus group discussion, workshop	Done (for survey)	Singh et al. (2017)
Chilapata reserve forest	Purposive sampling method	Questionnaire-based personal in-depth interviews (Purposive and random sampling method; 400) Group discussions	–	Dey et al. (2018)
Himalayan Foothills of West Bengal	Multistage simple random sampling method	Structured questionnaire survey (random 'walk' methodology; 384) Focus group discussion (FGD; 3)	Done	Ghosh and Ghoshal (2020)
Himalayan Foothills of West Bengal	Multistage purposive sampling	Structured questionnaire survey (random 'walk methodology; 100)	Done (for both interview and FGD)	Ghosh and Ghoshal (2020)
Garhwal Himalaya	Random selection	Household-level questionnaire (128)	Done (for interview)	Gupta et al. (2019)
Almora and TehriGarhwal (Uttarakhand, India) Bajhang, Kavre, and Terhathum (Nepal)	–	Selection depending on age, sex, and general well-being Discussions in focus groups (FGDs; 40) Semi-structured interviews at the household level (144) Participant observation	–	Macchi et al. (2014)
Uttarakhand, Western Himlaya	Random selection	Group discussions Individual interviews (Random selection; 1080 households)	–	Negi et al. (2017)
Nainital, Western Himalaya	–	Paper-based survey (Random selection, 90 households)		Pandey et al. (2018)

(continued)

Table 15.1 (continued)

Location	Site selection	Sampling strategy	Pilot study	Reference
Uttarkashi and Chammoli (Uttarakhand)	Purposive sampling	Semi-structured questionnaire-based survey (871)		Rautela and Karki (2015)
ChhotaBhangal area (Western Himalaya)	–	Stratified random sampling (240 individual interviews), 430 household survey Focus group discussion	–	Sharma et al. (2020)
Chakrata and Bhikiyasain tehsil (Uttarakhand)	based on state-level inherent vulnerability assessment	Household-level survey (Purposive sampling; 241)	Done	Shukla et al. (2019)
Sikkim	–	Semi-structured interviews (Random sampling, 228)	–	Sharma and Shreshtha (2016)
Chakrata and Kalsi blocks (Dehradun, Uttarakhand)	Random selection	Household-level survey (Random sampling; 95)	–	Pratap (2015)
(1) Senchal Wildlife Sanctuary, (2) Northern belt of Singalila (3) Ilam District (Nepal)	–	Household-level survey using a semi-structured questionnaire (Random sampling; 576: $N_1 = 326$; $N_2 = 121$; $N_3 = 129$) Focus Group Discussions (10)	–	Chaudhary et al. (2011)
Singalila National Park; Ilam District (Nepal)	–	Household-level survey using a semi-structured questionnaire (250) Focus Group Discussions (10)	–	Chaudhary and Bawa (2011)
Kullu and Solan (Himachal Pradesh, Northwestern Himalaya)	–	Household-level survey (Random sampling; 275)	–	Kimani and Bhardwaj (2015)
Tribal localities of Tones Valley, Uttarkashi (Uttarakhand)	–	Household-level survey (100) Focus Group Discussions	–	Rawat (2013)

be conducted by several possible methods that include personal interviews, questionnaires posted by mail directly to people or via institutions; computer-assisted questionnaires; email questionnaires; online questionnaires; telephone interviews, or a combination of these modes. There are benefits and detriments of each mode but the most appropriate method depends on the objective of the survey and the target audience (Leung 2001). The next step is drafting a clear, unambiguous, and bias-free survey questionnaire. A good question, as per Fowler and Cosenza (2009), would generate reliable answers and ensure accurate measurements as to what is being described. Questions must be designed in such a way that they are relevant to the survey objectives, phrased correctly and precisely, arranged in a smooth logical flow, and consistently understood by the respondents (Ho 2005; Fowler and Cosenza 2009). A survey can have open-ended (a statement that requires a longer response) or closed-ended questions (pre-defined answers to choose from) or even a combination of both. At times, it becomes important to set up meetings and group discussions with the members of the target group to identify key problems which further can be translated into question categories. Therefore, before surveying on a large scale, it is advisable to do pretesting or pilot testing of the survey with a small number of individuals to ensure its content, clarity, and consistency (Lau 2017). A pilot study is a preliminary feasibility research process that is "a small-scale experiment or set of observations undertaken to determine how and whether to launch a full-scale project" (Collins English Dictionary 2014, para 1; Fraser et al. 2018). It also helps to decide the type of method to be used to survey following the objective of the survey and ease of the respondents. When pilot testing, it is essential to anticipate the actual circumstances in which the survey will be conducted and choose respondents similar to the target population. Thus, such studies not only enhance the reliability and validity of a survey but also act as a tool to modify the questionnaire and survey design appropriately along with cutting down preventable temporal and monetary expenses for the main survey. Following this, is the selection of the types of methods used to conduct the survey. The right sampling method is the one that serves the purpose of the study most appropriately. The efficacy of a sampling method is based on the number of complex variables. To simplify that, the aforementioned approach can be used to obtain an appropriate sampling method using the following steps:

- Draw up a checklist of the study aims and objectives along with other factors like cost, convenience, accuracy, and precision.
- Identify the sampling method that would most likely be effective in achieving the goals.
- Test all the sampling methods one-by-one or even in combinations (assessing advantages and disadvantages of each method).
- Assess each method's potential of achieving the specified objective.
- Select the method that's most suitable for that objective.

The methodologies used for a survey are discussed in brief in the following section. After the selection of mode and type of methodology for data collection comes conducting the survey. While conducting a survey, the response rate is one of the most crucial parameters to be considered. Towards this, satisfactory information through publicity of the survey study purpose and usefulness of results are efficient ways of gaining respondents' cooperation. Another way to encourage survey response is by providing incentives for respondents to take the survey. Incentives can be monetary or nonmonetary and need not be substantial (as small rewards are often just as effective as large ones). Ensuring the anonymity of the respondent and confidentiality of the data obtained would also enhance the response rates (Ho 2005). In the case of field surveys, the strategy is to have a prior idea about the location if possible and map the fieldwork according to the best contact times (for example, planning household visits after office hours or during weekends). Another important factor in extracting maximum responses is interviewing skills of the researcher. The interviewer or researcher must be sufficiently trained, articulate, and must have detailed knowledge about the survey topic. One can use more than one mode (or method) to improve the response rate (Glasow 2005). Progress of the survey needs to be carefully monitored to get instantaneous appraisal/feedback on the challenges encountered while conducting the survey (Ho 2005). The collection of the survey data must be done ethically and rigorously. The ground information (data) should be checked and edited for accuracy before any statistical stimulation. It is also necessary that researchers attempt to eliminate or rectify any form of survey bias. Correction for missing values or obvious wrong data can be done by using imputation. Following this, is the analysis and representation of the data (and results) which are discussed in a separate section (Fig. 15.1).

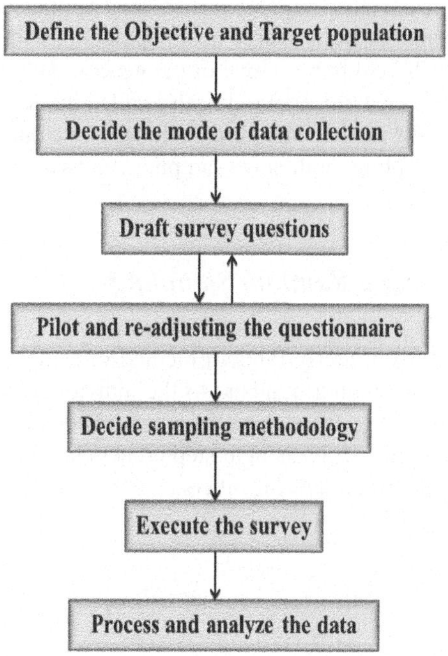

Fig. 15.1 Seven steps for conducting a survey. After organisation for economic cooperation and development (OECD) (2012), Leung (2001), and Taherdoost (2016)

Methodologies of Sample Collection

When doing a questionnaire (written) survey, it is practically impossible to study the entire population because of feasibility and cost constraints. Therefore, a relatively smaller percentage of individuals (a subset), considered as the representatives from a pre-defined population, are selected for observation and analysis. The process through which these representative samples are extracted from a population is called sampling. It should be done in such a way that making inferences and drawing conclusions from the target population (representative sample) can be generalized back to the full population of an area.

The methodologies used to collect information from the respondents are determined by the size and variance of the population, objectives of the study, nature of the sampling universe, i.e., homogeneity or heterogeneity in the constituent units as well as the mode of data collection. Sampling methodologies can be broadly categorized as probability (random) sampling and non-probability (non-random) sampling.

Probability (Random) Sampling

With probability sampling methods, each individual in the population has an equal (or at least quantifiable) possibility of being selected. To ensure a generalisability of the study results, this method applies statistics to select a handful of people at random from a large host population, with the expectation that all of their answers will be representative of the general demographic. Since the samples are randomly selected, this method is also called the 'random sampling' method. Types of probability random sampling include simple random sampling, stratified sampling, cluster sampling, multistage sampling, and systematic random sampling.

Simple Random Sampling

It is an utterly random method of sample selection. This sampling method works by assigning numbers to the individuals in a given population and then randomly choosing from those numbers. These randomly chosen numbers are considered samples. It is a single-step process in which the sample selection of each element is independent of the others.

Stratified Sampling

It is also defined as proportional sampling in which the population is heterogeneous and is split into non-overlapping groups called 'strata', based on some characteristic. This is followed by stratum-specific sample selection. If a simple random sampling method is employed for the selection of a sample from each stratum, i.e., stratification followed by simple random sampling, then the method is referred to as stratified random sampling. The size of the sample in each stratum is taken proportional to the size of the stratum.

Cluster Sampling

According to Ofer Abarbanel (2020) cluster sampling is used when mutually homogeneous yet internally heterogeneous groupings are apparent in a statistical population. When the population is huge, this method is used as an alternative for simple random sampling. The population is classified (known as clusters) and sampled clusters are selected using simple random sampling (Wilson 2014). From the selected clusters, the elements are then sampled either by a 'one-stage' or 'two-stage' cluster sampling plan.

Multistage Sampling

This method employs different probability sampling methods at different levels of research. The sample selection becomes a multistage process involving fewer sampling units in subsequent stages. By using a combination of sampling methods on a given population, it becomes possible to minimize selection biases.

Systematic Random Sampling

For systematic random sampling, a list of all the samples in the population is generated. After that, the first sample element, from the population list's first n elements, is chosen at random. Subsequently, every nth element on the list is then chosen. This implies that a random starting point is used to pick a member from a wider group, followed by a periodic sampling selection. In contrast to simple random sampling, the likelihood of sample selection is not equally likely in this process.

Non-probability (Non-Random) Sampling

Non-probability sampling methods are those in which the probability of sample selection is unknown or sometimes none. In this method, the samples are not selected randomly but rather selected based on their availability, convenience, or some other pre-defined criteria. Unlike the probability sampling method, this method does not give the most accurate representation of the population since the sample selection is non-random. It also does not facilitate the accurate statistical analysis, estimation of sampling errors, and may often have some kind of sampling-related bias. However, these methods come with their own set of advantages namely cost and convenience. The types of non-probability sampling methods are as follows.

Voluntary Sampling

A voluntary sample constitutes self-selected participants; those who volunteer to take the research survey. Such people may have a particular interest or opinion regarding the subject of the survey, may have spare time to join the survey, or might be a part of the survey for incentives or ethical reasons.

Convenience Sampling

This type of sampling includes people who are easy to reach. The sampling is based on factors such as convenient access, geographic proximity, availability at a specific time, and willingness to participate (Dörnyei and Griffee 2010). It is a quick, cost-effective, and convenient technique and requires very little planning. It is relatively casual and an easy way of sampling.

Other Non-probability Sampling Methods

Purposive Sampling

This technique is also known as judgment (or deliberate) sampling. The sample (individual) is deliberately chosen as a participant based on his/her qualities. The process of sample selection aims at finding people or groups of people that are conscious and well-aware about the topic of interest (Creswell and Clark 2011).

Quota Sampling

With the quota sampling procedure, the subgroups of the population of interest are decided based on the topic/area of research. The sample is chosen based on the frequency distribution of characteristics that is the reflection of the population of interest. It is however important to know the percentage of members of each subset to tally the percentages in the samples. In each subset, the individuals are not randomly chosen rather are selected based on their availability.

Expert Sampling

Interviewees are selected in a non-random manner based on their experience and expertise on the subject under investigation in this process. It is used when the opinions or assessment of people with high proficiency and dexterity about the study area (or phenomenon) are required. Expert sampling, at times, can be considered as a subset of purposive sampling. When used in this manner, expert sampling is reduced to a simple subset of purposive sampling. The strategy of this method is simply to define the "expert" and select the people who meet the "expert" criteria.

Discussion

Survey research involves the collection of information through answers to questions from a population and is aimed at finding solutions and solving problems faced. To answer a research question, it is not always possible to collect data from all cases or the whole population. Thus, there is a need to select a sample. Since it is not economically and temporally feasible to analyze the entire population, the above-mentioned sampling methodologies are applied to reduce the number of samples (individuals) and find a representative of the whole population. To produce scientifically accurate and statistically sound results, the probability sampling method seems more practical. Nevertheless, even when probabilities of sample selection are unequal, it is important to pre-determine them so that proper weighting methods can be applied to obtain unbiased population estimates. Another important advantage in this is the lack of bias. Because the individuals are chosen randomly, each person has an equal chance of being selected from a set of populations. Each probability sampling method above has various advantages and disadvantages which should be taken into consideration while selecting the appropriate sampling strategy.

In the case of a simple random method, sample collection is much straightforward than other methods. It is a one-step procedure where the individuals in the subset are selected randomly. Since the selection of individuals is random, every member of the population is equally likely to be selected. As a result, a balanced subset is formed that has the best chance of representing the larger group as a whole. Although using a simple random sample has distinct advantages, it also has inherent disadvantages. One of these disadvantages is the difficulty in obtaining a comprehensive list of a selected demographic. Furthermore, when a sample set of a larger population is not comprehensive enough, the representation of the entire population is altered, and the results can be biased, necessitating the use of additional sampling techniques.

Stratified random sampling enables to find out the best representative of the entire population by making sure that each subgroup of interest is represented. This method provides better coverage of the population since there is a robust control over the subgroups. However, stratified random sampling is disadvantageous when every member of the population cannot be classified into a subgroup. Another issue that comes while using this sampling method is the problem of overlapping by which the result can be misinterpreted or can inaccurately represent the population.

Cluster sampling entails categorizing the study's population into groups. Because there are far fewer operational and transport charges, this method is less expensive than simple random or stratified sampling. Major drawbacks of this method are its biases and sampling error. If the cluster formation was biased, the interpretations and conclusions of the target population would be biased too. Furthermore, the cluster method is highly susceptible to sampling error than other sampling methods. Cluster sampling differs from stratified sampling in a way that the sample for stratified sampling includes elements from each stratum, whereas the sample for cluster sampling includes only elements from sampled clusters. A more compound form of cluster sampling is multistage sampling. In this case, the population is first divided into large clusters. These large clusters are further divided into smaller clusters into multiple stages only to make primary data collection more manageable. One of the main drawbacks of this method is the subjectivity of the formation of the subgroups and their selection. The sample might not fully represent the entire population, and thus ensuing potential for biases. To form the sampling frames for multistage random sampling, information about groups on various levels is required which sometimes is not feasible. Nevertheless, it is a useful sampling technique due to its effective and flexible data collection strategy along with the cost and time effectiveness.

Systematic sampling is less complicated, more precise, and easier to understand than random sampling. Unlike simple random sampling, there isn't a need to number each member of the sample, because the objective is to create representative data of the entire group without specific individualized identifiers. Since the selection method is at a fixed distance between each participant, the probability of data contamination automatically becomes low. However, this method comes with its own set of disadvantages. These include a reasonable approximation of the demographic (which cannot be precise at all times), the fractional chance of sample selection (rather than equal), and less random when compared to simple random sampling.

Data Analysis and Presentation

Data on its own is meaningless without proper analysis and interpretation. Thus, to produce meaningful inferences, it is important to thoroughly examine and accurately interpret the data and results. In perception studies, data analysis is important for understanding the results of pilot studies, surveys, group discussions, finding gaps in the data, redesigning surveys, planning new statistical activities, and justifying the objectives. Qualitative data should be analyzed using established methods such as content analysis (Field and Morse 1985), whereas, for quantitative data, appropriate statistical tests can be applied. The analysis must be assisted by an effective evaluation scheme that considers the accuracy of estimates. Data analysis includes: cross-tabulating and filtering results, benchmarking, trending, and comparative data crunching, taking into consideration causality vs. correlation and comparison with the past data (if available). For the climate perception survey, it is important to compare the perception data with the meteorological data, to understand the coherence between the two and understand the impact of changing climate on people.

Data and results can be presented in multiple formats (such as diagrams, maps, graphs, tables). Organizing and displaying data in graphical formats aids in the identification of patterns and future projections. The data must be arranged in a logical order and order of relevance or importance and must be easy to read. Information about the source of the data, the type of analytical method used, and standard errors must be mentioned. Along with the above, it is equally important to mention any shortcomings in the data that may have affected the analysis. It has to be ensured that the intentions stated in the introduction/objective are fulfilled by the rest of the study and the conclusions are consistent with the evidence and possible recommendations are proposed. Hence, a comprehensive survey should include research objectives, mode of the survey, type of methodology, population coverage, sample design, sample size, and quality metrics (sampling error, response rate, and expected sources of non-sampling error). This would help the readers to evaluate the authenticity of the survey conclusions (Ho 2005).

Conclusions

The paper attempts to describe the stages of a systematic survey process. The study lists the steps of phases to be followed via. defining objectives and targeting a specific population, sample collection, drafting unbiased survey questions, a pilot study for any readjustment, sampling methodology, survey execution, and data analysis. Data analysis, correlations, graphical representations, and reporting of the findings is an important phase of the whole process and should be done carefully. The results should provide relevant and easily understandable and accessible information. It is also advised to add a section suggesting possible recommendations that discuss the ways to overcome mitigate or adapt to any given process.

Acknowledgements The authors are thankful to the Director, Birbal Sahni Institute of Palaeosciences, Lucknow, Uttar Pradesh, India for constant support and infrastructural facilities to carry out the present study.

References

Agho K, Stevens G, Taylor M et al (2010) Population risk perceptions of global warming in Australia. Environ Res 110(8):756–763

Ali SN, Dubey J, Ghosh R et al (2018) High frequency abrupt shifts in the Indian summer monsoon since Younger Dryas in the Himalaya. Sci Rep 8(1):1–8

Aryal S, Cockfield G, Maraseni TN (2015) Perceived changes in climatic variables and impacts on the transhumance system in the Himalayas. Clim Dev 8(5):435–446

Barua A, Katyaini S, Mili B et al (2014) Climate change and poverty: building resilience of rural mountain communities in South Sikkim, Eastern Himalaya, India. Reg Environ Change 14(1):267–280

Bhutiyani MR, Kale VS, Pawar NJ (2007). Long-term trends in maximum, minimum and mean annual air temperatures across the Northwestern Himalaya during the twentieth century. Clim Change 85(1), 159–177

Brody SD, Zahran S, Vedlitz A et al (2008) Examining the relationship between physical vulnerability and public perceptions of global climate change in the United States. Environ Behav 40(1):72–95

Chaudhary P, Bawa KS (2011) Local perceptions of climate change validated by scientific evidence in the Himalayas. Biol Lett 7(5):767–770

Chaudhary P, Rai S, Wangdi S et al (2011) Consistency of local perceptions of climate change in the Kangchenjunga Himalaya landscape. Curr Sci 504–513

Choi BC, Pang T, Lin V et al (2005) Can scientists and policy makers work together? J Epidemiol Community Health 59(8):632–7

Collins English Dictionary (2014) Pilot study. Retrieved from https://www.constantcontact.com/index.jsp

Creswell JW, Plano Clark VL (2011) Choosing a mixed methods design. Design Conduc Mixed Methods Res 2:53–106

Crona B, Wutich A, Brewis A et al (2013) Perceptions of climate change: linking local and global perceptions through a cultural knowledge approach. Clim Change 119(2):519–531

Data OA (2020) Cluster sampling. https://www.oferabarbanel-data.com/cluster-sampling-2/

Dey T, Pala NA, Shukla G et al (2018) Climate change perceptions and response strategies of forest fringe communities in Indian Eastern Himalaya. Environ Dev Sustain 20(2):925–938

Dörnyei Z, Griffee D (2010) Research Methods in applied linguistics. TESOL J 1.https://doi.org/10.5054/tj.2010.215611

Field PA, Morse JM (1985) Nursing research: the application of qualitative approaches. Chapman Hall, London

Fowler Jr FJ, Cosenza C (2009) Design and evaluation of survey questions. In: The SAGE handbook of applied social research methods, pp 375–412

Fraser J, Fahlman DW, Arscott J et al (2018) Pilot testing for feasibility in a study of student retention and attrition in online undergraduate programs. Int Rev Res Open Distrib Learn 19(1)

Gadenne D, Sharma B, Kerr D et al (2011) The influence of consumers' environmental beliefs and attitudes on energy saving behaviours. Energ Pol 39(12):7684–7694

Gentle P, Thwaites R, Race D et al (2014) Differential impacts of climate change on communities in the middle hills region of Nepal. Nat Hazard 74(2):815–836

Ghosh M, Ghosal S (2020) Determinants of household livelihood vulnerabilities to climate change in the Himalayan foothills of West Bengal, India. Int J Disaster Risk Reduct 50:101706

Ghosh M, Ghosal S (2020) Climate change vulnerability of rural households in flood-prone areas of Himalayan foothills, West Bengal, India. Environ Dev Sustain 1–26

Glasow PA (2005) Fundamentals of survey research methodology. Mitre Department 2013

Gupta AK, Negi M, Nandy S et al (2019) Assessing the vulnerability of socio-environmental systems to climate change along an altitude gradient in the Indian Himalayas. Ecol Indic 106:105512

Hansen J, Sato M, Ruedy R (2012) Perception of climate change. PNAS 109(37):E2415–E2423

Ho FW (2005) Survey as a source of statistics and factors affecting the quality of survey statistics. Int Stat Rev 73(2):245–248. https://doi.org/10.17226/4982
Kimani NC, Bhardwaj SK (2015) Assessment of people's perceptions and adaptations to climate change and variability in mid-hills of Himachal Pradesh, India. Int J Curr Microbiol Appl Sci 4(8):47–60
Kraemer KL (ed) (1991) The information systems research challenge survey research methods, vol III. Harvard University Graduate School of Business Administration
Lau F (2017) Chapter 13: Methods for survey studies. In: Handbook of e-health evaluation: an evidence-based approach [Internet]. Available online also at: https://www.ncbi.nlm.nih.gov/books/NBK481602/
Leiserowitz A (2006) Climate change risk perception and policy preferences: the role of affect, imagery, and values. Clim Change 77(1):45–72
Leiserowitz A (2005) American risk perceptions: is climate change dangerous? Risk Anal 25(6):1433–1442
Leung WC (2001) Conducting a survey. BMJ 322(Suppl S5). Accesssed on 3/3/2021
Lorenzoni I, Pidgeon IL (2006) Public views on climate change: European and USA perspective. Clim Change 73–95
Macchi M, Gurung AM, Hoermann B (2014) Community perceptions and responses to climate variability and change in the Himalayas. Clim Dev 7(5):414–425
Maurer JM, Schaefer JM, Rupper S et al (2019) Acceleration of ice loss across the Himalayas over the past 40 years. Sci Adv 5(6):7266
Morgan MG (1997) Public perception, understanding, and values. National Academy of Engineering. The Industrial Green Game: Implications for Environmental Design and Management. The National Academies Press, Washington, DC
Nawrotzki R, Kadatska P (2010) Addressing climate change with indigenous knowledge. Int J Clim Chang Impacts Responses 2(1):33–48
Negi VS, Maikhuri RK, Pharswan D et al (2017) Climate change impact in the Western Himalaya: people's perception and adaptive strategies. J Mountain Sci 14(2):403–416
O'Connor RE, Bard RJ, Fisher A (1999) Risk perceptions, general environmental beliefs, and willingness to address climate change. Risk Anal 19(3):461–471
Organisation for Economic Co-operation and Development Staff, 202. Measuring regulatory performance: a practitioner's guide to perception surveys. OECD
Pandey R, Kumar P, Archie KM et al (2018) Climate change adaptation in the western-Himalayas: household level perspectives on impacts and barriers. Ecol Indic 84:27–37
Pratap D (2015) Climate change impacts and adaptation by communities in a tribal region of central Himalaya: a study from Uttarakhand Himalaya India, commons amidst complexity and change. In: The fifteenth biennial conference of the international association for the study of the commons, Edmonton, Alberta, pp 25–29
Rautela P, Karki B (2015) Impact of climate change on life and livelihood of indigenous people of higher Himalaya in Uttarakhand, India. Am J Environ Prot 3(4):112–124
Rawat VS (2013) People perception on climate change and their influence on various aspects around Tones valley of Garhwal Himalaya. Environ Ecol Res 1(3):150–154
Raygorodetsky G (2011) Why traditional knowledge holds the key to climate change. United Nations University, p 13
Salick J, Byg A (eds) (2007) Indigenous peoples and climate change. Tyndall Centre for Climate Change Research, Oxford
Sharma A, Batish DR, Uniyal SK (2020) Documentation and validation of climate change perception of an ethnic community of the western Himalaya. Environ Monit Assess 192(8):1–22
Sharma RK, Shrestha DG (2016) Climate perceptions of local communities validated through scientific signals in Sikkim Himalaya, India. Environ Monit Assess 188(10):1–1
Shi J, Visschers VHM, Siegrist M (2015) Public perception of climate change: the importance of knowledge and cultural worldviews. Risk Anal 35(12):2183–2201. https://doi.org/10.1111/risa.12406

Shukla R, Agarwal A, Sachdeva K et al (2019) Climate change perception: an analysis of climate change and risk perceptions among farmer types of Indian Western Himalayas. Clim Change 152(1):103–119

Singh RK, Zander KK, Kumar S et al (2017) Perceptions of climate variability and livelihood adaptations relating to gender and wealth among the Adi community of the Eastern Indian Himalayas. Appl Geogr 86:41–52

Sullivan A, White DD (2019) An assessment of public perceptions of climate change risk in three western US cities. Weather Clim Soc 11(2):449–463

Taherdoost H (2016) Sampling methods in research methodology; how to choose a sampling technique for research. In: How to choose a sampling technique for research

Turner WR, Oppenheimer M, Wilcove DS (2009) A force to fight global warming. Nature 462(7271):278–279

Von Borgstede C, Andersson M, Johnsson F (2013) Public attitudes to climate change and carbon mitigation—implications for energy-associated behaviours. Energ Pol 57:182–193

Whitmarsh L, Capstick S (2018) Perceptions of climate change. In: Psychology and climate change. Academic Press, pp 13–33

Wilson J (2014) Essentials of business research: a guide to doing your research project. Sage

Chapter 16
Climate Change-Related Governance and Policies in Indian Himalayas

Supinder Kaur, Damanjeet Kaur, Anupreet Singh Tiwana, and Saurabh Gupta

Abstract Climate change is a serious concern that is posing a major threat to the environment and society throughout the globe. Some unexpected changes in climate, such as increased temperature, inconsistent rainfall, and other extreme meteorological conditions, are common. These eventually lead to prolonged periods of drought and floods which are deteriorating the natural resources. Consequently, the Government of India (GoI) is functioning at different spatial scales to reduce greenhouse gases emissions, reforestation, prohibiting the use of polythene, and use of renewable resources. These policies will ultimately prevent harmful anthropogenic intrusion with the climatic conditions and ensure economic development to proceed sustainably. Besides this, the Indian Ministry of Environment, Forests, and Climate Change (MoEFCC) is working to formulate methods for dealing with the inevitable impacts of climatic change. The GoI has made numerous initiatives of National Action Plan on Climate Change (NAPCC), National Adaptation Fund on Climate Change (NAFCC), Climate Change Action Programme (CCAP), and State Action Plan on Climate Change (SAPCC) to deal with climate change impacts. These plans address the climate concerns of the entire nation. The National Mission for Sustaining the Himalayan Ecosystem (NMSHE) under NAPCC is an innovative idea to deal with ecological issues in the Himalayas. The GoI has taken different initiatives such as Governance for Sustaining Himalayan Ecosystem (G-SHE), Hill Area Development Programme (HADP), Indian Himalayas Climate Adaptation Programme (IHCAP), Climate change adaptation project (CCAP), and National Mission on Himalayan Studies (NMHS) for Himalayan ecosystem sustainability. These are

---

S. Kaur
University School of Law, Sri Guru Granth Sahib World University, Fatehgarh Sahib, Punjab, India

D. Kaur · S. Gupta (✉)
Department of Microbiology, Mata Gujri College (Autonomous), Fatehgarh Sahib, Punjab, India

D. Kaur
Department of Biotechnology, Punjabi University, Patiala, Punjab, India

A. S. Tiwana
Department of Geography, Mata Gujri College (Autonomous), Fatehgarh Sahib, Punjab, India

School of Development Studies, Tata Institute of Social Sciences, Mumbai, India

S. Rani and R. Kumar (eds.), *Climate Change*, Springer Climate,
https://doi.org/10.1007/978-3-030-92782-0_16

providing various actions towards controlling and adapting to climate change. This chapter is a compilation of Indian policies, governance, and its role to minimize the climate change impacts and improve sustainability.

Keywords Indian Himalayas · Climate change · Adaptation · Policies · Sustainability

Introduction

Climate change is a crucial ecological concern grabbing global attention. Nowadays, humanity faces three interlinked challenges of reducing worldwide poverty, empowering development, and avoiding climate change (Edenhofer et al. 2012). Worldwide biological diversity, water, energy, human well-being, wildlife, agriculture, and forests are affected by extreme climate changes (UNFCCC 1992; National Research Council 2013). Despite the cumulative initiative taken by India at the country and international level towards climate challenge, its impacts are major concerns for Indian and international governance. India faces an apprehensive ecological challenge socially and economically to safeguard the rapid depletion of natural resources (Lolaksha and Anand 2017). Governance on climate change deals with mitigating and adapting while formulating effective methods of managing these measures and consequences across the different fields (Knieling and Filho 2013). Variations in climate saturate our lives in multifarious ways distressing the social, economic, and political system.

Climate change is a defining realm for the twenty-first century (Boykoff 2010). Global warming, natural or anthropogenic changes in climate, is repeatedly used in climate concerns-related literature (Cohen and Waddell 2009). Traditionally, climate change is defined 'as a significant change in climate state in the course of the certain time interval, where the means are taken over decades or longer' (Cohen and Waddell 2009). Literally, 'climate change' signifies a continuous change in the statistical distributions of atmospheric conditions, e.g. precipitation temperature, etc., over decennium to millions of years. Thus, different organizations have demonstrated changes in modern climate in the environmental discourse augmented by adverse anthropogenic activities. For example, deforestation and burning fossil fuels are often considered to change our climate (Rahman 2013). For more than the past 150 years, GHG emissions and air temperature have been profoundly augmented due to large-scale production by enterprises. Geological research studies highlight that fluctuations in global temperature have been noticed in the past, and these fluctuations associate with CO_2 content in the air. However, the evidence-based studies indicate that the CO_2 content of the air at no time for the duration of the past 650,000 years has been as high as it is today (Giddens 2009). Some scientists in the nineteenth century had postulated that GHGs could impact the climate globally. However, in the late 1950s, extensive international scientific discussion on global warming took off when atmospheric CO_2 concentrations data became available in a better way (Kreienkamp 2019). In 1958, highly accurate measurements of atmospheric CO_2 concentration were done, which further conveyed the atmosphere's changing composition through leading time-series documentation (Keeling 1960). This data in climate change science indicated the effects of human activities on the global chemical composition of the atmosphere. A true record of the global carbon cycle and fossil fuel burning is found in measurements on Mauna Loa in Hawaii (Le Treut et al. 2007).

Climate change will intensely impact human environments in the coming years through variants in rainfall patterns and high temperatures, etc. The chief perpetrator for these climatic conditions is human interference (Santer et al. 2013; Kumar et al. 2020). To tackle the climate change problem as a whole, its governance is a great issue that needs to be considered with all scientific and practical technicalities. The United Nations Framework Convention on Climate Change (UNFCCC) and the Kyoto Protocol (KP) are the fundamental structures of existing global climate governance from a legal perspective (Bernauer and Schaffer 2010). India's prevailing climate change policies and praxis are concerned with making decisions and actions through different missions. Significant alterations are found in the present climate change policy towards its actions in a concrete manner at the national and international levels (Saryal 2018). The NAPCC was the Indian government's officially recognized document on climate change concerns drafted by the Ministry of Environment and Forests (MoEF) in 2008. The core of the NAPCC approach had created eight national missions, 'representing multi-pronged, long-term and integrated strategies for attaining key goals in the context of climate change' (GoI 2008). The Indian political discourse is a seriously centralized quasi-federal system (Lijphart 1996) or nominal federalism (Parikh and Weingast 1997). The central government takes the lead in policymaking and has also been confirmed for environmental policies (Reich and Bowonder 1992; Jasanoff 1993). Consequently, the Union Ministry encouraged the state governments to set up their SAPCC by focusing on the broad objectives of the NAPCC. The states and UTs had formulated their policies to control climate changes. The objective of 'anti-poverty and rural development programmes is to reduce susceptibility to climate danger (Narain et al. 2009). Considering the importance of emerging climate change issues in the country, this study is an effort to review the research work on the emerging level of climatic changes and its influence on policies in the Indian Himalayan region.

Impacts of Climate Change

Due to the speedy emission of GHGs, global climate change is negatively impacting humanity and the ecologies on Earth (Gupta and Bhatt 2019). In the fifth assessment report, the Intergovernmental Panel on Climate Change (IPCC) provides that climate change will affect billions of people worldwide and the ecosystems, natural resources, social and physical infrastructure, and live (IPCC 2014). It is causing severe drought, which has triggered food and water insecurity in many regions worldwide. The world now faces a risk to all of Earth's systems and leaves future generations with a beleaguered planet fighting for its continuous existence (Abate 2019). The world's vision is at risk and not science fiction, futurist speculation, or a theoretical construct. It is reflected in universally degraded ecosystem functioning, including rising temperature, changed hydrological conditions, lowered agricultural yield, and instant displacement of some communities due to rise in sea level, desertification, catastrophe caused by some highest occurrences such as cyclone and rapidly spreading fire, etc. (Burger and Gundlach 2018). India is one of the utmost susceptible nations to climatic alterations (Cruz et al. 2007; NAPCC 2008; INCCA 2010). The influences of changing climate are manifested in every area. Finding adaptation and mitigation policies to face such variations is a significant task for India with a rapidly increasing population relatively large agriculture-based communities, large and fragile ecology, and numerous socio-economic and political causes in executing adaptation procedures. Maintaining water, food, and energy securities has become one of the complex issues in day-to-day life. As the economic system of India merely depends upon industrial and agricultural practices to provide subsistence to millions of people, any hindrance in these will affect all aspects of living beings (Nair 2009). Approximately, fifty per cent of the Indian population is agriculture-based or depends on other climate-sensitized fields (Bureau of Labour Statistics 2010). Production yield in terms of quality and quantity is negatively affected due to climatic changes like rising temperature, alterations in rainfall and monsoon, etc. (MoEF 2012). During 1968–2008 in India, the annual mean relative humidity, recorded over 244 stations, increased from 63 to 66% (Rajeevan and Nayak 2017). These changes indicate that extreme rainfall has been increased, whereas moderate rainfall has been decreased in the past 50 years. These trends have increased in both floods and droughts in India.

The Himalayas have experienced widespread climate changes in the twentieth century. During the period 1962–2004, in the Chenab basin (Himachal Pradesh), the size of small glaciers has been decreased by 38%, while the large glaciers had decreased (greater than ten sq. km) by 12% during the same period (Kulkarni et al. 2007). Hence, many small glaciers in the Himalayas may disappear entirely in the next 50 years (Srinivasan 2019). Shifts in the species' range are happening on land as well, particularly in the Himalayas. Average temperatures across the Hindu Kush Himalaya (HKH) have risen by 0.3 °C in 1901–2014, with 'dramatic warming' after the 1970s (Ren et al. 2017). It has contributed to an upward shift in the ranges of fruits, vegetables, oak trees, reptiles, birds, and other fauna across the Himalayan states, as these species find temperatures to which these are accustomed higher up (Adve 2020).

Further warming in the HKH will increase the temperature from 2.6–4.6 °C by the end of the twenty-first century. The increased temperature will further aggravate the snowfall and glacier decline resulting in intense hydrological and agricultural influences in the region (Sabin et al. 2020). The unusual melting of the glaciers in the Himalayas will soon affect wetlands, major canals, deltas, and millions of people, and the prosperous biological diversity. Sea-level rise is a significant concern since most people reside within 50 km of the coastal area of the sea. In India, the coastal states, such as West Bengal and Gujarat, are most susceptible to a rise in sea level. Therefore, it results in enhanced damage during windstorm surges and cyclone landfall (Srinivasan 2019). Consequently, extreme weather actions pose threats to human well-being, security of foodstuffs and availability of water, etc. (Cruz et al. 2007). These constant impacts of climate change affecting worldwide governance together at national and regional levels.

Climate Change Governance

Considering the emerging climate-related changes and their impacts, governance on climate change deliberates on describing and clarifying the policies on climate change. Governance on climate change necessitates a harmonized approach at every level (Ostrom 2010). Meadowcroft stated that 'climate change governance requires governments to take an active role in bringing about shifts in interest perceptions so that stable societal majorities in favour of deploying an active mitigation and adaptation policy regime can be maintained' (Meadowcroft 2009). Likewise, governance is the distinguishing characteristic of an institutionalism approach dealing with monitoring structures conjoining public and private, hierarchical and network forms of action management (Renate 2004). Governance on climate change is a rising area, which is directly linked with public administrative systems of the state and the attitude of private actors, inclusive of a commercial segment, along with non-governmental organizations and civil society. Governance on climate change provides for both mitigation and adaptation procedures (Knieling and Filho 2013). In the 1950s, environmental safety discourse commenced (Wilde 2008), whereas the climate change discourse initiated approximately after twenty years. Some noticeable developments in the twenty-first century have heightened attention towards climate security. 'The Political Economy of Climate Change: Stern Review Report' examined the imminent impact of climate change on the economy (Stern 2007). The United Nations Security Council (UNSC) had deliberated that climate changes are a major threat to the maintenance of world peace and security. The IPCC, from its inception, disseminates knowledge on climate change. The Human Development Report, 2007–2008 unambiguously underpins the severe task of 'fighting climate change' (Sahu 2019). The UNFCCC and the KP had founded the basic structure of the prevailing international governance on climate change (Bernauer and Schaffer 2010). In 1988, the IPCC was instituted by the World Meteorological Organization (WMO) and the United Nations Environment Programme (UNEP) to offer policymakers regular scientific assessments regarding climate changes, its implications, possible future risks, and also to put forward adaptation and mitigation strategies. It was mandated to evaluate scientifically, technically, and socio-economically data pertinent to conceptualizing the scientific base of anthropogenic persuaded climate change threat, its probable effects, and preferences for alteration and mitigation on a wide-ranging, exposed, and translucent basis. Although the IPCC was the outcome of the UNEP and WMO, it put forward all areas which are distinct and beyond their specific expertise (Drexhage 2008).

Presently, India is a subject of maximum exposure to climate impacts. It has genuine negotiations for attaining a significant result and a mounting attentiveness of its prospective role in achieving such goals. The GoI recognizes that now it is requisite to give much weightage to these objectives other than domestic emergencies, mainly to accomplish extraordinary heights of socio-economic progress and poverty alleviation. Recently, a shift has been seen in Indian strategy towards consultations surrounded by the UNFCCC and extra innovative climate policy action in the nationwide and subnational areas. This tendency towards 'multilevel governance' with a more independent subnational dimension shaped the policy at each level (Atteridge et al. 2012). In India, the MoEFCC collaborates with the ministries of new and renewable energy (MNRE), urban development, science and technology, water resources, and agriculture to minimize climate-related issues and concerns. The main feature of the NMSHE is the governance to provide sustenance to the Indian Himalayas. The focus of this mission is to conduct studies and research for the formulation of Himalayan region policies. The Department of Science and Technology (DST) is working on a varied range of goals in this mission. The government had engaged six task force institutions to administrate this mission, including geological wealth, water, forest resources, traditional knowledge, and Himalayan agriculture (http://www.knowledgeportal-nmshe.in/).

Global Climate Change Negotiations and Indian Governance

Climate change has developed as a matter of significant discussions in transnational politics and diplomacy over the last few decades. In 1972 by proclaiming the 'UN Conference on the Human Environment in Stockholm, the 'UN General Assembly' specified that the principal aim behind convening the conference was to motivate and arrange strategies for actions to be taken by governments and transnational organizations intended to safeguard and make a better environment for humanity (Brisman 2011). 'The Brundtland Commission Report 1987' familiarized the idea of 'sustainable development' (Adams 1990; Carter 2001). It was pursued by the 'Rio Earth Summit in 1992' and further adopted by the UNFCCC. The IPCC had provided scientific reports with all technicalities and socio-economic guidance to the international community, mainly to the parties of the UNFCCC through regular evaluations (Saryal 2018). In the first stage, the Indian policy framework was appeared commonly principled and conceptual, declaring fundamentally that India is an emerging country, and climate change alleviation was apparent to be developed countries task (Messner 2017; Saryal 2018). The existing stage climate change policy of India consequently looks categorized by realism. A significant revision is initiated in India's present climate change policy, which tries towards solid engagements at both national and international levels (Saryal 2018) (Table 16.1).

Table 16.1 Governance on climate change in Indian Himalayas, and at international level

Year	International	Indian
1972	United Nations Conference on Human Environment (Stockholm Declaration)	Development of a legal framework to protect the environment
		'National Council for Environmental Policy and Planning'(NCEPP) established
		The Wildlife (Protection) Act enacted
1985	–	MoEF established
1986	–	The Environment (Protection) Act enacted
1987	'The Brundtland Commission Report 1987' 'Our Common Future' acquainted with the idea of sustainable development	The Factories Act, 1948 amended in 1987 which provided environmental security by guaranteeing the welfare and health of the workers
1988	Establishment of IPCC	National Forest Policy (NFP) The MoEF has set up G.B. Pant Institute of Himalayan Environment & Development (GBPIHED)
1989	By resolution of the 'United Nations General Assembly' 44/207 Negotiations started for the 'Framework Convention'	MoEF establishes 'Expert Advisory Committee on worldwide ecological concerns
1990	IPCC prepared its report (Ist) on causes, impacts of climate change 'Intergovernmental Negotiating Committee (INC)' convened for UNFCC negotiations	'Conference of Select Developing Countries on Global Environmental Issues' assembled in India and the Nation prospered in acquiring the wide-ranging support of the developing world and secured its place globally on climate change discussions
1991	INC negotiations procedure was supported beneath the umbrella of the General Assembly	–
1992	UNFCCC signed at Rio De Janeiro	UNFCCC signed by India
	The Convention on Biological Diversity (CBD)	
1993	–	India ratified UNFCCC
1994	UNFCCC came into force	–
1995	IPCC prepared Assessment Report (IInd)	–
	Conference of Parties (COP) 1 adopted the 'Berlin Mandate'	
1997	KP accepted at COP 3	–
2001	IPCC prepared assessment report (IIIrd)	–
	At COP 7, 'Marrakesh Accords' accepted	
2002	COP 8 arranged for 'climate adaptation'	India ratified KP
		India hosted COP 8 in Delhi

(continued)

Table 16.1 (continued)

Year	International	Indian
		The Biological Diversity Act National Wildlife Action Plan
2003	–	India instituted 'National Clean Development Mechanism Authority (CDM)'
2005	KP enforced Ad-Hoc Working Group on KP (AWG-KP) constituted at COP 11 to scrutinize 'second commitment period' targets 'Dialogue' initiated at COP 11 on long-term cooperative action (LCA)	–
2006		National Environment Policy (NEP)
2007	'G8+5 Summit' held in Heiligendamm, Germany IPCC prepared assessment report (IVth) 'Bali Action Plan' accepted at COP 13 Ad-Hoc Working Group on LCA (AWG-LCA) constituted at COP 13	Establishment of Prime Minister Council on Climate Change (PMCCC)
2008	AWG-KP and AWG-LCA held meetings at COP 14	GoI adopted NAPCC and launched eight missions to deal with climate change issues in India and also comprises of specific Mission for Himalayas (NMSHE) National Biodiversity Action Plan (NBAP)
2009	L'Aquila Summit of 'Major Economies Forum on Energy and Climate (MEF)' considered '2 degrees C' limit COP 15 'takes note of Copenhagen Accord' AWG-KP and AWG-LCA obligations extended to COP 16	India signed MEF Covenant and proclaimed an intended 'emissions intensity' cut of 20–5% by 2020
2010	'Cancun Agreements' adopted at COP 16 AWG-KP and AWG-LCA obligations projected till COP 17	Planning Commission launched Expert Group on 'low carbon economy'
2011	At COP 17, 'Durban Platform for Enhanced Action' adopted and Ad-Hoc Working Group on Durban Platform for Enhanced Action (ADP) constituted	–
2012	At COP 18, 'Doha Amendment to KP' agreed, and ADP meetings organized	–
2013	COP 19 held in Warsaw requested parties to formulate and inform Intended Nationally Determined Contributions (INDCs)	–

(continued)

Table 16.1 (continued)

Year	International	Indian
2014	IPCC prepared Assessment Report (V^{th})	MoEF was renamed as MoEFCC to highlight agendas on climate change High-Level Committee on Forest and Environment was constituted to review the following laws: (i) Environment (Protection) Act, 1986 (ii) Forest (Conservation) Act, 1980 (iii) Wildlife (Protection) Act, 1972 (iv) The Water (Prevention and Control of Pollution) Act, 1974 (v) The Air (Prevention and Control of Pollution) Act, 1981 (vi) The Indian Forests Act, 1927
	COP 20 interpreted the 'principle of common again, but differentiated, responsibilities and respective capabilities' (CBDR&RC) as 'CBDR&RC in the light of different national circumstances' (CBDR&RC-NC)	
2015	At COP 21 'Paris Agreement' adopted	India increased its solar power aims by fivefold
		India communicates INDC promised 'emissions intensity' cut of 33–5% by 2030
		India introduced the 'International Solar Alliance' with France
		National Mission on Himalayan Studies (NMHS)
2016	'Paris Agreement' enforced At COP 22 Ad-Hoc Working Group on Paris Agreement (APA) constituted to discuss the Paris Agreement guidelines	'Paris Agreement' ratified by India
2017	Pronouncement of the US departure from 'Paris Agreement' Meetings of APA at COP 23	India repeated provision for 'Paris Agreement'
		Working groups established by NITI Aayog of GoI for 'sustainable development in mountains of Indian Himalayan Region (IHR)'
2018	COP 24, held in Katowice, Poland, provided guidelines to carry out obligations of the 'Paris Agreement'	India organized around 20 sessions on issues important for climate change adaptation and mitigation, including sustenance of the Himalayan ecosystem
		NITI Aayog established 'Himalayan State Regional Council'
2019	COP 25 intended to proceed for implementation of guidelines of the Paris Agreement at COP 24 and the subsequent important steps on climate change	India played a leading role in the financial demands of developing nations for climate concerns and also strongly criticized developed nations approach towards climate threat

(continued)

Table 16.1 (continued)

Year	International	Indian
		National Clean Air Programme launched
2021	UN Climate Change conference in Glasgow in 2021(COP26)	–

Source http://moef.gov.in; https://www.downtoearth.org.in; https://unfccc.int; https://legal.un.org; http://www.knowledgeportal-nmshe.in; https://www.ipcc.ch; https://sustainabledevelopment.un.org; http://www.nihfw.org; http://dst.gov.in

Climate Change Policies and Institutions

NAPCC (NAPCC 2008)

The NAPCC, issued by the MoEF in 2008, is the Indian government's managerial accreditation of climate change and associated concerns and is committed to tackling climate change congruously. 'The NAPCC addresses the country's urgent and critical concerns through a change in the development pathway. The NAPCC classifies measures that promote the development objectives and compliant co-benefits for addressing climate change effectively' (GoI 2008). The NAPCC and national climate policy endured exhaustive consideration in the 12th Five Year Plan (2012–17), which emphasized that climate change worry should infuse all procedures of strategy for any assignment to thrive in the long term; it should have different objectives, enthusiastic execution technology, and sufficient financial support (GoI 2013; Kumar and Naik 2019). The core of the NAPCC includes eight national missions that symbolize 'multi-pronged, long-term and integrated strategies to achieve key goals in the perspective of climate change' (GoI 2008). The NAPCC comprises of following missions on different aspects:

National Solar Mission (NSM)
National Mission for Enhanced Energy Efficiency (NMEEE)
National Mission on Sustainable Habitat (NMSH)
National Water Mission (NWM)
National Mission for Sustaining the Himalayan Ecosystem (NMSHE)
National Mission for Green India (GIM)
National Mission for Sustainable Agriculture (NMSA)
National Mission on Strategic Knowledge for Climate Change (NMSKCC)

NSM and NMEEE

The NSM under the NAPCC intended to encourage the enhancement and utilization of solar energy to generate power and electricity, with the eventual purpose of creating competition with fossil-dependent energy resources. The mission embraces reinstituting a solar research centre, developing global alliances on research and technology, enhancing domestic manufacturing competence, and raising funds and international assistance. The NMEEE strives to reinforce energy competence by the functioning of creative business ideas in the energy efficiency field. The NMEEE has undertaken four programmes, including Perform Archive and Trade (PAT), Energy Efficiency Financing Platform (EEFP), Market Transformation for Energy Efficiency (MTEE), Framework for Energy Efficient Economic Development (FEEED), to increase energy efficiency in energy-intensive industries (www.mnre.gov.in).

The GoI at COP 21 was enacted dynamically to place India as fully alert about worldwide accountability on concerns associated with climate change. For universal obligation concerning the climate change issue, the Indian government framed a new mission, 'International Solar Alliance' with France. The mission intended to expand the worldwide reception of solar energy, particularly by the tropics. In addition to this, the Indian government has decided to raise the nation's ability to generate solar power by fivefold, from 20 gigawatts (GW) to 100 by 2022, to execute the commitments set up by the NAPCC (Navroz 2020). The Indian policy concerning the nexus between climate and energy has been multi-dimensional and budding; different policies and Nationally Determined Contributions (NDCs) gave eminence to renewables and energy efficiency. This is generally evident that NDCs emphasize diminution of the strength of the emissions of the GDPs by 33–35% from 2005 to 2030 with a goal of 175 GW of renewables by 2022 (NITI Aayog 2015).

NMSH

Sustainable habitat mission planed for present and future strategies to address climate change mitigation. The mission has also outlined policies to deal with different aspects of climate problems, enhance preservation of natural resources, energy efficiency, enhance preservation of natural resources, energy efficiency, uphold better planning in urban areas, ecological habitation standards, and concurrently tackle climate-related issues. The GoI ensured the implementation of this mission by four leading missions of the Ministry of Urban Development: 'Atal Mission on Rejuvenation and Urban Transformation (AMRUT), Swachh Bharat Mission, Smart Cities Mission, and Urban Transport Programme'. The purpose behind this is to acclimate and lessen emissions of GHGs and to manage sustainability. The report for the COP21 at Paris drafted by the energy and resources institute under the Ministry of Urban Development projected that effective enforcement of these four missions under NMSH could mitigate GHG emission up to 133 million tonnes CO_2 eq by 2021 and 270 million tonnes till 2031 (http://cpheeo.gov.in).

NWM

Water being nature's bequest is essential for aquatic as well as terrestrial livelihoods and sustainable development. The current situation of water sources and their management has given augmentation to numerous concerns. The NWM under NAPCC concentrates on safeguarding the water, appropriate re-use, and inspiring humanity to adopt eco-friendly modes of water harvesting and protection approaches. The major objective of NWM includes 'conservation of water, minimizing wastage and ensuring its more just supply both across and within States through integrated water resources management'. The five acknowledged goals of the mission are:

> "Comprehensive water database in public sphere and assessment of the effects of climatic changes on the water resource
>
> Promotion of individuals and state action for water conservation, expansion, and preservation
>
> Consideration to vulnerable areas, including over-exploited areas
>
> Increasing water use efficiency by 20%
>
> Promotion of basin level integrated water resources management."

The Ministry of Water Resources is accountable for framing policies and programmes to regulate the country's water resources. The major programmes under this ministry include 'Ganga Rejuvenation; Inter Linking of Rivers; Command Area Development and Water Management; Flood Management Wing Programmes; Research and Development Programme in Water Sector; Dam Rehabilitation and Improvement Programme' (www.jalshakti-dowr.gov.in).

GIM and NMSHE

The GIM under the NAPCC intends to protect and enhance Indian forest cover to equilibrate the menace caused by climate change. The mission envisions an exhaustive prospect of greening and emphasizes various ecological amenities, sequestration of carbon, and emissions reduction as co-benefit. The goals of this mission are to expand forest cover to the extent of 5 million hectares (mha) and improve the quality of forest cover on extra five mha of forest/non-forest lands to uplift ecological community features like carbon sequestration and sustainment (in forests and other biomes), hydrological services and biological diversity; together with the provision of resources like fuel, food, and forest produce and to upsurge the forest-dependent subsistence of approximately 3 million households (www.naeb.nic.in). The national executive council of GIM has endorsed four states' purposive plans, including Sikkim, Maharashtra, Madhya Pradesh, and Himachal Pradesh. The GIM has initiated the Ecosystems Services Improvement Project (ESIP), funded by World Bank on 13 July 2018, is being executed in Chhattisgarh and Madhya Pradesh. Indian Council of Forestry Research and Education (ICFRE) established a project implementing unit and ESIP enforcement activities (www.naeb.nic.in). The NMSHE under

NAPCC attempts to deal with serious concerns regarding melting glaciers, biodiversity, wildlife, people's livelihood, and planning to sustain the Himalayan ecosystem. The main objective of the NMSHE is to deal with governance for sustaining the Himalayan ecosystem (http://www.knowledgeportal-nmshe.in).

NMSA and NMSKCC

The NMSA emphasizes the promotion of crop breeding to develop abiotic stress-resistance crop varieties and generate better farming practices. Prospective approaches to acclimatize agri-business to climate change also correlate with other plans relating to inherent reserves like water, land, and forest (www.nmsa.dac.gov.in). The NMSKCC is predominantly expected to support studies on climate change in academics by setting up universities, institutes, and organizations to enhance commercialized sectors to formulate techniques for acclimatization and mitigation (Aryal et al. 2019). Moreover, the NAPCC defines the process for executing the eight missions through different institutions in collaboration with other ministries and specialists from industry, the academic world, and organizations. The structure of these institutions is varied as per the targets that need to be achieved.

SAPCC

After the NAPCC developments to deal with grave issues regarding climate change, GoI directed states to prepare plans based on the policy framework of the eight NAPCC missions. India being a quasi-federation, decentralization is required because various areas of these missions, like agriculture and water, are state subjects (Dubash and Jogesh 2020). Therefore, it necessitated accomplishing unity amongst the policies and actions at the central and state levels. Hence, the central ministry encouraged the governments of the states to formulate their SAPCC by accomplishing the goals of NAPCC. Consequently, states/UTs had framed their SAPCCs to mainstream climate change distresses in their policy procedure. The states/UTs have been directed to review their SAPCCs and focus on their responsibilities to enable their post-2020 goals (MoEFCC 2019). The Indian Network for Climate Change Assessment (INCCA) had set up institutions to scrutinize climatic conditions, make GHG inventories, and prepare a mechanism to harmonize research studies. The government has also made the NAFCC 'assist particularly vulnerable States, grounded on the requirements and priorities recognized under the SAPCC and the related missions under NAPCC' (MoEFCC 2016).

The National Bank for Agriculture and Rural Development (NABARD) is a body for financial assistance in climate-concerned projects under the NAFCC (Dubash and Ghosh 2020). Therefore, GoI and respective state governments under the NAPCC and

SAPCC taking varied actions through missions to eradicate climate-related complications. Constitution under Article 21 of the Indian Constitution, the right to life is of wider connotation, includes a variety of rights. The right to a pollution-free environment is also protected under this provision. The legal system of India expanded the scope of the right to life by declaring a pollution-free environment as a fundamental right. Despite the role played by the executive and judiciary, environment-related problems are increasing day by day due to the lack of implementation of these policies. Its immediate victim is not only the present but also the future generation. Indian environmental policy and laws have thus needed to be in harmony with international commitments. This may necessitate structural modifications in the environmental policies and laws of India. The further utmost requirement is proper enforcement of these policies.

Indian Governance and Policies for Sustainability of Himalayas

The Himalayan Mountains surround the Hindu Kush Mountain region and the Tibetan autonomous region of China. The mountains are extended over Afghanistan, Pakistan, China, India, Nepal, Bangladesh, Myanmar, and Bhutan and cover 34.4 Lac km^2. The IHR is extended over an area of around 5.37 Lac km^2, between 21°57′–37°5′ N latitudes and 72°40′–97°25′ E longitudes (about 2500 km in length, 250–300 km in width and rise from low lying plains to above 8000 m AMSL). The sustainability of the Himalayas is of major concern, so different policies and action plans have been framed for mitigation and adaptation to climate change. In 1988, the MoEF set up the GBPIHED with a commitment to achieving sustainable development and ecological preservation goals in the IHR. The institution prepared 'Action Plan for Himalaya' in 1992 to recommend environmental and economic development of Himalaya. Different centres are functioning to develop actions plans for natural resource management and ecological sustainability in the IHR. The Centre for Land and Water Recourse Management, Centre for Socio-Economic Development, Centre for Biodiversity Conservation and Management, and Centre for Environmental Assessment and Climate Change are working to devise environmental concerns for mitigation and adaptation strategies to deal with climate change risks in the IHR. (http://gbpihed.gov.in). India's contributions towards multidimensional negotiations in the UNFCCC are substantial, and it endures to promote functional, complaisant, and reasonable universal methodologies based on the principle of 'CBDR&RC'. To execute these commitments, India has established the NAPCC in 2008 with eight missions. The NMSHE is a mission to evolve nationwide determinations aimed at sustaining a fragile Himalayan ecology. After that, Indian states, including Himalayan states, prepared the SAPCCs for the sustainability of the Himalayas. The GoI has initiated different programmes for safeguarding the

IHR, including G-SHE, HADP, IHCAP and CCAP, etc. (http://www.knowledgeportal-nmshe.in).

NMSHE

The NMSHE under NAPCC is an endeavour to deal with critical issues regarding melting glaciers, biodiversity, wildlife, people's livelihood, and planning to sustain the Himalayan ecosystem. The main objective of NMSHE is to deal with governance for sustaining the Himalayan ecosystem. This emphasis on research and actions is concerned with formulating policies for the Himalayan region to control climate-related issues and attain sustainable development goals. The DST is working in collaboration with MoEFCC for proper enforcement of goals under NMSHE (http://www.knowledgeportal-nmshe.in). The primary objectives of the mission are:

> "To conserve biodiversity, forest cover, and other ecological values in the Himalayan region
>
> Sustainable development of the country by enhancing the understanding of climate change, its likely impacts, and adaptation actions required for the Himalayas
>
> To facilitate the formulation of appropriate policy measures and time-bound action programs to sustain ecological resilience and ensure the continued provisions of key ecosystem services in the Himalayas
>
> To evolve suitable management and policy measures for sustaining and safeguarding the Himalayan ecosystem along with developing capacities at the national level to assess its health status."

Under the principles of cooperative federalism, the NMSHE involved twelve states in the Himalayas, strengthening their capacity to formulate and implement climate change alteration plans and undertake susceptibility evaluation and spread consciousness amongst the common people on climate change effects (http://www.knowledgeportal-nmshe.in). In 2017, NITI Aayog of the GoI established working groups associated with institutes for sustainable development in the mountains of the IHR to prepare reports on every aspect. On recommendations through reports by working groups, NITI Aayog had established 'Himalayan State Regional Council' to attain sustainable development goals in the IHR and review the reports (https://niti.gov.in).

G-SHE

The G-SHE is an integral part of a comprehensive climate change adaptation strategy and provides guidelines regarding the governance and management of the Himalayan ecosystem. Further, it deals with major ecological challenges of the IHR, including water, energy, urbanization, forest management, and tourism (http://moef.gov.in). It aims to provide good infrastructure, healthy, decent quality of life to the population residing in the IHR with a clean and sustainable environment. Therefore, it provides for improvement in the general quality of life in the urban and rural areas by promoting cleanliness (http://gbpihed.gov.in). The G-SHE is multi-dimensional covers the following programmes:

Smart City Mission

Swachh Bharat Mission

Door-to-Door Garbage Collection

Ban on Plastic

Community-Based Ecotourism

Spring-shed Development Programme etc.

NMHS

The Indian government has introduced the NMHS to conduct studies on the IHR. The NMHS aims to create a knowledge network of experts and institutions to find out strategies for the sustainable development of the IHR. This mission aims to improve people's livelihoods in the IHR, in consonance with the NEP, 2006 (www.nmhs.org.in).

The objectives of this mission are:

"Fostering conservation and sustainable management of natural resources

Enhancing supplementary and alternative livelihoods of IHR peoples and overall economic well-being of the region

Controlling and preventing pollution in the region

Fostering increased/ augmented human and institutional capacities and the knowledge and policy environments in the region

Strengthening, greening, and fostering the development of climate-resilient core infrastructure and basic service assets."

HADP

The HADP has been set up with the commencement of the 5th Five Year Plan to deal with hill area concerns. The primary purpose of this programme is the socio-economic upliftment of the population residing in the hills along with ecological developments. Therefore, the programmes executed under the HADP have aimed at stimulating the basic life support systems for residents of hill areas with justifiable use of the natural resources in those areas protected by the programme (https://niti.gov.in).

IHCAP

The IHCAP is a project under the Global Programme Climate Change of the Swiss Agency for Development and Cooperation associated with the DST. The IHCAP focused on enriching knowledge regarding climate change concerns in the IHR and building capabilities of research institutions, planners, experts, and other stakeholders to find out strategies and execute climate change adaptation procedures. Therefore, the IHCAP has taken initiatives to develop an awareness of climate change issues in the IHR and to form strategic associations to support and facilitate the enforcement of the NMSHE and related action plans in the hilly states at their level. Its purpose is to strengthen the capabilities of Indian institutions and also Himalayan states institutions towards climatology, with a particular emphasis on glaciology and related issues along with policy implementation (http://dest.hp.gov.in/). The programme focused on the following goals:

> "Strengthening capacities for adaptation planning and implementation in HP through research, training and capacity building
>
> Scientific capacity building in the field of Glaciology and related areas Facilitating dialogues between Himalayan states and key stakeholders for mainstreaming climate change concerns into development planning."

NAFCC

The Indian government has established NAFCC intending to assist the adaptation projects and programmes to diminish the effects of climate change facing communities and sectors. The focus of the NAFCC is to support states and UTs susceptible to unfavourable impacts of climate change to meet adaptation expenses. The NABARD has been appointed as a national implementing entity liable to implement adaptation schemes under the NAFCC. A climate change adaptation project entitled 'Sustainable Livelihoods of Agriculture-Dependent Rural Communities in Drought Prone District of Himachal Pradesh through Climate Smart Solutions (SLADRC)' has been sanctioned funds by the MoEFCC, GoI under NAFCC in three blocks of district Sirmaur of Himachal Pradesh (HP). The major objectives and actions under NAFCC are:

> "Identification of sectoral adaptation strategies to assist rural communities for implementation
>
> Development of long term activity-wise action plan
>
> Assessment of community-level vulnerability with exposure, sensitivity, and adaptive capacity different from conventional planning process on Agriculture, Water-irrigation, Crop diversification & livelihood practices
>
> Documentation of best practices being adopted by the farmers
>
> Development of GIS-based information systems to represent impacts of climate change in district Sirmour of HP
>
> Create an enabling framework for climate change adaptive capacity
>
> Training module development on climate-smart approaches

Training / Orientation of target farmers on climate-resilient agriculture / horticulture
Extension services and handholding support to target farmers from time to time
Demonstration of different packages of practices, adaptive to climate variability
Organizing dissemination workshops on project learning."

CCAP

The CCAP is framed under the NAFCC in HP to decrease the climate-associated susceptibility and to improve the adaptive competence of rural small/marginal farmers, including rural women, by establishing a union of climate-smart farming technologies together with mandatory social engineering and capability-building processes leading to enhanced food safety and livelihood to increase resilience. For improvement of capacities of rural marginal farmers, particularly women, the following seven training modules have been prepared as per training need assessment (http://sladrc.in/):

"Drought resilient varieties and cropping systems training
Efficient water management systems including micro-irrigation for water use efficiency
Management of soil nutrition, including practices to enhance soil organic carbon
Efficient pest control through integrated pest management
Agro-met advisory plan as per local weather conditions
Governance aspects of community institutions and convergence
Farmers producers' organization and climate change adaptation."

International Initiatives

The International Centre for Integrated Mountain Development (ICIMOD) has been working for the Himalayan countries (Bhutan, India, Myanmar, and Nepal). MoEF arranged significant support to the ICIMOD. The GBPIHED serves under the MoEF as the lead institution (https://www.icimod.org). The International Climate Initiative (IKI) is one of the effective mechanisms of 'the German Federal Ministry for the Environment, Nature Conservation, and Nuclear Safety for the worldwide funding of climate change mitigation and biodiversity. The IKI functions under the agenda of the UNFCCC and the CBD, funding for climate change mitigation and biodiversity conservation worldwide. The IKI approved more than 750 climate and biodiversity projects and financed over more than 60 countries between 2008 and 2020. These actions of the IKI are towards putting into practice the United Nations Agendas for Sustainable Development Goals (https://www.international-climate-initiative.com).

Conclusions

Climate change is an emerging global environmental issue and undoubtedly affected human well-being, thus resulting in a great impact on climate change governance. The UNFCCC and the KP, from a legal perspective, are the pillars of present global climate governance. The present status of Indian climate change policies, therefore, appears to be characterized by pragmatism. A significant amendment has been found in the Indian current climate change policy, which now revolves around decisive ventures at the country and international levels. The NAPCC formulated eight national missions representing 'multi-pronged, long-term integrated strategies' to achieve key goals in the context of climate change. The MoEFCC collaborates with the different ministries involving power, renewable energy, urban development, science and technology, water resources, and agriculture to combat these continuously growing climate-related issues. The GoI is formulating and implementing different policies to increase the nation's solar power generation capacity. Further, the GoI has also organized some missions, including Atal, Swachh Bharat, and smart cities for sustainable habitat. The TERI prepared reports regarding the successful implementation of the missions. The GIM under NAPCC is working for the afforestation of our country. The NMSHE engaged all the twelve states in planning to adapt and mitigate changes in climate in the Himalayan region. The NMSHE G-SHE, NMHS, IHCAP, etc., are working for the sustainability of the IHR. The objective behind all these missions is to mitigate the emission of GHGs. Despite numerous efforts of the MoEFCC and other ministries through NAPCC, SAPCC, etc., India faces tremendous challenges in responding sufficiently to climate change. So, it is essential to develop highly effective strategies to adapt and mitigate climate change.

References

Abate RS (2019) Climate change and the voiceless: protecting future generations, wildlife, and natural resources. Cambridge University Press

Adams WM (1990) Green development: environment and sustainability in the third world. London, Routledge

Adve N (2020) Impacts of global warming in India. In: Navroz DK (ed) India in a warming World: integrating climate change and development, Oxford University Press, pp 64–78

Aryal JP, Sapkota TB, Khurana R, Khatri-Chhetri A, Rahut DB, Jat ML (2019) Climate change and agriculture in South Asia: adaptation options in smallholder production systems. Environ Dev Sustain 22:5045–5075

Atteridge A, Shrivastava MK, Pahuja N, Upadhyay H (2012) Climate policy in India: what shapes international, national and State Policy? Ambio 41:68–77

Bernauer T, Schaffer L (2010) Climate change Governance, ETH Zurich centre for comparative and international studies. University of Zurich, pp 1–29

Boykoff MT (2010) Politics of Climate Change: A Survey. Routledge, London

Brisman A (2011) Stockholm Conference, 1972. In: Chatterjee DK (eds) Encyclopedia of Global Justice. Springer, Dordrecht. https://doi.org/10.1007/978-1-4020-9160-5_655

Burger M, Gundlach J (2018) Public Health and the Law, Cambridge UK

Carter A (2001) Presumptive benefits and political obligation. J Appl Philos 229–243
Cohen SJ, Waddell MW (2009) Climate change in the 21st century. McGill-Queen's University Press
Cruz RO, Harasawa H, Lal M, Wu S, Anokhin Y, Punsalmaa B, Honda Y, Jafari M, Li C, Ninh N (2007) Asia. Climate change 2007: impacts, adaptation and vulnerability. Contribution of Working Group II to the Fourth Assessment Report of the Intergovernmental Panel on Climate Change, pp 469–506
Drexhage J (2008) Climate change and global governance. International Institute for Sustainable Development (IISD)
Dubash NK, Ghosh S (2020) National climate policies and institutions. In: Navroz DK (ed) India in a warming World: integrating climate change and development. Oxford University Press, pp 329–348
Dubash NK, Jogesh A (2020) State climate change planning in india from margins to mainstream. In: Navroz DK (ed) India in a warming World: integrating climate change and development. Oxford University Press, pp 349–369
Edenhofer O, Wallacher J, Lotze-Campen H, Reder M, Knopf B, Muller J (2012) Climate change, justice and sustainability: linking climate and development. Springer, Netherland
Giddens A (2009) Politics of climate change. Polity Press, Cambridge, UK
GoI (2008) National action plan on climate change. In: Prime ministers advisory council on climate change, New Delhi, Ministry of Water Resources, Government of India
GoI (2013) Water. In: Twelfth Five Year Plan (2012–2017), Faster, more inclusive and sustainable growth, vol 1. New Delhi, Planning Commission, Government of India, pp 144–189
Gupta DK, Bhatt RK (2019) Carbon sequestration potential of Hardwickia binata Roxb. Based agroforestry in hot semi-arid environment of India: an assessment of tree density impact. Curr Sci 116:23–31
Himalayas Climate Change Portal, An Initiative under National Mission for Sustaining the Himalayan Ecosystem, (Under Climate Change Program) Department of Science and Technology, Ministry of Science and Technology, Government of India http://www.knowledgeportal-nmshe.in/Home.aspx
INCCA (2010) Climate change and India: A 4x4 assessment. Ministry of Forest and Environment, Government of India
IPCC (2014) Climate change 2014: synthesis report. In: Contribution of working groups I, II and III to the fifth assessment report of the intergovernmental panel on climate change [Core Writing Team, Pachauri RK, Meyer LA (ed)], IPCC, Geneva, Switzerland
Jasanoff S (1993) Bridging the two cultures of risk analysis 1, 2. Risk Anal 13(2):123–129
Keeling CD (1960) The concentration and isotopic abundances of carbon Dioxide in the Atmosphere. Tellus 12:200–203
Knieling J, Filho WL (2013) Climate change governance: the challenge for politics and public administration, enterprises and civil society. In: Knieling J, Filho W L (eds) Climate change governance. Climate Change Management. Springer, Berlin, Heidelberg, pp 1–5
Kreienkamp J (2019) The long road to paris: the history of the global climate change regime. Global governance institute policy brief series. University College London, London
Kulkarni A, Bahuguna I, Rathore B, Singh S, Randhawa S, Sood R, Dhar S (2007) Glacial retreat in Himalayas using Indian Remote Sensing Satellite data. Curr Sci 92:69–74
Kumar G, Kumari R, Kishore BSPC, Saikia P, Kumar A, Khan ML (2020) Climate change impacts and implications: an Indian perspective. In: Roy N, Roychoudhury S, Nautiyal S, Agarwal SK, Baksi S (eds) Socio-economic and eco-biological dimensions in resource use and conservation strategies for sustainability. Springer Nature, Switzerland
Kumar P, Naik A (2019) India's domestic climate policy is fragmented and lacks clarity, economic & political weekly, vol 54
Le Treut H, Somerville R, Cubasch U, Ding Y, Mauritzen C, Mokssit A, Peterson T, Prather M (2007) Historical Overview of Climate Change. In: Solomon S, Qin D, Manning M, Chen Z, Marquis M, Averyt KB, Tignor M, Miller HL (eds) Climate change 2007: the physical science

basis. Contribution of working group I to the Fourth Assessment Report of the Intergovernmental Panel on Climate Change Cambridge University Press, Cambridge, United Kingdom and New York, NY, USA

Lijphart A (1996) The puzzle of Indian democracy: a consociational interpretation. Am Polit Sci Rev 90:258–268

Lolaksha NP, Anand A (2017) Climate change and its impact on India: a comment. NLUO Law J 4:81–97

Meadowcroft J (2009) Climate change governance. Policy Research Working Papers. Washington DC

Messner PD (2017) Game Theory in International Climate Change Negotiations: Why does the US plan to withdraw from the Paris Agreement and what are the consequences for countries and future climate protection?

Ministry of Environment and Forest, Annual Report 2008–2009; 2012–2013. Available at http:moef.gov.in

MoEFCC (2016) Ministry of Environment, Forest and Climate Change, Government of India, Annual Report 2015- 2016. Available at http://moef.gov.in

MoEFCC (2019) Ministry of Environment, Forest and Climate Change, Government of India, Annual Report 2018- 2019. Available at http://moef.gov.in

Nair S (2009) Climate change: global risks. In: Challenges and decisions IOP Publishing IOP Conferences Series: Earth and Environmental Science vol 6, p 292050

NAPCC GOI (2008) http://pmindia.nic.in/Pg01-52.pdf

Narain S, Ghosh P, Saxena NC, Parikh J, Soni P (2009) Climate change perspectives from India. United Nations Development Programme (UNDP), India

National Mission on Sustainable Habitat Central Public Health & Environmental Engineering Organisation (CPHEEO), Ministry of Housing and Urban Affairs, Government of India, http://cpheeo.gov.in/cms/national-mission-on-sustainable-habitat.php

National Research Council (2013) Climate and social stress: implications for security analysis. National Academies Press

Navroz DK (2020) India in a warming World: integrating climate change and development. Oxford University Press

NITI Aayog, GOI (2020) Sustainable Development. In: The Indian Himalayan Region, assessed on 27th June 2020, https://niti.gov.in/sustainable-development-indian-himalayan-region

Ostrom E (2010) A multi-scale approach to coping with climate change and other collective action problems. Solutions 1(2):27–36

Parikh S, Weingast BR (1997) A comparative theory of federalism: India. Virginia Law Rev 83:1593. https://doi.org/10.2307/1073770

Rahman MI (2013), Climate change: a theoretical review. Interdisc Description Complex Syst 11:1–13

Rajeevan MN, Nayak S (eds) (2017) Observed climate variability and change over the Indian region. Springer

Reich MR, Bowonder B (1992) Environmental policy in India. Policy Stud J 20(4):643–661

Ren YY, Ren GY, Sun XB, Shrestha AB, You QL, Zhan YJ, Rajbhandari R, Zhang PF, Wen KM (2017) Observed changes in surface air temperature and precipitation in the Hindu Kush Himalayan region over the last 100-plus years. Adv Clim Chang Res 8(3):148–156

Renate M (2004) Governance theory as an advanced control theory? Max planck institute for the study of societies, Cologne.

Reports of Intergovernmental Panel on Climate Change, 1990; 1995; 2001; 2007; 2014, https://www.ipcc.ch/

Sabin TP et al. (2020) Climate change over the Himalayas. In: Krishnan R, Sanjay J, Gnanaseelan C, Mujumdar M, Kulkarni A, Chakraborty S (eds) Assessment of climate change over the Indian Region. Springer, Singapore

Sahu AK (2019) The democratic securitization of climate change in India. Asian Politics Policy 11:438–460

Santer BD, Painter JF, Mears CA, Doutriaux C, Caldwell P, Arblaster JM, Cameron-Smith PJ, Gillett NP, Gleckler PJ, Lanzante J, Perlwitz J, Solomon S, Stott PA, Taylor KE, Terray LP, Thorne W, Wehner MF, Wentz FJ, Wigley TML, Wilcox LJ, Zou CZ (2013) Identifying human influences on atmospheric temperature. Proc Natl Acad Sci USA 110:26–33

Saryal R (2018) Climate change policy of India: modifying the environment. South Asia Res 38:1–19

Srinivasan J (2019) Impact of climate change on India. In India in a Warming World, Oxford University Press, pp 31–44

Stern N (2007) The economics of climate change: the stern review (Cambridge University Press)

United Framework Convention on Climate Change (1992), https://unfccc.int/resource/docs/convkp/conveng.pdf

Wilde JH (2008) Environmental security deconstructed. In: Brauch HG, Spring UO, Mesjasz HC, Grin J, Dunay P, Behera NC, Chourou B, Kameri-Mbote P, Liotta PH (eds) Globalization and environmental challenges, hexagon Series on human and environmental security and peace. Springer, Berlin Heidelberg, pp 595–602

The manufacturer's authorised representative in the EU is Springer Nature Customer Service Centre GmbH, Europaplatz 3, 69115 Heidelberg, Germany. If you have any concerns regarding our products, please contact ProductSafety@springernature.com

Printed and bound by CPI Group (UK) Ltd, Croydon, CR0 4YY
28/09/2026
02224093-0010